Marco the Great and the Mystery of Phaseville

By

SK Bennett

Illustrations by Rylee Heavner

Marco the Great and the Mystery of Phaseville

Published in the United States by MathBait LLC, Albuquerque
www.MathBait.com

Library of Congress Control Number:
2024907719

ISBN: 979-8-9880861-2-3

Editorial: Carla DuPont
Cover Design: Garrett Myers

TO A AND R. MY TWO NUMBER WIZARDS WITH UNBELIEVABLE POWERS.

Warning

No one ever learned magic from a spell book. They had to cast incantations, levitate and move objects, or shrink and stretch formations with a flick of their wrist to harness their powers. You too will need to practice.

Enroll in *The Kryptografima*, the premier institute for Saints, to embark on missions and strengthen your abilities.

Ahead lies puzzles, mysteries, and secrets. You'll want to figure them out, so be prepared. To get started, carefully read the next page. As you embark on your journey, solve the clues and mark your findings. You will be rewarded handsomely if you can complete this task. Watch in amazement as the hidden is revealed.

Lastly, this is the second book in a series. If you haven't read *The History of Numberville*, you'll do okay. To master the spells in this book you need only to be an excellent hunter. Flip to the Glossary in the back to get caught up.

The trek in front of you is exciting, thrilling, and dangerous. You will be provided with powerful skills that you must use responsibly and wisely. Be careful. This book may cause you to think. Allow it. Question every assumption and convince yourself.

You have been warned.

UNCOVER THE SECRET

This book contains a mystery. To reveal the hidden, you must utilize the powers you gain on your journey.

1. Applying what you learn, solve the clues at the base of each page.
2. The number line on the right acts as center stage: the y-axis. Use it to identify the solution to each clue.
3. Carefully "dog ear" each page to match your findings by creating a fold.
4. You can use this page to aid your quest by removing it to act as your horizontal number line.
5. Can you uncover the hidden message and image?

Visit www.MathBait.com/folds for hints and tips.

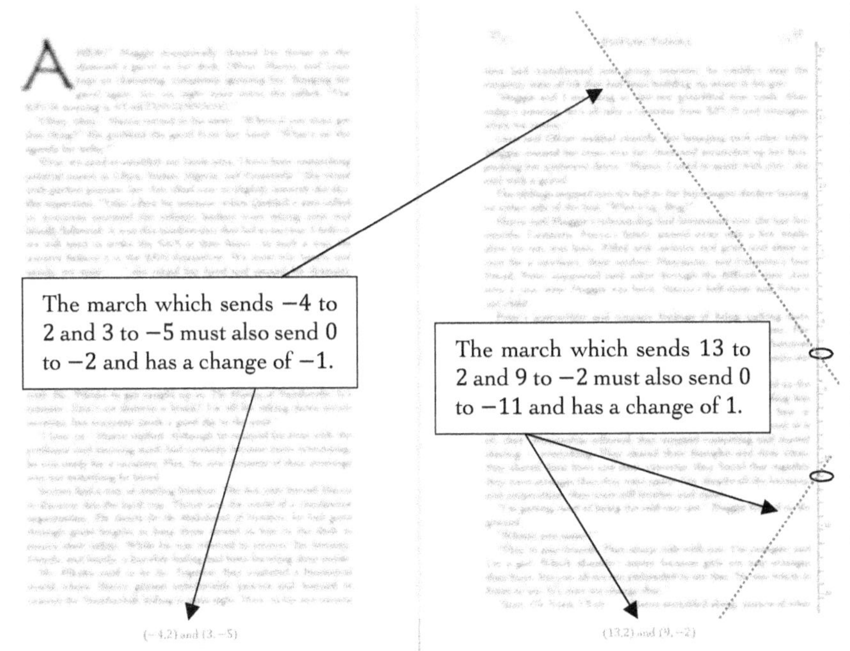

The march which sends −4 to 2 and 3 to −5 must also send 0 to −2 and has a change of −1.
The march which sends 13 to 2 and 9 to −2 must also send 0 to −11 and has a change of 1.
(−4,2) and (3,−5)
(13,2) and (9,−2)

20
15
10
5
0
5
10
15
20

CONTENTS

MARCO THE GREAT

AND THE MYSTERY OF PHASEVILLE

Preface

> "WHAT WE CALL CHAOS IS JUST PATTERNS WE HAVEN'T RECOGNIZED. WHAT WE CALL RANDOM IS JUST PATTERNS WE CAN'T DECIPHER."
>
> -CHUCK PALAHNIUK

The man stepped out of the shadows and into the sunlight. He took a moment to allow the warm rays to penetrate his skin as if he were harnessing the energy of the star directly into his veins.

Dark sunspots clouded his vision as he slowly opened his eyes. Allowing the vast field that lay in front of him to come into view, he inhaled sharply. The landscape seemed infinite. He considered for a moment that the three colors he saw were the only ones in the world. The bright green clovers contrasted sharply against the forest green (so deep and burnt, it was almost black) of the tree line to his far left. Above it all was the crisp powder blue sky without a cloud in sight.

He took a step. Hesitated. Something felt off. It was the sky. Wasn't it? It was . . . different.

Gazing up, he realized he had taken for granted the beauty of the familiar bright blue ceiling. Now that it was gone, he missed it. The sky was a mild teal as if the green from the field had mixed into the air ever so slightly like the neighboring paint blobs on an artist's palette. He shook the thought away.

As he knelt down, the man brushed his hand over a patch of clovers. They bent, then immediately sprung back as a demonstration of their resilience. The soft leaves tickled his palm. His fingers clasped together as he gave the ground a sharp tug, punishing the flora for their disobedience. The clovers lay lifeless in his hand. He reached into his jacket and revealed a small monocle.

Studying the now departed plants he shouted, "NO!" His voice carried far across the field with nothing to block or hamper it. He leapt to another area. Another jerk. Another scream. Again and again. He moved, ripped out a patch, investigated the clovers, and yelled in frustration.

"You okay dear?" The soft feminine voice from behind startled him. He turned to see an elderly lady holding a woven basket. She pushed onto her tiptoes to get a glimpse of what he held.

"Oh! How spectacular!" she gasped. "You know, they say four-leaf clovers are lucky? This is nothing to be upset about dear . . . this is surely a sign of great things to come in your future!"

The man's body relaxed, and he let out a soft chuckle. "I think they say that because four-leaf clovers are *rare,* or at least they used to be."

Scrunching her nose, the woman shifted her gaze to her feet. Her eyes widened as the reflection of the green field invaded her light gray irises.

"You see," the man continued, "the *entire* field contains four-leaf clovers." He unclenched his fist and began to walk away, allowing the dead plants to float to the ground. As he left, he heard the woman yelling, "This is a miracle! A miracle! We gotta call the local news, I'll be on TV! Gabriella will have nothing on me!" [*]

"If only you knew," he muttered to himself. "The Pattern cannot be denied. The laws of nature are changing."

As he made one last glance towards the woman, he considered warning her, telling her she should be afraid, very afraid. He thought about insisting she run, hide. But where would she go? As she danced around the field, her bright smile growing with excitement, he stopped. *Let her have this. Soon enough, no one, no one at all will be smiling.*

---

[*] Gabriella Gerhardt (at the time of printing) holds the world record for the largest collection of four-leaf clovers with 118,791 of these good luck charms.

HOW TO HUNT

Who is hiding behind the numbermask?

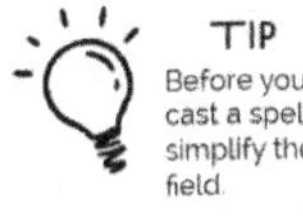

TIP
Before you cast a spell, simplify the field.

A CAPTURED NUMBERFOLK

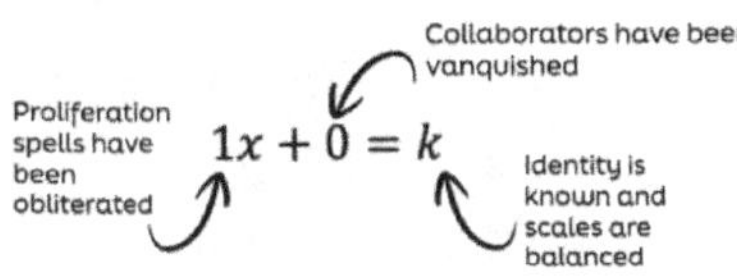

VANQUISH

Cast a duel with an evil twin to vanquish a Numberfolk and create a zero.

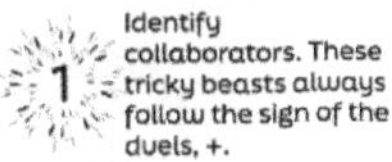

1
Identify collaborators. These tricky beasts always follow the sign of the duels, +.

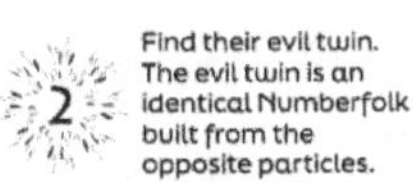

2
Find their evil twin. The evil twin is an identical Numberfolk built from the opposite particles.

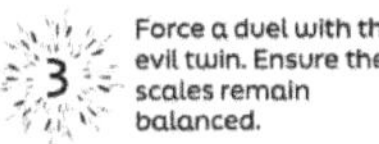

3
Force a duel with the evil twin. Ensure the scales remain balanced.

OBLITERATE

Cast a proliferation spell to shrink or enlarge to obliterate a coefficient and create a one.

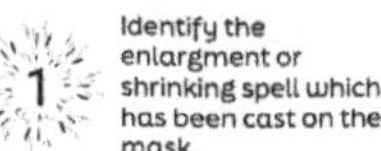

1
Identify the enlargment or shrinking spell which has been cast on the mask.

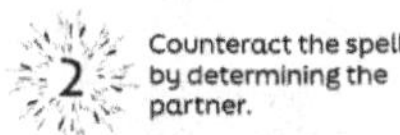

2
Counteract the spell by determining the partner.

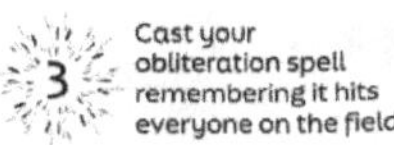

3
Cast your obliteration spell remembering it hits everyone on the field.

SIDE BY SIDE

There are no rules in hunting. You can force any duel and cast any proliferation spell. However, the skilled hunter will know how to quickly zone in on their prey.

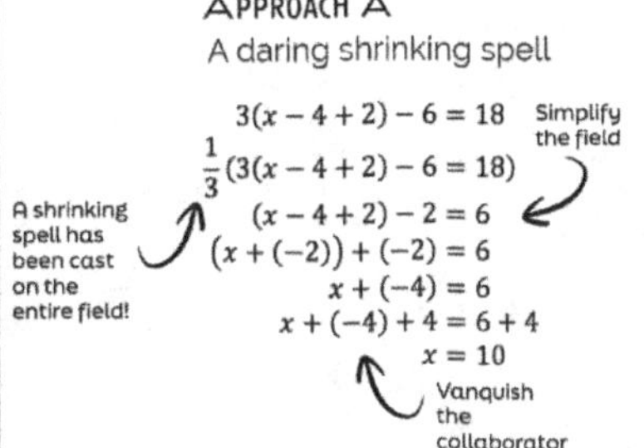

APPROACH A
A daring shrinking spell
$3(x - 4 + 2) - 6 = 18$
Simplify the field
$\frac{1}{3}(3(x - 4 + 2) - 6 = 18)$
A shrinking spell has been cast on the entire field!
$(x - 4 + 2) - 2 = 6$
$(x + (-2)) + (-2) = 6$
$x + (-4) = 6$
$x + (-4) + 4 = 6 + 4$
$x = 10$
Vanquish the collaborator

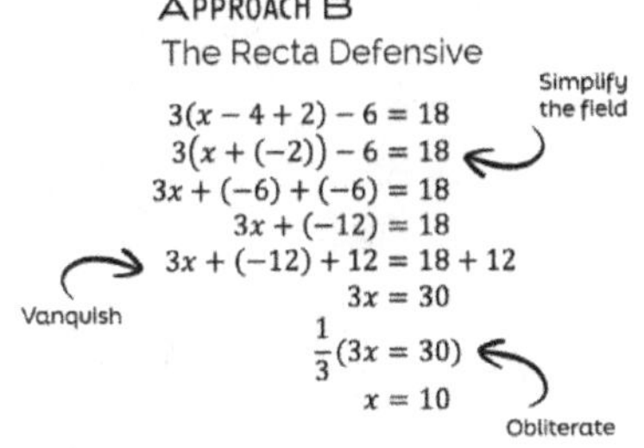

RECTA DEFENSIVE

When a Numberfolk is hiding in a march, the Recta Defensive is a combination of spells that is very effective.

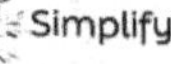

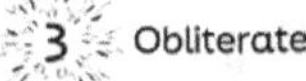

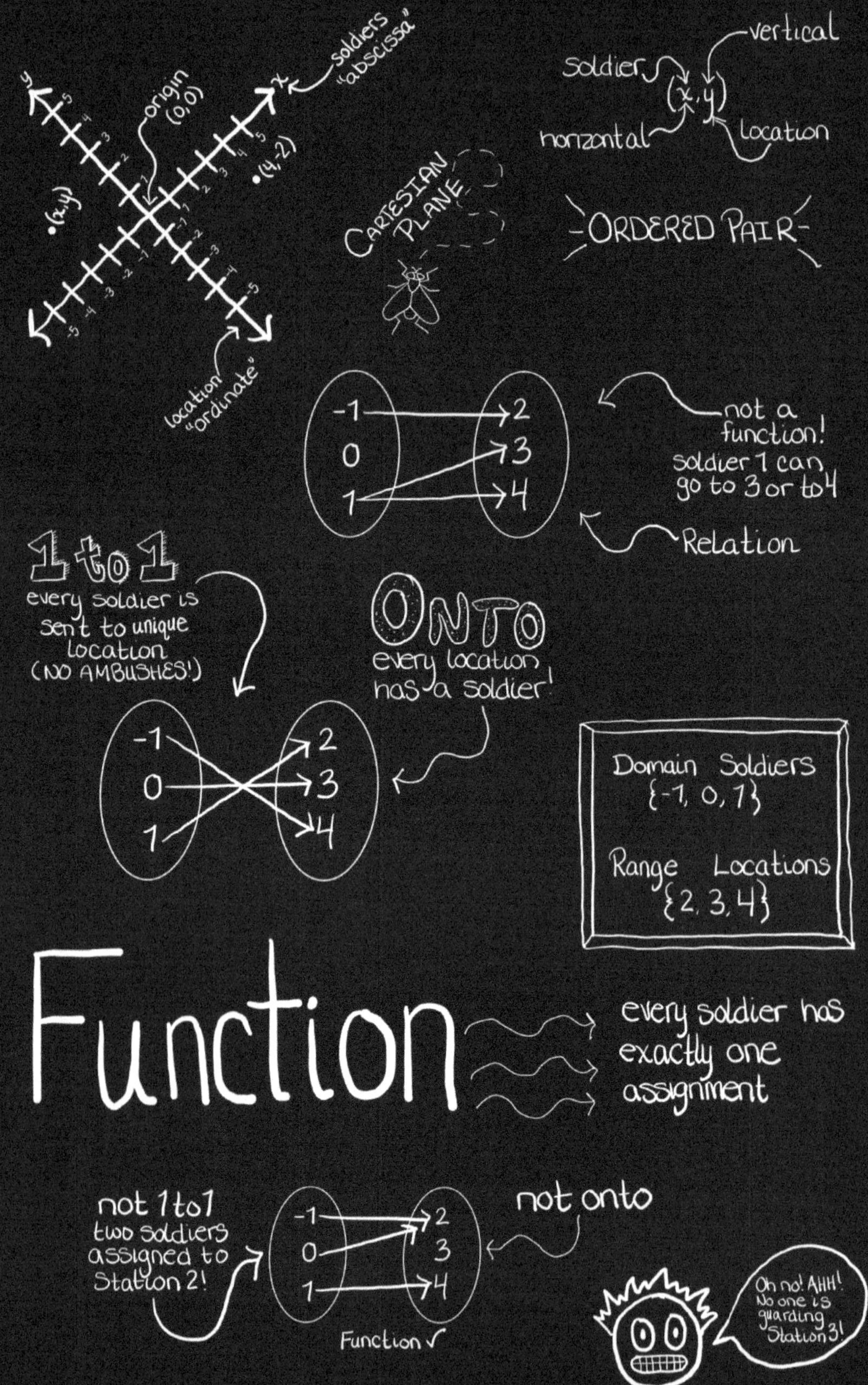

y
x
origin (0,0)
soldiers "abscissa"
(x,y)
(4,-2)
location "ordinate"
CARTESIAN PLANE
soldier
vertical
(x, y)
horizontal
location
ORDERED PAIR
not a function! soldier 1 can go to 3 or to 4
Relation
1 to 1
every soldier is sent to unique location (NO AMBUSHES!)
ONTO
every location has a soldier!
Domain Soldiers {-1, 0, 1}
Range Locations {2, 3, 4}
Function
every soldier has exactly one assignment
not 1 to 1 two soldiers assigned to Station 2!
not onto
Function ✓
Oh no! AHH! No one is guarding Station 3!

1

DYSFUNCTIONAL

"THE GREAT ARCHITECT OF THE UNIVERSE NOW BEGINS TO APPEAR AS A PURE MATHEMATICIAN."

-J.H. JEANS

"AHEM!" Maggie dramatically cleared her throat as she slammed a gavel on her desk. Oliver, Marco, and Liam kept on chattering, completely ignoring her. Banging the gavel again, six, no, eight more times she yelled, "The KFUN meeting is STARTIIIIIINNNNG!"

"Okay, okay," Marco turned to his sister, "Where'd you even get that thing?" He grabbed the gavel from her hand. "What's on the agenda for today?"

"First, we need to establish our battle plan. I have been researching political unrest in Libya, Sudan, Algeria, *and* Venezuela." She stood with perfect posture, her chin tilted ever so slightly towards the sky, like superman. "Take Libya for instance, when Qaddafi's men rolled in, everyone assumed the military leaders were taking over and blindly followed. It was this misdirection that led to success. I believe we will need to strike the SAN at their heart—in such a way the masters believe it is the SAN themselves. We must win hearts and minds, we must . . ." she raised her hand and paused for dramatic effect.

Marco saw this as possibly his only moment to interject. "Maggie, we are four kids. How do you expect us to overtake a centuries-old institution?"

Oliver started to chuckle. Maggie's head whipped around and gave him a glare that quickly turned his look of glee into one of fear. "Your sister's scary, bruh," he whispered.

"Look." Liam chirped, "We've spent the last three months working with Mr. Pikake to get caught up on *The History of Numberville*. It's summer. Don't we deserve a break? I'm all for taking down secret societies, but everyone needs a good dip in the pool."

"I hear ya," Marco replied. Although he enjoyed his time with the professor and learning math had certainly become more interesting, he was ready for a vacation. Plus, the new dynamic of their meetings was not something he loved.

Secrets had a way of stealing freedom. The last year forced Marco to discover this the hard way. Thrust into the world of a clandestine organization, *The Society for the Abolishment of Numbers*, he had gone through great lengths to keep those closest to him in the dark to ensure their safety. While he was relieved to recover his honesty, friends, and family, a horrible feeling had been festering deep inside.

Mr. Pikake used to be *his*. Together they explored a fantastical world where Marco gained unbelievable powers and learned to control the Numberfolk hiding in plain sight. Now, as his one-on-one

time had transformed into group sessions, he couldn't stop the runaway train of ick that had been building up steam in his gut.

"Maggie and I are going to visit our granddad next week. After today's meeting, let's all take a vacation from KFUN and strategize when we return."

Liam and Oliver nodded cheerily, fist bumping each other while Maggie crossed her arms over her chest and scrunched up her face, pushing her eyebrows down. "Marco. I need to speak with you," she said with a growl.

The siblings stepped into the hall as the boys began shadow boxing on either side of the bed. "What's up, Bug?"

Marco and Maggie's relationship had blossomed over the last few months. Cameron, Marco's father, passed away only a few weeks after his son was born. Filled with sadness and grief, and alone to care for a newborn, their mother, Maryanne, and Cameron's best friend, Peter, supported each other through the difficult time. Just over a year later Maggie was born, Marco's half-sister and Peter's *real* child.

Peter's insecurities and constant feelings of being nothing more than a consolation prize translated into bullying his stepson. The events of the last year carried everything to the surface and shattered their family. In a twist of good fortune, Maggie and Marco came out stronger than ever before.

Being nearly the same age as her brother and yet branded as the "little" sister, Maggie took after her father, consistently feeling less than which bubbled into an irrational need to be the best at everything. Marco went to extremes to protect her. At the end of it all, their relationship softened; they stopped competing and started sharing — everything. They shared their thoughts and their ideas, they shared their fears and their concerns, they found that together they were stronger than they were apart. Yet, despite all the harmony and cooperation, they were still brother and sister. . . .

"I'm getting tired of being the odd-one out." Maggie kicked at the ground.

"Whatta you mean?"

"They're *your* friends. They *always* side with you. I'm younger, and I'm a girl. Which shouldn't matter because girls are way stronger than boys, but you all are too pinheaded to see that. No one wants to listen to me. It's time we change that."

"Sure. *Uh*. Yeah. I'll *uh* . . ." Marco stumbled along, unsure of what he was trying to say and absolutely sure he had no clue how to appease her.

"Look. I want to bring someone new into KFUN." Maggie's confidence had magically returned as she looked her brother dead in the eyes.

"I don't know about that. . . . That doesn't sound like a good idea. You know it's dangerous. I already feel responsible enough for you. I can't handle somebody else."

"I don't need *your* protection!" she snarled. "I'm not some little defenseless kid. I'm barely even younger than you and I am *way* better with numbers than you are! You brought *your* friends in. I should be able to bring someone in too."

At that moment, Liam and Oliver burst out of the room, still boxing. They surrounded Marco throwing pretend jabs in all directions filling the hallway with laughter.

"I . . . I'll think about it okay?" Marco yelled back as his friends pulled him down the stairs. Maggie stomped five feet behind them, arms crossed, the entire way to the library.

* * *

His long legs dangled from the chair like a giant in a dollhouse. Even in the dead of summer, a record-breaking 98° didn't stop Mr. Pikake from fashioning his habitual tweed suit. Dark blue pants, an olive-green vest, and a khaki blazer made Marco hot just looking at his tutor. Mr. Pikake flipped over his hand, palm facing up, to reveal a large, oversized watch face on his inner wrist.

The tutor was right out of a cartoon. A cross between *Doctor Who* and Ms. Frizzle, he was the Enderman of numbers able to masterfully take and rearrange, uncovering hidden patterns and invisible truths. He led Marco out of the cave of shadows, training him to be a master hunter. Along the way, Mr. Pikake became like a father to him. The professor's perfect pronunciation that caused sounds to explode as he spoke, the way he flopped like he somehow had less bones than the rest of humankind, and his larger-than-life smile backed with genuine care and compassion were only some of the lovable traits that gained Marco's admiration and trust. But there was a catch. Mr. Pikake was part of the very group working to abolish Numberfolk control and destroy anyone who got in their way.

Determined to discover the true agenda of the Numberfolk, Marco and his friends created their own secret society, KFUN: *Kids for the*

Understanding of Numberfolk. They couldn't do it alone. Working as a double-agent, the professor was now guiding the group, helping them build the skills they needed. Unfortunately, this put everyone in danger.

"Perfectly punctual," Mr. Pikake popped as his Cheshire cat grin slowly nestled onto his face. Like Jack Skellington, he jumped up and in two large steps had crossed the room to the board.

"Hurry, hurry, gather in. No time to dawdle." The professor ushered the students to their seats.

With one smooth movement, he drew a horizontal line across the board in green marker. "Let me illuminate the illustrious line," he purred. Marking the line with a small vertical dash, he wrote the number 0.

"Liam!" he called out, "Please punctuate with another Numberfolk."

The boy walked to the board, his gait somewhere between a sasquatch and a teeny bopper. His long arms floated back and forth while his body bounced happily along. Liam was tall and skinny. His light brown hair seemed impossibly straight. Marco always thought the strands were tiny unbendable arrows that refused to do anything other than point directly down. Adopting a 2009 Bieber-style, Liam tried to force the locks to "swoop", but they resisted and large chunks always fell to his face giving off an unintended emo look.

"You want me to put a number here? Like this?" Liam made a short and stubby vertical mark to the right of Mr. Pikake's 0 and labeled it with the number 1.

"Sensational!" the professor beamed. Maggie rolled her eyes and slouched down in her chair. The tutor pointed at Marco with the marker. Bouncing up, thinking quite highly of himself, he made a sharp vertical line to the *left* of the 0, adding −1 beneath.

"Very Nice!" Mr. Pikake burst, "We have created a Numberfolk line. We order them from smallest to biggest moving left to right."

"We learned this in like 2nd grade!" Maggie spit. "How exactly is this going to help us take down the SAN???"

"Alright, alright. Calm down little one."

Maggie's blood started to boil but before she could erupt, Mr. Pikake continued.

"You know now how to hunt the Numberfolk. You can cast proliferation spells to shrink or enlarge them in size, force duels, vanquish, and obliterate, but!" The familiar pop rang through the room, "Imagine the power to command them — to order them to bend to your will."

Marco was in awe. As a teenager, it felt like nothing was ever in his control. Someone was always telling him where to go, what to do. Worse, he didn't even have dominion over his own emotions. They too had a mind of their own. The kind of power Mr. Pikake was describing would be astronomical, he'd have an entire army to wield.

Oliver chuckled, "Yeah Maggie, it's like lining up kindergarteners from shortest to tallest to walk to the lunchroom. Didn't you do that last week?"

She growled at her brother's friend.

"Now!" With one word, Mr. Pikake altered the energy in the room. "Who says we need to make this line horizontally?" Another swift wave and a long vertical line was added to the board, this time in blue.

"You're talking about the coordinate plane." Maggie said matter-of-factly. She bounced up from her seat and quickly slashed at the board before moving to the side to reveal her work. "The coordinate plane is when we combine two number lines. We draw one horizontally and one vertically. On the horizontal line, we label the marks, or 'ticks' as mathematicians call them, from least to greatest, from left to right. On the vertical line, we label the ticks from least to greatest from the bottom to the top. The place where the two lines meet is zero on both lines. It's called the origin."

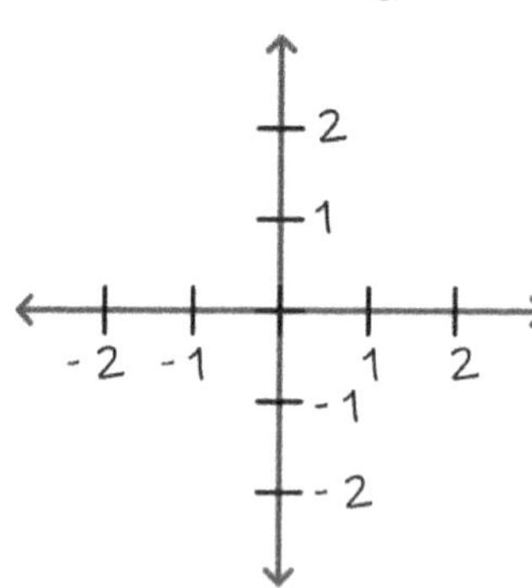

The room was quiet. No one knew what to do. No one knew what to say.

"Yes . . . *ah*, very valuable input Maggie," Mr. Pikake whispered.

A hot embarrassment burned on Marco's cheeks as he tried to force himself into a ball. The table blocked like a knee-goalie, only allowing him to achieve a strange squatting position. His sister was acting like a brat, a dog off her leash, in front of his friends *and* his mentor. As much as he longed to just disappear, he felt a heavy responsibility as the big brother to manage her.

Trying to shift the awkwardness he started, "*Um*. Okay. So. We can put two number lines together. What's the point?"

A large smile invaded Mr. Pikake's face. "The point *is* the point!" With a surge of excitement, the tutor began frantically making dots, points, all over the board. He turned back to the children, "Have you ever played *Battleship*?"

"A superior test of military strategy!" Liam beamed.

"Precisely!" Mr. Pikake boomed back. In only a few seconds he revealed a remarkably perfect grid on the library whiteboard.

"But *Battleship* has letters too. That way, when you say something like A7, you know where to go. Without the letters, you can't be sure what 2-4 even means," Liam whined.

"Outstanding observation!" Mr. Pikake was dancing around the room now, tripping over every chair and backpack and pair of legs that got in his way. "How to decide if we intend 2 horizontally and 4 vertically, or 4 horizontally and 2 vertically, unless . . ." he held his "s" like a snake, ". . . we all agree on whom we shall call first and whom we shall call second." He paused to catch his breath. He was speaking with an impressive speed that reminded Marco of the *Cat in the Hat*. "Obviously, we originate with the original! The abscissa."[*]

Maggie perked up. This was new and therefore interesting. "What's an abscissa?"

"*Ah*, that is the question!" Mr. Pikake's hand flew into the air. "Imagine I told you to walk 8 miles. What would you do?"

Maggie grabbed her right elbow with her left hand and wrapped her forefinger and thumb around her chin indicating she was thinking. She was clearly unsure, which was a foreign feeling for her. It felt uncomfortable. But it also felt *not* boring, which was, for the moment, exhilarating.

Before she had made up her mind Oliver shouted out, "She'd walk 8 miles!" The boys giggled; Maggie glared angrily.

"Which way? Left, right, up, down, north, south?" Mr. Pikake was clearly enjoying this line of questioning. "Abscissa means a cut-off line. Which was fine when we had but only one line! Walking 8 miles from your position simply meant to travel on the line to the right 8 units."

---

[*] If you are wondering how to say this word it is ab-sis-uh. I am not sure why someone decided to throw in that random silent "c."

"Why to the right?" Liam looked lost. He was unsure where the conversation turned from *Battleship*. Was abscissa one of the boats?

"Because the right is positive," Oliver chimed in. "If they wanted you to walk to the left, they'd tell you to walk −8 miles."

Liam was transported to Bizzarro world. He imagined he was lost in the streets and had to ask for directions.

"How do I get to the House of the Seven Gables?"

"You're gonna go −2 on Essex, then +1 on Bentley. Finally, take a $-1\frac{1}{2}$ on Derby to Turner. Can't miss it! It's right there on the positive."

He shook the thought from his head. His tiny strands of hair flew back and forth united, refusing to change their shape.

"As Liam poignantly pointed out, in *Battleship* we require two pieces of information, just as we do here. The second piece is the ordinate. With the pair, we can navigate to any space on the board." Mr. Pikake waved his right hand in front of the grid like a game show model presenting a new car.

"So which way do we go?" Marco understood two pieces of information were needed, and there was a preset order of announcing them, but which way was the abscissa and which way was the ordinate?

"First, you go left or right. Then, you go up or down." Maggie spoke in crisp, slow words. Marco couldn't tell if she was being patronizing or trying to mimic their tutor. Whatever it was, he didn't like it.

Mr. Pikake dove into his brown messenger bag, like Mary Poppins his arm seemed to reach farther than humanly possible before he reeled it back up with a small glass jar in hand.

Marco instantly recognized the contents. "Is that a *fly*?"

"Ahh, that it is!" Mr. Pikake looked at the insect endearingly, as you would a pet. With his free hand, he motioned upwards. The children all tilted their heads back to see what he was pointing to.

The ceiling of the library room was constructed with small fiberglass square tiles. There were 7 across and 15 up and down. Each tile had a thin aluminum strip bordering it and keeping it in place.

"I shall release the fly I have in this jar. Predict where it will land. We can observe how this system of navigation is part and parcel to our existence."

(−4, 7) and (4, 1)

"You know, flies went extinct in the 80s?" Liam began. "The flies you see today are really tiny video devices the government uses to watch us and collect data."

"Bruh, if that were true then why are flies always hovering around poop?" Oliver retorted.

Maggie giggled, "Let it free! Let it free!"

Mr. Pikake popped open the jar and the fly began to buzz around the room. "Call out your predictions!" he screamed, "hurry!"

"Wait!" Marco shrieked. His internal rollercoaster had picked up speed. It was about to flip upside down when he realized his seatbelt was unbuckled. Picking a location was like attempting to get the *click* before he plummeted to his death. "There aren't any numbers on the ceiling!"

"Well done! When we use this method for navigation, we must identify a key point. Your question is really: where do we originate?"

"The origin. The starting point. Zero, zero. But I don't know where that would be on the ceiling," Maggie chimed in. The fly was buzzing around the room, gaining altitude.

"This method was first described by the exalted Saint, René Descartes. While he was sick in bed, he examined how a fly moved across his ceiling which led to his description of the Cartesian Plane, named, of course, after himself. Descartes placed his origin at the bottom left corner of the room, we may do the same." Mr. Pikake spoke quickly but clearly.

"Four-eight," Liam called out.

"Ten-ten!" Marco screamed wildly. He wasn't sure if his seatbelt was securely fastened.

"Ten-ten isn't even possible, *Marco.*" Maggie said her brother's name with a whine. "There are only seven tiles along the edge."

"Five-ten!" he screamed back.

"Why are you yelling?" she said calmly before adding in her guess, "two-nine."

"Oliver?" Mr. Pikake motioned to the last student.

"Uh. Seven-eight."

The fly bounced back and forth along the ceiling, before finally settling on the nice piece of aluminum at the convergence of six and twelve.

"I guess no one won," Marco said in defeat.

"Couldn't we figure out who was closest?" Liam suggested.

"That we can! But that is for another day. Our primary purpose presently is to comprehend relationships." He snapped twice, and the fly completed a nosedive with a loop-de-loop before landing back in the jar. The tutor popped the lid back on. Liam's eyes widened in amazement. Marco felt nauseous and thankful it was over.

"To find any location, we require one Numberfolk per dimension. The ceiling is two-dimensions . . ." He placed his hand on the table and in one fluid move slid his long legs across the surface before laying back. ". . . horizontal and vertical, and thus two Numberfolk are required. The full room is three-dimensions. To find a place in the space, we'd need three Numberfolk."

In a deep and brooding voice Liam bellowed, "Beyond is another dimension. You are moving into a land of both shadow and substance, of things and ideas. Time is the fourth dimension."

They all stared dumbfounded. "*Twilight Zone*." He whispered with a nod and a wink, "But I added in that last part."

"Certainly!" Mr. Pikake broke the silence springing up like a reanimated corpse and jumping back to the ground. "If we wanted to observe the fly's movements over time, we could add an additional dimension. But!" A pop. "We're getting ahead of ourselves. Let us begin with a tame two: a soldier and where we command them to travel."

He pranced to the board, improving his skill of dodging and diving around the students and chairs that now filled the tiny library room, and motioned to his grid. "We can describe any place on this board with two Numberfolk. The first, the abscissa, commonly called the x-coordinate dictates our soldier positioned on the horizontal line. The ordinate, alternatively the y-coordinate, announces their vertical placement. Together they form," he paused and looked over the room before lowering his voice and concluding, "an ordered pair."

Maggie gulped. "Are ordered pairs . . . *dangerous*?"

"Of course not!" The professor perked up in an instant. "But they are powerful!"

Oliver nudged Marco and spun his notebook for his friend to see. He had drawn an ordered pair, (x, y), each boasting outrageous muscles. The two quietly giggled.

"Ordered pairs form a map. If the Numberfolk are plotting something, we can use our information, our

intelligence, to create a blueprint and then determine their every move." He turned to the board and began scribbling.

$$A = \{-1,0,1,2,3\},$$
$$B = \{2,3,4,5,6\}.$$

"These Numberfolk represent two teams. Marco and Oliver, you'll command Team A, Liam and Maggie, yours is Team B."

"We have some traitors in our midst." Oliver tilted back in his chair.

"Whatta you mean?" Marco asked.

"Look! The 2 and the 3 are playing for both teams."

Marco imagined the numbers on a field. The −1 passed the ball to 2, who proceeded to throw it in the air, catch it again, and change jerseys in the blink of an eye before scoring on its own goal.

"No, no. Same name, different player," Mr. Pikake clarified. "There are many Olivers in the world, are you all the same person?" He lifted his left eyebrow in inquiry.

Marco updated his mental image to include two twos in different colored jerseys both with the name *Steve* on the back. From the sidelines the coach was yelling, "Pass it to Steve! To STEEEVE!" The confused −1 scanned the field, found a Steve, and threw the ball. Wrong Steve. As hard as Marco pushed, he couldn't keep the smile from taking over his face.

"We are playing tag. Each team will select a member of the other team to 'tag'. I'll transcribe the turns. Begin."

Oliver and Marco huddled, whispering, before Oliver declared, "−1 tags 4."

Mr. Pikake jotted the play, $-1 \to 4$.

Liam shouted out, "6 tags 0!" without consulting Maggie which earned him a sharp elbow-jab to the side.

Marco's turn. "0 tags 3."

Before Liam could open his mouth Maggie burst out, "6 tags 3" then turned to Liam and twisted her neck in defiance.

Oliver called out the next play, "−1 to 6."

Marco questioned his partner's move, "Negative one already tagged someone?"

"Bruh, he's our best player. Gotta use him well."

"Alright, alright. Very good, let's examine your exercise," Mr. Pikake interrupted. He drew out three ovals on the board and connected them with lines to represent the plays.

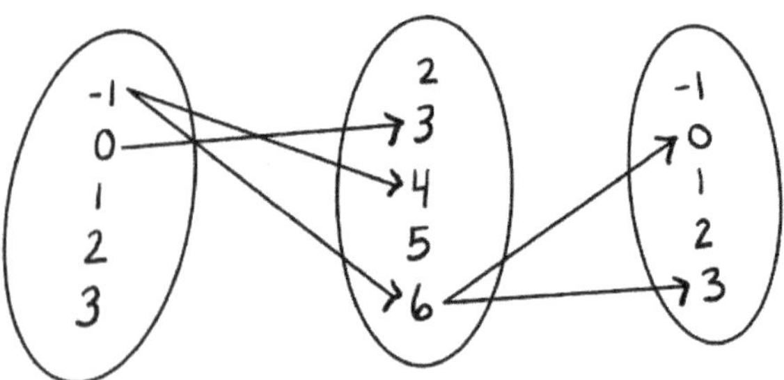

"Why does their team get two bubbles?!" Maggie whined.

Ignoring her, the tutor continued. "Now, if we wanted to make this our battleplan, instructions on what to do, whenever we went onto the field, the −1 on Team A could pick to tag Team B's 4 or 6, while 0 would always tag 3. The remaining players are there but did not complete any tags."

"Our plays suck," Liam sighed. "We set it up so 6 is our only tagger and they can tag 0 or 3 but no one else."

"Excellent examination," the professor exclaimed. "We call this the domain. It is your dominion, those you command. The domain of a relation is all the x-coordinates, taggers, soldiers. A larger domain allows for more variety of plays."

"So, our domain is −1 and 0?" Marco asked.

"Precisely! And predictably, your range of motion is the players your team can tag. Where they are allowed to go."

"Our range is 0 and 3," Maggie said defeatedly.

"And ours is 3, 4, and 6," Oliver followed. "In all fairness, we did get an extra play."

Team A	**Team B**
$-1 \to 4$	$6 \to 0$
$0 \to 3$	$6 \to 3$
$-1 \to 6$	

"True! Team B, you have one more play. Can you create a mapping to ensure the size of your domain and range is equal to that of your opponent's?"

Maggie and Liam huddled up and began whispering. "We will have 4 tag 1," she responded proudly.

Mr. Pikake scribbled, drawing in an additional arrow. "Your domain now?"

"Our domain is 4 and 6 and our range is 0, 1, and 3." Liam was

$(0, 7)$ and $(4, 3)$

either intentionally attempting to sound like Maggie or unintentionally had picked up on her "I'm better than everyone else" tone.

"Very well, you have created two relations." He turned his back to jot more letters and numbers on the board.

$$T1 = \{(-1,4),(-1,6),(0,3)\},$$
$$T2 = \{(6,0),(6,3),(4,1)\}.$$

"Where . . ." he continued to add,

Domain T1:	$\{-1, 0\}$	Range T1:	$\{3, 4, 6\}$
Domain T2:	$\{4, 6\}$	Range T2:	$\{0, 1, 3\}$

"I fail to see the relevance of this," Maggie said sophisticatedly.

"Total domination not good enough for you?" Oliver chuckled.

"Relations, mappings, are good for many things," Mr. Pikake responded. "Suppose instead of teams, these represent bus stops. If I get on at the bus depot, stop 0, it takes me to stop 3. No buses leave from stop 3, so hopefully that is near my destination."

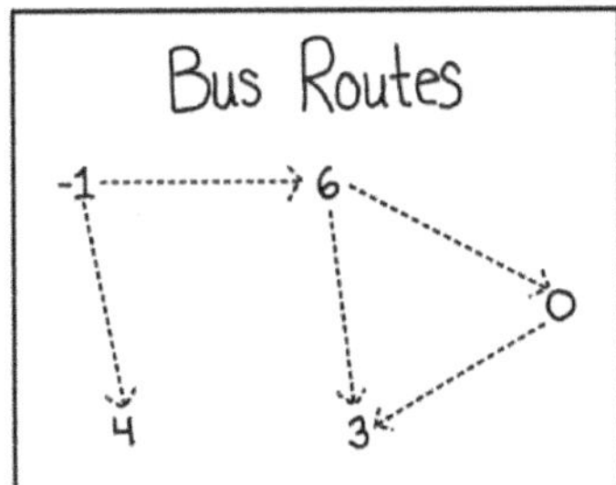

"Oh, I see," Oliver started. "If I wanted to waste the day riding around, I'd want to take the bus from stop −1. That would take me to stop 4 or stop 6. If I got off at stop 6, I could then ride to the bus depot at stop 0 and from there to the end of line at stop 3."

"That you could! And while relations are certainly useful, they can track family lines, city of residence, teachers and students . . . they can be horrible for battle plans. That is, when they allow for too much choice."

"Choice?" Maggie asked.

"We don't want our soldiers making choices. If they do, the battlefield becomes dysfunctional. If I sent Oliver to stop −1, this directive would allow him to select his next move and I'd have no idea where he ended up. At 4? Or perhaps at 6? Since stop 6 also has choices, he could be anywhere in the city by now!" He paused and held still like a statue before suddenly continuing as if someone hit the pause button on their remote. "A function is a special type of relation, one where there is no choice. A soldier has their orders. Each Numberfolk is assigned to one, and only one place."

"*Cogito, ergo sum,*"[*] Liam stated coldly. No one knew what he was talking about, so they ignored him.

"So, some of the soldiers just sit there and do nothing?" Oliver interjected. "Like the players without anyone to tag?"

"Insightful!" Mr. Pikake lowered himself into the chair at the front of the room, placed both hands on the table, and leaned in. "To construct a functional system, we have a second requirement. You must command everyone in your dominion! We cannot leave soldiers without an order."

"Alright, let's work this out." Maggie took the lead. Predictably. "Say our bus stations are $-2, -1, 0, 1,$ and 2. For this to be functional, we have to have a route leaving each station."

"And!" Liam joined in, "No station can go two different places. They all need one specific order."

"Wait, do all the stations they go to have to be the same? Like we could tag 6 even though 6 wasn't on our team?" Oliver looked to Mr. Pikake.

"Certainly not! You can send your soldiers to any station . . . that exists."

"Okay then, let's make the destinations $0, 1/2, 1,$ and 3."

The group worked together to create a set of bus routes by all picking a station and assigning its destination. They looked over their work.

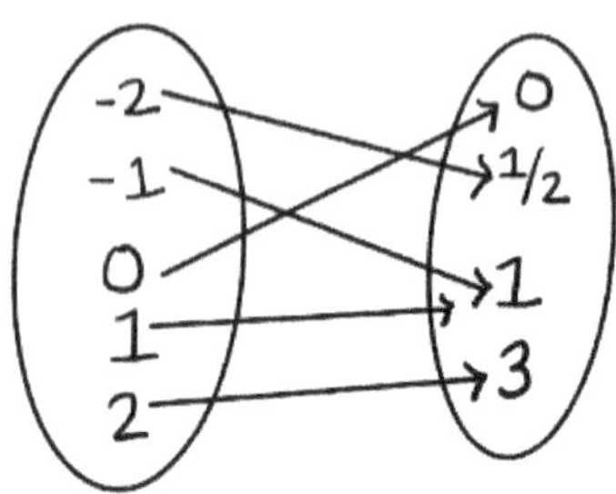

[*] This famous saying was from none other than Descartes, the inventor of the Cartesian Plane. It means "I think, therefore I am" and was René's way of convincing himself he actually existed. Liam is commenting on the idea of not having a choice or not being able to think for yourself.

"Liam, you dummy. You made bus stop 0 go to itself!" Oliver admonished.

"It's the Hotel California." Liam replied happily, "You can never leave!"

"That's not even the biggest problem." Maggie pointed to their diagram. "We made −1 and 1 both go to 1. That isn't allowed."

"Why not?" Marco, for once, felt ahead of his sister. "The rule was each soldier needed one order and every soldier had *an* order."

"Perfection!" Mr. Pikake jumped up. "Marco is correct. You have indeed defined a functional command. Every soldier has been assigned *and* every soldier has only one assignment."

Maggie threw her arms across her chest and slouched in her chair.

"But!" a pop, "Maggie's observation is optimal. In this battle command, you have sent two soldiers to the same location."

"What if there's a huge assault at 1?" Liam argued. "Wouldn't you want to send more of your soldiers there?"

"Riveting reflection," the tutor responded. "Where we send our soldiers, the bus routes we select, depends on the situation. Sometimes, we want to spread them out." He stretched open his fingers revealing large gaps between each bony digit. "And other times we may want to rush a single location." He violently clenched his hands together in fists. "Our situation will dictate our design."

The professor swiveled on one foot like a general to face the board and wrote 1 – 1.

"Zero!" Liam called out eagerly. Everyone stared.

"Not one *minus* one. One-to-one. I suppose I could write . . ." his voice trailed off. Using his sleeve to erase the markings, he replaced it with 1: 1.

Oliver rolled his eyes. "Ratios. *Great.*"

Marco didn't love ratios, but he did love LEGO. He recognized the familiar 1:1 from the instructions. They always included that when the picture of the piece was the same size as the actual piece. He knew Mr. Pikake was saying that something matched up exactly.

"One-to-one simply means one soldier — one place. Your battle plan is not one-to-one since two different soldiers have been sent to the same place."

"But we can't fix it," Maggie protested. "There are fewer places than soldiers so two soldiers *have* to go to the same place."

"We can add more stops, right? We picked the stops, so we can just pick some more. Let's add stops at −1 and 2," Marco suggested.

The children leaned in and updated their diagram.

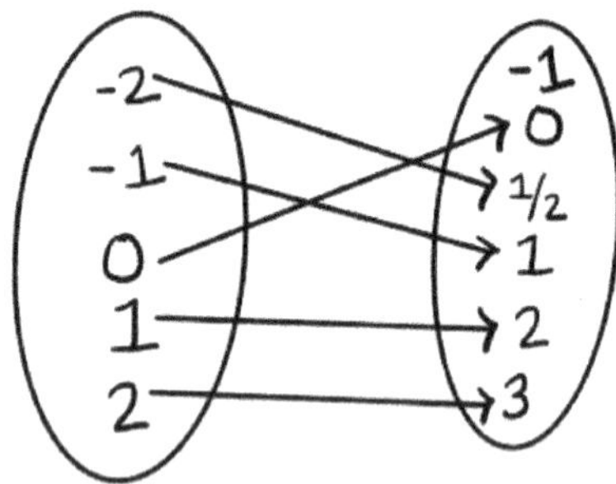

"Look! Now every soldier is going to their own unique place. It's one-to-one!" Maggie shivered with excitement.

"We left the −1 without a soldier!" Liam shrieked. He was taking this *very* seriously.

"And you have discovered the second type of function! Onto!" Mr. Pikake called out. "An onto function is when the range is equivalent to the destinations. What is your range?"

"The range is the part of the battlefield we have guarded, so that is 0, 1/2, 1, 2, 3. We have an unattended station," Maggie sighed.

"Forget the fuss! A functional command doesn't *need* to be one-to-one and onto, or one-to-one or onto. It is the choice of the commander. But," the pop bounced off the walls of the small room. "Knowing is essential." Mr. Pikake began to shovel items back into his bag.

"One!" The tutor's first skeletal pointer sprung to life. "We have before lined up the Numberfolk on a horizontal line; these are our soldiers. Our domain, dominion, the ones we command."

"The abscissa!" Maggie shouted excitedly, relishing in her new vocabulary word.

"Indeed! We have now added a second dimension, a vertical representation which dictates where each soldier is sent. The range of our formation."

"The ordinate!" Oliver shouted in a high-pitched tone. Maggie scoffed.

"Two!" his finger popped up to make a duo. "Together, a soldier and their location form an ordered pair (x, y), horizontal and vertical in harmony. A relationship."

"x and y sitting in a tree . . ." Liam started, everyone groaned.

"Relations are a group of ordered pairs that dictate their connections," the professor continued.

"But a relation, like a family, can be dysfunctional!" Marco shouted out. Everyone else had contributed and the invisible weight of peer pressure was pulling him down. He needed to make sure he

didn't disappear into the crowd and Mr. Pikake knew that he was still there.

After each session, Marco always made sure to recap everything in his journal. The professor had not gifted journals to the rest of KFUN, so it remained a tiny secret bond that was just theirs. He couldn't help the nugget of jealousy that burned in his throat. His special world had been bulldozed and replaced with a chaotic mosh pit where everyone was constantly talking over each other.

"Precisely!" The tutor flashed Marco a warm smile. Giving in to the chaos he started, "A functional relationship must be such that . . ." he paused allowing the children to chime in.

"Every soldier has exactly one order." They blurted out unenthusiastically and not at all in sync. Liam said "assignment" rather than order and Maggie said the whole thing backward. It sounded more like "every soldier mur mum mah fas lah moo."

"Excellent, that makes three." Mr. Pikake threw his thumb to the side. Liam couldn't stop staring at it, considering the odd choice. "For a functional command, you must assign every soldier in your group an order, and each soldier can have only one place they are assigned to go."

Another finger jumped from the professor's hand, it was the tiniest, the pinky. "Finally, four! Should we wish to spread out our soldiers, we can create a 1:1 order, a command such that every soldier is sent to a unique location, no ambushes. And to ensure every position on the field is guarded, we can create a command that is onto the entire front, nowhere left unattended."

He placed a soft, brown, explorer hat on his head, pushing down his salt and pepper hair and bowed towards the students. "Good day." With that he ducked out the door and disappeared.

"Did you see his fingers?" Liam burst. "Who counts like that?! Oh! And the fly! I mean, you can't train a fly. That *had* to be a government spy-fly."

They all giggled and chattered as they exited the library to a bright, warm, and sunny day. One would think they would have noticed the man. The man dressed in all black in the dead of summer. The man who was tall and thin and loomed over everyone that passed by. The man who stood, staring. But they didn't. Not one of them noticed a thing.

2

MIRROR, MIRROR, AND A WALL

"IT'S NOT A BORING PLACE TO BE, THE MATHEMATICAL WORLD. IT'S AN EXTRAORDINARY PLACE; IT'S WORTH SPENDING TIME THERE."

- MARCUS DU SAUTOY

Birthed by the charmed ones, Natural was created in their image. The Great Scale had other ideas. Knowing the universe required balance, for each being the charmed ones created, it ordered the Mirror of Wonders to make another, their twin. The two groups together, Natural made up of positive Numberfolk $(1, 2, 3, \ldots)$, TNZ consisting of their evil twins $(\ldots, -3, -2, -1)$, and 0, the only resident of Whole County outside the Village of Natural, made up the state of Integer.

For eons, the Numberfolk existed within Integer. As time went on and societal struggles erupted, a massive stadium was constructed to appease and quiet the masses. The arena was raised outside the state of Integer, in Rational country. There, they engaged in their own outrageous and entertaining sport: The Vinculum Games.

The vinculum was a large bar that determined, with the help of the Great Scale, the winner of each game. Unbeknownst to the Numberfolk, they found that with each match they formed a new and interesting being. These creatures were the result of the destruction, the pulling apart, of the magical particles that gave the Numberfolk their abilities: Spogs. The Vinculum Games created some of the most well-known Numberfolk today like one-half, and two-thirds.

All things eventually come to an end. It was a combination of the disasters of the 53rd Games and a greater understanding of the vinculum and Logos that caused the Numberfolk to abandon the stadium. * And there it sat.

While the Numberfolk had friends and families, they had enemies and rivals, and they had feelings, they were all very rational. They made decisions based on facts. They followed logic and they followed rules. This caused Numberville to grow into a remarkably fascinating place.

The Village of Natural, home to the "original" Numberfolk was by far the nicest area in all the land. The Negative Zone, or TNZ as it was called, continued to decay away. The Negatives were considered bad, dirty, and evil, although they certainly didn't see themselves that way. Without question, the worst area to reside was in the outskirts of Rational Country, home of the Logos.

The Logos didn't differentiate between "negatives" and "positives,"

---

* A *Logos* is what Numberfolk call a Rational citizen who is not an Integer. All Rationals can be disguised with Integers and a vinculum. For instance, two is none other than $4/2$ or $8/4$ or the infinite other possible costumes. A Logos is a Numberfolk such as $1/2$ or $8/7$. They are the torn Numberfolk made from parts. They are those that will never alone be whole.

they were all simply discarded, not worthy of entering Integer state. The outskirts of Rational was home to the one-halves and the two-thirds, but it was also home to the negative seven-eighths and the negative twenty-four twenty-firsts. It was home to the broken. All of them, in some way, had been ripped apart and glued back together the best they could.

It wasn't long before the Naturals decided it was best to separate themselves from the remainder of Numberville. Nothing good came from negative and broken beings. They enjoyed the comfort and tranquility of their pristine Village and wanted to keep it that way. In an attempt to insulate themselves, force the past behind them, Natural erected large walls that stood tall for centuries. The Village was a perfect place. Everyone was whole. The streets were clean and peaceful, and everything was safe.

Rational, outside the Village, was a different story. This next chapter of Numberville history was wrought with strife. It would bring about the rise of a maleficent ruler, an outstanding army, and a shocking revelation. Too often we are unaware of the warning signs, the factors that allow these dangerous powers to take control. Thus, first we must journey down the road that led Numberfolk to such a devastating outcome. It all began with the identity crisis.

The Identity Crisis

One was special. Constructed as the image of Un, ones possessed a remarkable power gifted to their ancestor by the Mirror of Wonders: the power of reflection.

Most ones didn't perceive this power as a gift, but rather a curse. Ones could never see themselves in their own children, their numberbabes were always the spitting image of their mate. And while every other Numberfolk could cast a proliferation spell, shrinking or enlarging their neighbor, One's reflection meant such a charm had no visible effect. A particularly innovative One (having been born and lived her entire life in the outskirts of Rational) was determined to harness her power into something new, something creative.

As a Natural living in the outskirts, One was already a bit of an oddity. While she tried to find comfort amongst the other Naturals, she never could. She had something in common with all of them, but it wasn't much, and it was never enough. To make matters worse, the zeros used her power to taunt her. They'd scream, "You're rubber, we're glue!

Numbers stick to us and bounce off of you!" *

Being a loner gave One a lot of time to herself. She'd use this time to go on very long hikes traveling further than any Numberfolk before her. Because of this, she was the first to ever set eyes on the volcano.

The volcano was an amazing sight. One had seen mountains before. There was a gigantic ridge between the walls of the Village and the boundary of TNZ that could be spotted for miles and miles in any direction. This mountain was different. At first glance, it was nothing more than a tall hill constructed of solid rock. One probably wouldn't have been able to distinguish it from the surrounding landscape if not for the smoke.

Beautiful, fluffy clouds plumed from its head, and she had to get a closer look. Climbing it was nearly impossible. The terrain was like sand and as hard as One pushed, she slid back down again and again. She persisted, and it was worth it. When she finally reached the ridge, One was taken aback by the beauty of the black beach that lay at her feet.

A sea of obsidian, so steep it was almost vertical, stood in front of her. She had never seen such a thing! No one had. She picked up the black, shiny rock and gazed into its face. She looked closer. She polished it off to make sure her eyes weren't playing tricks. No. It was really there.

One saw another in the rock, trapped in the hard stone. She cradled it all the way home. As she walked, she told Rock-One her life story. She shared her ideas and her ambitions. One wanted desperately to make an impact on the world, and Rock-One heard all about it.

The more time she spent with Rock-One, the more she was sure it was a demon. It never responded, so it was clearly cursed to live a silent existence. One also noticed Rock-One would mock her. When she was sad, Rock-One would plaster on a face of sorrow. When she was happy, Rock-One would smile back at her, but in a twisted, evil way. One had to know if more demons lay at the foot of the volcano. Gathering her things, she began the long trek.

When she reached the black beach, she began foraging, picking up a new rock, then another. They all contained the trapped soul of another one. Convinced the universe was playing a horrible trick on her, she let out a loud and miserable yell. The mountain screamed back. Suddenly, it began to cry tears of thick, slow, blazing drops of orange and red.

---

* This was of course because given any number, a, $1 \times a = a$ (One bounces) while $0 \times a = 0$ (Zero sticks).

As the mountain's tears poured over the black sand, a pool emerged. The rocks rippled like water. Scared, One gently poked at the puddle. It was warm and thick, pulling her into its embrace. Dying to know how deep these inky ponds were, she reached farther and farther into its mouth allowing the liquid to suck her in.

She awoke on the sea of rocks. Looking down, she couldn't believe her eyes. The black water had changed her into something new. She was 10/10. The universe wasn't tricking her, it had blessed her. Slowly standing, she struggled to gain her balance as she adjusted to her new body. This was it. This was what she had been searching for her entire life. One was about to make a huge mark on the world.

Quickly, she gathered the demon rocks. Dipping them into the mountain's tears she fused them together before dragging her creations back home.

The news of One's remarkable new ability spread like wildfire. She called the demon rocks *Mirare* or *to look at* and allowed Rationalites (by appointment only) to peer into the demon's face to receive an instant make over.

Numberfolk could now test out all sorts of outrageous outfits they never would have dreamed of. A 2/3 left her shop as 1/3/2 (two vinculums!), while an older 3/7 tried on dozens of looks before settling on 51/119. They finally felt they could show the world their true selves.

Naysayers who doubted the power of One began to sing a different tune. Her reflection gave any and every Numberfolk the ability to see themselves in a new light, from a fresh perspective.

What started out as a high-end boutique soon became retail outlets throughout Rational country as the fancy *Mirare* was replaced with the slang *mirror*. Rational had enough problems and mirrors became the tipping point. Crime ran rampant. With everyone in disguises that changed from day-to-day, witnesses were never really sure who they saw.*

Yet in all this disorder, beauty blossomed. As Logos, their children, and the discarded families of TNZ grew, a neighborhood formed which they named, New Addition.

New Addition was bustling with diversity. It was a haven for art and

* If you are thinking that in a world of Rational beings crimes would not exist, you would be very wrong. The detectives were great, strong deductive reasoning, and the crimes were always sensical. Unfortunately, with so many in need, it was nearly impossible to catch the culprit and then the reasoning became, "if we aren't going to get caught, the benefit far exceeds the risk."

culture. While only whole Numberfolk and their opposites $(\ldots, -2, -1, 0, 1, 2, \ldots)$ were allowed in Integer State, anyone could live in the outskirts of Rational. It was home to Numberfolk of all types. Together they erected amazing structures. An outstanding community center with 8 pillars on the left and 5 on the right left visitors in awe. (Although some claimed that something wasn't quite right about it.) They learned to vibrate and found that different Numberfolk vibrations created new and interesting sounds. Together they began to compose beautiful music. For all its problems, New Addition was a vivacious place. Meanwhile, Naturals lived a very different existence.

The Fable of Six

The Village of Natural was perfect. It was orderly. It was clean. And it was dreadfully boring. No one knew this more than Six. Afterall, Six was known (along with Twenty-eight and Four-hundred Ninety-six, and Eight-thousand One-hundred Twenty-eight and . . . you get the picture) as being a *perfect* Natural. This placed a heavy responsibility on his shoulders. Everyone expected him to do great things for the Village. Everyone expected him to do everything right, to never make a mistake.

It was lonely at the top. Having fans and having friends were two very different beasts. Six felt an overwhelming sense of displacement, like he didn't belong. A celebrity, when Six walked through the town square he was bombarded with smiles and waves. The little Natural girls would giggle when he waved back and run off in groups to whisper and point.

Longing to get away, Six found a quiet little spot next to the wall where he'd spend most of his days. One afternoon, everything changed. As Six relaxed in his special place, he noticed a small piece of the wall had tumbled to the ground. It was tiny, the size of a pebble, but it was enough to spark his curiosity. Six began to scratch at the wall. To his surprise, more mortar crumbled to his feet.

Six became enamored with the stone barrier. It was an itch he couldn't get rid of — the more he scratched, the more it tickled. Scared someone would notice, he began placing large handfuls of debris in his pockets. He'd walk around the square waving and laughing, shaking hands and kissing babies, all while inconspicuously dropping tiny pieces as he walked. He scattered the wall all around the Village and no one noticed.

Every day he was back at work, chiseling, until finally the hole was large enough to look through. No bigger than a quarter, Six took a deep breath before peering through his creation. He saw nothing. He was heartbroken. Six gazed upon Whole. It was barren and desolate. He started to believe

his family was right. There was nothing good outside the walls. There was simply nothing.

Six released his boyhood ambitions and fell in line. He campaigned to become Mayor of the Village and won in a landslide. He stood a little straighter — lost his slouch, and although he had everything, the admiration of his peers, the love of his neighbors, the devotion of his followers, Six still felt empty.

He went on a campaign to convince all the villagers there was nothing else in Numberville. Leaning on the message, "For the charmed ones created us and nothing more," they followed him blindly. But secretly, he made a promise to himself to visit the hole once a year. His annual trek to his favorite boyhood hangout always ended in disappointment. Six would look through the void and see nothing. It infuriated him. Yet, it was a reminder that there was only the Village, where all was good, all was perfect.

Eventually Six settled down, married a 2, and had children. In time, he became old and frail. He lost his sturdiness and used a cane to hold himself up. In the end, when he knew his days were numbered, he looked to his granddaughter and asked her for a favor. She was a beautiful 36. While she wasn't perfect, Six always knew there was something special about her. She gleamed in a way he had rarely seen. Knowing his hole would be forgotten, just as he would, he wanted to pass on the reminder that there was nothing more. He saw in his granddaughter the same drive and ambition he saw in his young self. He worried she too would have outlandish dreams, and before he died, he needed to quell these outrageous thoughts.

She held onto her grandfather as they limped to the wall.

"Go. Go. Look. When I was a boy, I too had silly notions. I thought there was more outside Natural. Like you, I had a boisterous spirit, I had hope that Natural wasn't all there was. So, I carved out that hole. I came back to this very spot year after year, and I peered outside and there was nothing. There is nothing but Natural, nothing but home. I am an old man and will soon release my Spogs to join the charmed ones. I want to leave you with this gift. I see me in you. I see you tire of Natural. I see you long for something more. I see you too have a hole. This is my gift to you: truth. Put behind your childish notions, don't repeat my mistakes, begin living your full life here in Natural."

Six was tired. He pointed to the wall, "Look. Then you too will know."

Thirty-six allowed her grandfather to rest on a nearby rock as she moved towards the wall. She peered through the hole for the first time just as he had so many years ago. Like Six, she saw nothing, a vast uninhabited land, empty, depressing. But then, like a mirage in the hot desert, something caught her eye.

"Grandpa! Grandpa! There *is* something out here!" she yelled. Squinting her eyes trying to focus, she marveled at the sight. Like a child's first glimpse of a giraffe, she stared in awe. The creature was amazing, but what struck her was that it looked just like her grandfather. It was a 6, but one that stood on stilts!

While she didn't know it at the time, Thirty-six had gazed past Whole and into Rational Country spotting none other than 6/11, and what a strange creature it was.

"Come. Look! You won't believe this!" She turned back to the rock where her grandfather rested and saw six glowing orbs. The Spogs that were once his floated in the air surrounded in dust as his energy returned to the universe.*

Thirty-six wept.

She screamed.

For his whole life, her grandfather believed there was nothing outside Natural. The cosmos had squashed his truth, his hopes. It lied to him.

At that moment she hated the universe. She hated the charmed ones. She hated the Village. And she vowed to do something about it.

It took years, but Thirty-six followed in her grandfather's footsteps and became the Mayor of Natural. So much time had passed inside the wall, no one knew what lay outside, an entire history was lost. She couldn't stand the lies, everyone needed to know the truth. Her first order of business was to deploy expeditions to examine this strange land. What they found was shocking.

During the centuries of isolation, Rational had evolved into what some may consider the *wild west.* Duels were commonplace and everyone was always trying to obtain what someone else had.

With Natural closed off, many families were forced into Rational country. If two Negatives had a child, their numberbabe was a Natural. This caused a problem. The numberbabe, being made of positive particles was not permitted into TNZ, for only whole negative particles could reside there. With Natural walled off, that same child could never visit

---

* This is the *law of conservation of energy* which was part of the charmed ones' creation. Knowing the importance and the power of Spogs, the charmed ones decided that when a Numberfolk passed on, their Spogs could not be extinguished. Thus, this law states that energy cannot be created nor destroyed, only transferred.

their homeland. These families were forced out into Rational country as this was the only place they could live together in harmony.*

The same phenomena occurred within the Logos. The child of soul mates 2/3 and 3/2, was none other than a one as partners always completed each other creating a whole. But other pairs like 6/7 and 14/3 also gave birth to Natural numberbabes and none of these children were allowed inside the Village either.

This made Thirty-six sick. Not only was there a bustling world outside the walls, but their own people had not been allowed to come home for ages. Feeling there was no other choice, she enacted The Citizenship Program. Opening the gates raised a security issue, and thus the task was passed onto the great military leader, General Three. GT, as he was called, was Natural to his core. He was strong, whole, resolute, but most of all, GT was hungry for power. Not taking this assignment lightly, he set forth to create the most grueling citizenship test that could ever be conceived. With that, the inaugural games commenced.

The Citizenship Program

Although there was no denying that Naturals existed in Rational, General Three was particularly concerned about these long-lost cousins. They had lived their entire lives among negative and broken beings. It would be irresponsible to assume their environment hadn't taken some toll. GT determined the only way to ensure these Numberfolk truly belonged in Natural was through a trial.

Naturals were notoriously known for being well-ordered. You could gather any group of Naturals, finite or infinite, and it would always contain a smallest Numberfolk to act as the leader of the pack. This led General Three to believe that only true Naturals were capable of completing complex tasks. Broken and negative Numberfolk were not well-ordered, they could go on and on down the rabbit hole forever. GT was convinced non-Naturals were unable to ever truly finish what they started and create something

---

* For example, $-3 \cdot -2 = 6$. See *The History of Numberville* for a full accounting on Numberfolk genetics.

magnificent, and thus, such a task would be the perfect test.*

After many days and many nights, General Three looked over his citizenship test with pride. It was perfect. Not only did it ensure anyone who passed was able to follow rules like a true Natural, it also had the bonus of only allowing a small portion of candidates to progress. It was much too dangerous to simply throw open the flood gates. By inducting a small number of new citizens at a time, Three's army could keep a watchful eye on these interlopers.

The test was tournament style. Round one would separate the candidates into five groups. Round two would split them again, creating ten teams: a number both fitting and symbolic of the ten charmed ones. Finally, players would compete until only one was left standing, the ultimate Numberfolk from each team granted citizenship.

General Three named his game Operation. The rules were fairly simple, they were the same rules that governed the universe. In each round, players were given chips to arrange on a board to define an operation. All Numberfolk were familiar with the idea, it was the same concept that determined if a 2 dueled a 3, the result was $2 + 3 = 5$. The operations the candidates created would dictate what happened when any two chips were combined and must satisfy the following requirements:

1. **The Reflector:** Candidates must assign one and only one chip to be their reflector. When any chip operates with the reflector, the result would be themselves.
2. **Evil Twins:** Every chip must be assigned an evil twin. When evil twins operate, the result must be the reflector.
3. **Closure:** When any two chips operate, the result must be an existing chip. For instance, if the chips are red, blue, and green, the result cannot be yellow or tangerine.
4. **Associations:** The operation must produce the same result when carried out multiple times in different orders. That is, just as with dueling, $(4 + 1) + 5 = 4 + (1 + 5)$.

The final rule didn't have a direct impact on the players as they were only working with two pieces at a time, however it caused an interesting set of

* Consider a group where each member is doubled. In Natural, this creates the even Numberfolk $\{2, 4, 6, 8, \dots\}$ in which 2 is clearly the smallest. The same rule in TNZ gives us $\{\dots, -6, -4, -2\}$, no smallest element, they just keep getting littler and littler forever. Rationals are also not well-ordered.

$$\frac{-6}{8}x + \frac{64}{4}$$

dominoes to fall. General Three was both clever and devious in his design to ensure only the best, if any, could pass his test.

In the first two rounds, all candidates who constructed the same or equivalent boards were grouped together. In all following rounds, only the first player to create a unique operation would advance. General Three provided an example.

"Suppose you are given only two chips: ♥ and ♦. By the first requirement, one must be the reflector. I'll pick heart as my mirror and name my operation star. This of course means heart star diamond, ♥*♦, must produce a diamond, as the heart reflector reflects. Heart star heart yields a heart by the same token."

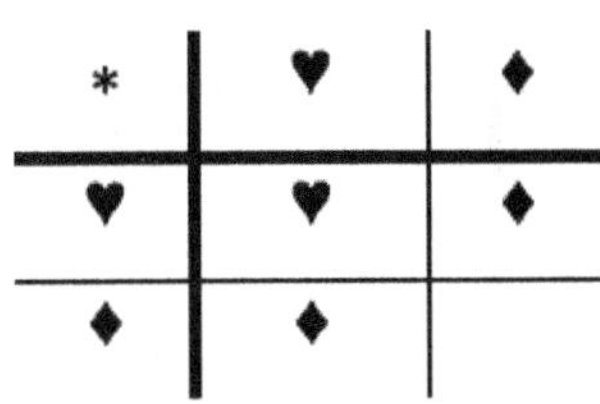

*	♥	♦
♥	♥	♦
♦	♦	

He demonstrated the construction of his board for all to see. "Notice I have but one piece remaining. By the second requirement, every chip must have an evil twin. This implies there must be a chip that operates with diamond to form the reflector. As ♦*♥ produces a diamond, which is not my reflector, heart and diamond cannot be twins. The only option is for diamond to be its own evil twin so that ♦*♦ produces the ♥."

Completing his board, the candidates stood in awe at both the beauty and the logic of the task. A young 17 raised their hand.

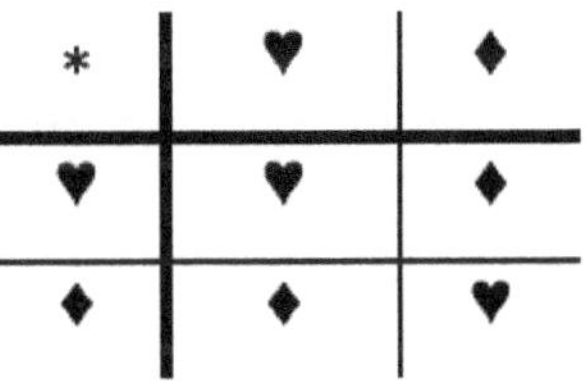

*	♥	♦
♥	♥	♦
♦	♦	♥

"So, the only way to make a different board is by making the diamond the reflector? There are only two possible arrangements?"

GT considered the question. He didn't want to give away any strategies, but also couldn't afford to have the candidates unclear on the rules allowing them a reason to dispute the results. Carefully selecting his words, he responded. "No. This board has only one arrangement. All others are equivalent." He named his new operation $, made diamond the reflector, and completed the lineup.

"This board is indistinguishable except for a name."[*]

"No, they aren't!" A pugnacious 5 argued. "In your first board ♥*♥=♥, in the second ♥$♥=♦."

[*] We call these isomorphic, meaning having the same structure. A way to test if the operation is in fact unique is by switching the names and rearranging.

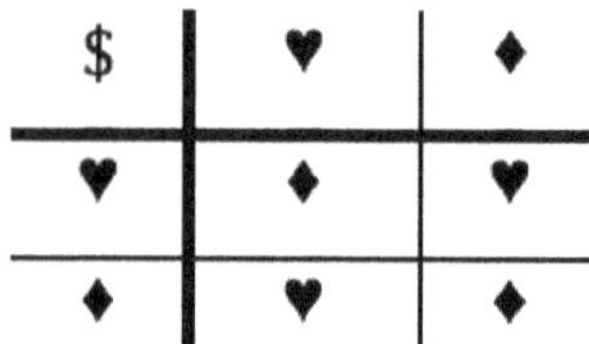

Annoyed, GT snapped, "I doubt you're a Natural at all if you cannot see these are the same! To simplify things for you, let me rearrange this board you claim to be 'different'."

First, General Three swapped the rows. Making sure to keep the operations unchanged, he placed the diamond row on top and the heart row below. "Do you agree *this* is the same board as that you created?"

$	♥	♦
♦	♥	♦
♥	♦	♥

The 5 gave a sharp nod noticing the general's new board had all the same results has his initial $ arrangement.

Next, keeping everything together, the general swapped the columns. He presented it to the group. "And what of this board? Would you agree it is too the same?" This time everyone furiously nodded in agreement, attempting to gain a spot on the leader's good side, assuming he had one.

"Wonderful! Then we all agree there is only one arrangement of two chips!" GT announced with pride.

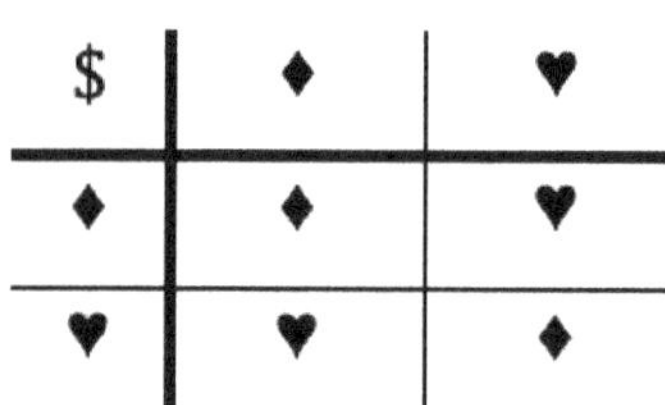

"Wait. What?" The 5 coughed. "You haven't done anything but rearrange the board. They are still just as different!" As the 5 twisted his neck back and forth between the general's first operation of * and his updated version of $, a feeling of doubt crept into his spine. They did look *awfully* similar. [*]

"Can you see *now*?" General Three pushed his nose right up to 5's. "All you have done is switched the names. If I call the reflector Tom and you call the reflector Jim it doesn't make a different board. The structure is identical. I will not fall for such tomfoolery!" GT stomped his foot. The candidates realized this citizenship test would be harder than it seemed. Many wondered if those already living in Natural could even pass the outlandish exam themselves. Not having any choice, the games began.

While most candidates focused on naming their reflector and building a board by adhering to the first three rules, a curious Four couldn't help

---

[*] Try it out. On the * board, color the reflector green and the second chip blue. On the $ board do the same remembering 5 made ♦ the reflector. What do you notice?

the obsession that was building in his gut around the fourth rule. Determined to understand why there was a fourth rule at all, he began to tumble down the rabbit hole.

Why would he give us a rule about three chips if our boards only use two at a time? The four pondered. As General Three began setting up the test, Four allowed his mind to play with the idea.

If I had three chips $a, b,$ *and* c*, could I make it so that* a *and* b *have the same outcome as* a *and* c*?* Combining all four rules, Four quickly identified the general's trick. He named his operation # and allowed $a\#b = a\#c$.

Since the second rule stated every chip must have an evil twin, and the third rule declared only existing chips are allowed, one of the chips must be e, the evil twin of a. Forcing the operation to each side, to keep the scales balanced, Four found,

$$e\#(a\#b) = e\#(a\#c).$$

Now he could examine the mysterious rule four, which allowed him to operate in whatever order he wanted. He decided to determine $e\#a$ first. Because e was the evil twin of a, Four knew $e\#a$ had to produce the reflector. The general had explained any time a chip operates with their evil twin, the reflector is the result. This led Four to,

$$\begin{aligned}(e\#a)\#b &= e\#(a\#c)\\ r\#b &= e\#(a\#c).\end{aligned}$$

He knew he was getting somewhere. The reflector reflects.

$$b = e\#(a\#c).$$

Applying the same rules to the right side he found,

$$b = e\#(a\#c) = (e\#a)\#c = r\#c = c.$$

That was the trick! General Three had left out, but also not left out, a key clue in winning his game. If $a\#b = a\#c$ it would allow the same chip to appear more than once in any row or column. The cunning fourth rule put a stop to that. While GT hadn't told them directly, his rules secretly dictated that every column and every row could have one, and only one, of each chip. Four was ready to play.

Round one gave each player eight chips. As it so happens, there are five unique ways to create an operation with eight items. General Three carefully studied the boards, placing candidates into groups who shared the same arrangement. Over half the hopeful Naturals failed to produce a valid board and were automatically eliminated. This included the mouthy

5 and Three took pleasure in personally removing him from the testing arena.

Next, the chips were lowered to six. Only two boards are possible with six chips, thus partitioning the players into ten groups and squashing the many hopefuls who could not form a valid arrangement in time.

#	r	b	g	y
r	r	b	g	y
b	b			
g	g			
y	y			

The head-to-head competition began. Round 3 would only allow for the first players with a unique board to advance. This cut down the numbers from hundreds of players to a mere twenty candidates. Everyone was provided a board and four chips. The witty and innovative four was the first to advance. The chips were red, blue, green, and yellow. He assigned red to be the reflector, named his operation #, and began building, easily completing the first row and column.

His next task was to determine opposites. Since the reflector was their own evil twin, Four recognized he had but two options: either create his board so that every chip was their own twin, or allow two chips to be their own twins and the remaining two would be each other's clones. Deciding the former was quick and easy, he placed his red chips along the main diagonal. Next, Four turned his attention to the second row. Only $b\#g$ and $b\#y$ remained. As he had already used up his blue and red chips for that row, the decision came down to assigning $b\#g = g$ or $b\#g = y$. Immediately noticing allowing $b\#g = g$ would break rule 1 making blue a second reflector, Four quickly completed his board and sprinted to turn it over to General Three.

#	r	b	g	y
r	r	b	g	y
b	b	r	y	g
g	g	y	r	b
y	y	g	b	r

With Four guaranteed a spot in the next round, there was only one opening left for a member of his group.

Eleven was a discarded Natural. Her parents, -1 and -11 opted to stay in TNZ and left her to fend for herself in Rational. It was a hard life. Making matters worse, 11 was prime. Rarely having anything in common with other Naturals, she was always getting herself into one argument or another. Thus, as a precaution, she became a master of the sword, making frequent trips to Whole to duel against zeros.

Her approach was to name her operation +, the sign of the duels. While

+	0	1	2	3
0	0	1	2	3
1	1	2	3	
2	2	3		
3	3			

she was forced to use red, blue, green, and yellow chips, Eleven decided to assign each to a number to recreate a duel. Red was 0, blue 1, green 2, and yellow 3. This allowed her to quickly fill in most of the board.

Now came the tricky part. In a traditional duel $2 + 2 = 4$, but that would break requirement three. Every result had to be a chip she already possessed. Homing in on the second row, she noticed the only chip missing was a red, a 0. This meant she must define $1 + 3 = 0$. Then it came to her. If she forced the duels to work on a circle by spiraling the number line onto itself, she could complete her task.

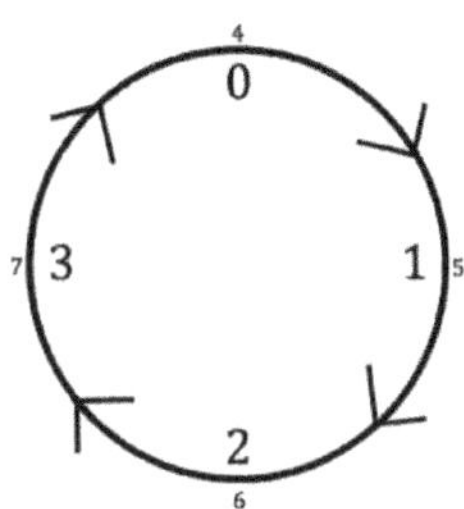

"Nicely done." GT nodded at Eleven before announcing the remainder of their group had been eliminated. Only Four and Eleven would progress having been the first to identify the two possible combinations.*

The twenty remaining players advanced onto the final round. This would leave only ten winners, ten new citizens to be welcomed into Natural.

General Three walked around collecting the colored chips and providing each candidate with three new markers adorning the symbols €, £, and ¥. "Three is a special Numberfolk!" he boasted. "We are powerful and strong thanks to our unwavering structure." Standing at full attention on center stage he concluded, "There is but a single board with three chips. The first from each team to construct this superior arrangement will be admitted through the gates and allowed into Natural! Begin."

+	€	£	¥
€	€	£	¥
£	£	¥	
¥	¥		

The players went to work. Eleven immediately employed her circle technique. She made € represent 0, £ was 1, which left ¥ assigned to 2.

* Four produced what is called the Klein Group V and Eleven constructed $\mathbb{Z}_4$. These are the only two unique groups with four elements. Eleven's method of using the duels in a circular form is known as modular arithmetic.

She knew all too well anyone who dueled with 0 came out unharmed. This forced € to play as the reflector, and Eleven had over half the board completed in seconds.

The middle square £+£, was easy, since she had decided £=1, £+£=2 and Eleven slid a ¥ chip in the center. She saw her path to freedom. Leaving the slums behind, Eleven was ready to enter paradise, ready to enter Natural. While she could determine by adding on her circle that £+¥ would take her back to zero, she instead noticed the second row and second column were missing only the € making the final place a £. Out of the corner of her eye she saw Four stand. Determined, Eleven jumped up and sprinted towards the general, placing her pieces as she ran. Four was close behind, but nothing was going to stop her from winning citizenship.

+	€	£	¥
€	€	£	¥
£	£	¥	€
¥	¥	€	£

General Three carefully studied the board, he grabbed Eleven's arm and pulled it into the air. "I am proud to announce, we have our first new citizen of Natural!"

Eleven joined the nine other victors for an elegant dinner with the mayor before being led to her new home. She took time to touch every surface, investigate every cabinet and drawer, and listen to the deafening silence of peace she had never heard before.

Her whole life, Eleven hated the wall. It was a gigantic reminder that she was somehow, through no fault of her own, less than. But as she snuggled under the soft blankets tucked into her plush new bed, she gained a new perspective. The wall kept out her enemy, it separated her from poverty, and now Eleven loved the wall, for inside the cage she was finally free.

OPERATION: ARE YOU A NATURAL?

The Citizenship Test . . .

General Three's citizenship test represents how we define an algebraic group.

DEFINITION

You are already familiar with operations such as + (addition) and x (multiplication). But you could define many operations, such as a*b=|ab| (an operation which takes the absolute value of the product of two elements), or define a#b as the larger of the two elements so that 8#3=8 and -2#-1=-1.

When you have a really big set, making a rule is the best way to define an operation. But for a small set like {a, b, c, d, e}, it can be easier to create a table that defines your operation.

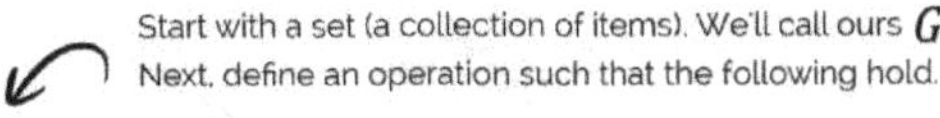

Start with a set (a collection of items). We'll call ours G.
Next, define an operation such that the following hold.

1

IDENTITY ELEMENT

We call this the reflector because . . . it reflects!

Let e be an element of G such that for all elements x in our set we have: $e * x = x * e = x$

Then e is our reflector i.e. the identity element.

In other words, under our operation, when anyone goes against e the result is themselves.

0 is the identity element for addition while 1 is the identity element for multiplication!

INVERSE

We call these evil twins because they are both the same and different.

Every element in your set must have an evil twin (inverse) also in the set.

For all elements a in G there must be an a' in G such that: $a * a' = a' * a = e$

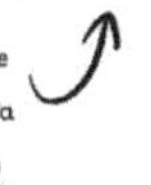

The additive inverse is -a since a+(-a)=0.
The multiplicative inverse is 1/a as (a)(1/a)=1.
Notice the problem? Division doesn't form a group because there is no inverse for 0!

CLOSURE

This just means the result of your operation must be an element of your group. For instance, the Integers are not closed under division. Notice for the Integers 2 and 3, the result of dividing 2÷3 is 2/3, who is not an Integer. . .

ASSOCIATIVITY

Order shouldn't matter. If you take any three elements of your set a, b, c then,

$$(a * b) * c = a * (b * c)$$

Can You Build It?

There are 5 valid operations you can create on a group of 8 items. Can you find all 5?

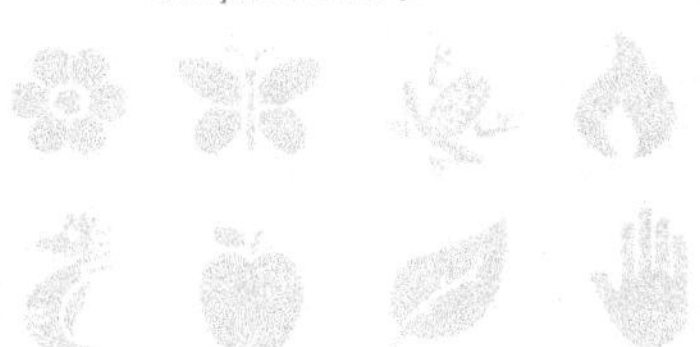

Draw items into the grid to define your operation.

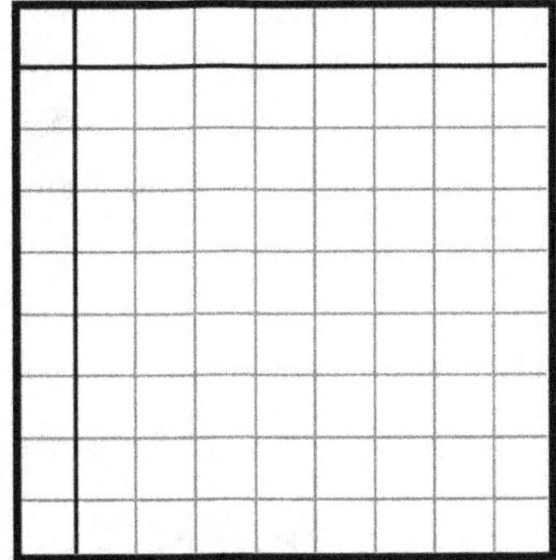

HINT

PICK A REFLECTOR AND FILL UP AS MUCH AS YOU CAN. THEN PAIR UP EVIL TWINS AND FILL UP THOSE SQUARES. FINALLY, USE THE RULES ABOVE TO DETERMINE HOW TO COMPLETE THE REMAINING BLANKS.

CLUES

* 4 digits
* each digit unique

Guess	Digits	Places
1234	2	1
5678	1	0
9140	2	0
4780	2	0
6294	2	1
8013	1	1

two in "1234" and one in "5678" means last digit must be 0 or 9!

5 or 6 or 7 or 8

one digit is 0 or 9, other must be 1 or 4

since 1 or 4 other digit must be 2 or 3

2054

6294 — 2 digits correct

* 1 or 4
* 2 or 3
* 0 or 9
* 5 or 6 or 7 or 8

6 and 2? No 9 ⇒ 0, No 4 ⇒ 1 — ¿6201? ✗ Breaks 4780

6 and 9? No 2 ⇒ 3, No 4 ⇒ 1 — ¿6931? ✗ Breaks 4780

Needs 2 of these digits

6 and 4? No 9 ⇒ 0, No 2 ⇒ 3 — ¿6403? ✗ Breaks 8013 ⟸ can't have 0 and 3

2 and 9? No 4 ⇒ 1, No 6 ⇒ 5, 7, or 8 — ¿291_? STOP No 4 or 0 no matter what will break 4780

2 and 4? No 9 ⇒ 0, No 6 ⇒ 5, 7, or 8 — ¿240_?

* 2405
* 2407 ✗ Breaks 4780 can't have 4, 0, and 7
* 2408 ✗ Breaks 4780 can't have 4, 0, and 8

9 and 4? No 2 ⇒ 3, No 6 ⇒ 5, 7, or 8 — ¿943_?

* 9435 ✗ only 4, needs 7, 8, or 0
* 9437
* 9438 ✗ can't have 8 and 3!

2405 or 9437 — what order?! → !IMPOSSIBLE!

1234: 4 is last

2 0 5 4

8013: 0 is 2nd

5678: 5 can't be 1st!

3

8013 says 3 is last

1234 says 3 can't be last!

MR. PIMRIDGE

"THERE IS STRENGTH IN NUMBERS, BUT ORGANIZING THOSE NUMBERS IS ONE OF THE GREAT CHALLENGES."

-JOHN C. MATHER

Surprisingly, the week flew by and before they knew it, the time had come for the family's annual summer pilgrimage to visit their mother's father. Residing in a nursing home for the last seven years, Marco's grandfather was a powerful man. Tall and slender, but muscular, he commanded any room he entered. While the other residents went by their first names, every attendant knew that he was "Mr. Pimridge" which was generally followed by a "sir."

Marco loved his grandfather. Never knowing his father meant Mr. Pimridge was Marco's only paternal rock, carving out an exceptional place in his grandson's heart. A veteran of WWII, he worked with the Navajo code talkers which ignited a fascination of puzzles that continued throughout his life.° Fond memories of crosswords and riddles danced around Marco's mind on the drive up.

When they arrived, the distinct smell of the nursing home was a repugnant, but familiar, reminder of their location. A mix of a sweet decay and fabric softener filled the air as Maggie and Marco raced down the hall towards their grandfather's room. Bursting through the door first, what Marco saw stopped him in his tracks. The comfortable living room where he remembered sitting on his grandfather's lap as they worked together on cryptograms was unrecognizable. The couch was replaced with a hospital bed and a frail and skinny man laid upon it.

"Marco," a soft voice whispered with more air than sound, "Maggie! Come here right away, I barely recognize you." Mr. Pimridge motioned to them. If Maggie noticed the change, she didn't show it. She pushed Marco aside and jumped up to squeeze her grandfather's neck. He let out a sharp welp of pain which sent Maggie recoiling in fear.

"Maggie doll, Dad is recovering. You know about his surgery. You have to be gentle." Their mother gracefully entered the room and gave her father a quick peck on his forehead. "Need anything?" she asked with a smile.

"No, no. They take good care of me here. Just wish I got to see these little ones more. Thought I saw a ghost when Marco came in, he is nearly grown and a spitting image of Cameron."

"That he is." Marco's mother's eyes glossed over a bit.

° If you are trying to do the math, which I hope you are, Mr. Pimridge was born in 1924. Marco's mother, Maryanne, was born in 1975 when her father was 51 years old. She then welcomed Marco at the age of 35.

$$\frac{\Delta y}{\Delta x} = -\frac{3}{4}, (8,8)$$

Although their grandfather looked fragile, his brain was as sharp as ever. He instructed Marco to retrieve his Go board, and the three of them played for over an hour. Maggie and Marco teamed up and together resisted his advances until ultimately their grandfather captured one piece after another declaring his victory.

The colors around them set with the sun as the room was enveloped in warm and muted hues. A nurse brought their grandfather his dinner on a plastic purple tray. "Probably best to allow Mr. Pimridge to get some rest, he's had an eventful day," she said before tilting her head to the side indicating she'd like to speak to their mother in the hall.

"Want my sherbet?" Mr. Pimridge asked Maggie. She enthusiastically nodded and plopped into an armchair in the far corner of the room scooping down the dessert.

Marco looked out toward the hall. The nurse was the only one talking. His mother hung her head and nodded slowly like a child being admonished by a principal.

"Marco?" Mr. Pimridge called out. He made his way to his grandfather's side. "I have something for you. Over there, top drawer, the box."

Following the directions, Marco pulled out a cedar chest from the dresser. It was small, about six inches wide, four inches deep, and four inches tall. Made from solid wood, it boasted tones of reddish-brown, each streak unique in both color and decoration. The grains swerved and curved in all directions, and if he looked at it just right, he could almost see them move. It reminded Marco of a stereogram: a picture that at first glance looked to be only a mix of patterns and colors but if you stared at it just right and for long enough, an amazing three-dimensional image popped from the flat plane. He carefully handed the chest to his grandfather.

"I received this puzzle box in 1950." Mr. Pimridge let out a cough. Then another.

"Puzzle box?" Marco squinted his eyes and raised his left eyebrow.

His grandfather slid open a small drawer at the base of the chest, an old folded up piece of stationery was the only thing inside.

"Dad," his mother interrupted them. Mr. Pimridge quickly threw his blanket over the box to hide it from view. "I'm going to stay here tonight." She was talking quickly, obviously upset. "I need to get the kids set up in the hotel."

"Yes. Yes." He responded in a stern and firm voice. For just a moment, Marco saw a glimpse of the man his grandfather used to be as he projected strength and authority. "Why don't you take Maggie and get things set up? Marco can stay here with me until you return."

Their mother nodded. Maggie gave her grandfather one more hug. Learning from her mistake, she carefully climbed up next to him and gently squeezed. "See you tomorrow!" Maggie chirped before bouncing along behind her mother, sherbet in hand.

Mr. Pimridge painted a kind smile on his face and kept it there. When the door clicked shut, his expression switched with it. "Hurry, we are running out of time."

Something about his grandfather's tone made Marco feel uneasy.

"This box contains a secret. To ensure its protection, two items were carved from the same tree." He motioned to the cedar chest, "This, and a clock. Long ago, I was entrusted with the box and," his grandfather hesitated for a moment, "another was given the clock. Splitting the items was a safety measure meant to protect its contents from the dangerous people who seek it."

Dangerous? Marco thought. *What could possibly be inside?*

"The box contains eight locks. Only those who can decipher all eight will be granted access to the treasure within." As his grandfather unfolded the note, Marco caught a glimpse of the small dials inside the drawer. There were four, each resembling a clock face but rather than twelve marks, they contained only ten.

"A unique digit." His grandfather answered his unspoken question. "Turning each to their correct placement will trigger a pin." Holding up the paper he continued, "My own foolish addition." Mr. Pimridge began to cough. Then again and again. Marco thought it would never stop. Starting to dismount the bed to locate a nurse, his grandfather firmly clasped onto his arm. "I'm fine." Another cough.

"I first stumbled upon the pair with a friend, a brother. We didn't understand and as pollyannaish young boys, all we saw was a puzzle to solve and a secret we desperately wanted to uncover. I wish we hadn't." He lowered his voice. "With the clock in hand we were able to discern how the two worked together. The clock would "tick" when we had a correct digit, and "tock" when a digit was in the correct place. That was all we needed."

His hand trembled as he held out the stationery like a runner ready to pass the baton. Marco carefully accepted, but Mr. Pimridge's grip

$$\frac{\Delta y}{\Delta x} = -1, (8,2)$$

remained firm. "Marco. Son. I am passing not only the secret, but the protection of the secret, to you. You must guard this with your life. I promised to burn this, but never did. I couldn't bring myself to do it. It's yours now."

Like a game of tug-of-war when one side suddenly released the rope sending the opposing team flying, his grandfather relinquished the paper. It was so light and yet carried a heavy responsibility that Marco could feel in his hand. Slowly unfolding the creases, he read through the first paragraph of text.

First Chamber

1234 Ticks: 1 Tocks: 0

7890 Ticks: 2 Tocks: 1

1598 Ticks: 4 Tocks: 1

A tsunami of questions flooded Marco's mind. Unsure where to begin, he looked at his grandfather ready to burst. He stopped. Mr. Pimridge lay peacefully on the bed, his eyes closed. Sharing his story had drained every last drop of strength and he had nodded off to sleep.

Marco slowly dismounted the bed so as to not disturb him. He found some scrap paper on the desk and plopped down in the armchair.

Ticks are digits, tocks are places. Marco thought. *The last code had four ticks, so the digits must be 1, 5, 9, and 8 in some order. But what order?* He began to study the previous clues. *Since 1234 had no tocks, the 1 couldn't go first.* He drew out four lines on his paper and placed the digits 1, 5, 9, 8 under each. He crossed the 1 from the first blank.

____	____	____	____
~~1~~ 5	1 5	1 5	1 5
9 8	9 8	9 8	9 8

The two ticks from the second clue have to be the 8 and the 9. There is only one tock, which means the 8 must be second or the 9 third.

Marco rubbed his eyes. It had been a long day, and this wasn't the easiest of puzzles.

Okay. Get it together. If I guess the 8 is in the right place in the second code, the 9 can't go third because that would be two tocks, so 9 must be first or last. But then what digit is in the right place in 1598? It can't be the 1, that isn't first, it can't be the 8, that's second. It can't be the 5, because it's where the 8 goes, and it can't be the 9 either. That's everything. It doesn't work.

$$\frac{\Delta y}{\Delta x} = 1, (8,2)$$

"The 8 isn't right in the second code," Marco said aloud. He looked at his grandfather making sure he hadn't woken him.

The 9 must be in the right spot in the second clue which makes it right in the third too. Marco scribbled the 9 in the third blank and crossed it from under the others.

What about the rest? Well, 8 can't be second or there would be two tocks for clue two, and 8 can't be last, because there is only one tock for clue three. That means the 8 must be first! He scribbled the 8 into the first blank.

8 ___ 9 ___

Wait. This is impossible. There is no way to know if the code is 8591 or 8195. He meticulously studied the page. An urge to just pick one and try it swelled through his body. He hesitated.

What if an incorrect code locks me out? He stared waiting for the answers to jump up and hit him over the head. They did. *The 5 can't be second! If it was second, the final clue would have two tocks, not just one. The code must be 8195!*

Marco slowly slid open the drawer. He carefully twisted the first dial to the eighth notch. Nothing happened. He wondered if he should wait, talk with his grandfather more when he woke up. He couldn't. He tried. The five minutes felt like five hours. He had to know if his code was right. Turning the second dial to 1 and the third to 9, he inhaled sharply before rotating the final dial to the fifth notch.

The box clicked.

Marco jumped.

Not knowing what to expect, the noise startled him and sent a lightning bolt of excitement through his veins. He was ecstatic. In his head, he jumped up and danced around the room, he ran around in circles and shook his grandfather awake to share what he had done. In reality, he didn't even flinch.

The apartment was quiet and dark. Everyone in the entire nursing home must've been asleep. Marco began to feel lightheaded and realized he had been holding his breath, too scared to make a peep. With a sharp inhalation, his eyes scanned to the next clue. It was much longer than the first.

Second Chamber

1234 Ticks: 2 Tocks: 0

5678 Ticks: 1 Tocks: 0

$$\frac{\Delta y}{\Delta x} = -\frac{4}{3}, \left(8, -\frac{2}{3}\right)$$

9012 Ticks: 2 Tocks: 1

2356 Ticks: 3 Tocks: 1

5302 Ticks: 3 Tocks: 3

The dead silence was broken by the sharp sound of heels on the hard tile floor. Marco quickly turned the dials back to 0000 hiding his progress. He stuffed the paper inside the box and closed the drawer just as his mother tiptoed into the room.

Grabbing his jacket from the desk, in one smooth and fluid motion, he threw it over his hand that held the cedar chest. He hoped his coordination, with the help of the lack of light, was enough to obscure his grandfather's gift from his mother.

Once in the hall, she spoke swiftly and softly. "Thanks Marco. I set your sister up in the hotel across the street, room 312. I ordered pizza for the two of you. I'm going to stay here with Dad. I'll see you in the morning." Marco could tell she wasn't interested in a conversation. Thankfully, he wasn't either. He gave her a quick, one-armed hug and hurried toward the exit.

The air outside was brisk. He wasn't expecting the cold to hit him in the middle of the summer. The nursing home was in the mountains north of their house. The thin air of the higher altitude was supposedly good for people like his grandfather. He remembered his mother describing *erythropoiesis*. She explained that higher altitudes caused the body to absorb less oxygen, which seemed like a bad thing. But, it actually also caused the body to produce more red blood cells to compensate, which was a good thing.

The mountain air made Marco disoriented. Shadows bounced along and created abstract shapes. He was nauseous. Worse, he felt paranoid. The box was surprisingly heavy for its size and its weight mirrored the weight on his shoulders. He had to protect this secret. Unfortunately, he had no clue what this confidential information even was.

Walking quickly to the hotel, by the time the double glass doors slid open he was out of breath.

"So much for erythropoiesis," he mumbled to himself.

When he arrived at room 312, Maggie had already dug into the pizza and claimed her bed. She was watching *Life on our Planet* on Netflix and didn't even acknowledge his existence. Marco saw his bag still on the luggage carrier in the entryway. Grabbing it, he made sure to keep the box hidden beneath his jacket as he disappeared into the bathroom.

He tugged out a loose t-shirt and a pair of shorts and changed. Carefully wrapping his dirty clothes around the cedar chest, he buried it in his bag before reappearing.

"Hey Bug." He tried as hard as possible to sound casual, but it came out awkward and forced.

"You won't *believe* what happened while you were gone," Maggie started in. "The pizza got here, and I was so excited, I grabbed a slice right away. It was COLD!" She paused for a moment, her face frozen, eyes wide, and mouth in the shape of an O. "I called them right up and they put me on hold. The voice kept saying 'you are caller number 4'. I waited and waited. She kept saying 'you are caller number 4', 'you are caller number 4'. Seriously?!? I never got to be caller number 3. I mean, what kind of service is this? I'm not just a number. I am a PAYING CUSTOMER!"

"But are you tho?" Marco asked, tilting his head to the side.

"Whatever. Enjoy your ice pizza." She rolled her eyes before fixing them back on the TV.

Marco grabbed a slice. He cringed as he realized Maggie was right. The pizza tasted like it had been sitting out all day. He threw it back in the box and crashed onto the bed. His hands cradled his head as he picked a kernel on the popcorn ceiling to stare at as he attempted to process what had been a strange and exciting day.

When her episode ended, his sister began to unpack her things. She had developed some weird obsession with hotel rooms after she read a study that claimed most housekeepers used the same rag to clean the toilet as they did the kitchenware. She dramatically threw the hotel blankets to the ground before remaking her bed with the sheets she packed from home. Her last step was to collect her small gray stuffed hippo with turquoise ears she'd had since she was a baby. Hugging it tightly, she snuggled in.

He hadn't realized it before, but in that moment, he noticed how young Maggie seemed. Only a year behind him, she managed to hold on to an innocence, a pure view of the world that Marco had lost somewhere along the way.

His thoughts began to pick up speed as he fell down a spiral hole of possibilities and the familiar game of table tennis formed in his mind.

$$\frac{\Delta y}{\Delta x} = -1, (8, -1)$$

On the one hand, he was sure he should tell Maggie about the cedar chest. She'd certainly come in handy solving the codes and he'd learned his lesson about keeping secrets from her. On the other hand, his grandfather had specifically mentioned danger and had their mother take Maggie away. No matter how much Maggie insisted she didn't need Marco's protection, as her big brother he felt a sense of duty to preserve her innocence and keep her safe.

Pros and Cons served back and forth, back and forth. Just when one team gained a point, the other team quickly struck back to tie the score. By the time he dozed off to sleep, he still hadn't made up his mind.

* * *

The bright rays of sunshine peeked through a hole in the curtains. Rubbing his eyes, Marco was surprised to see his mother curled up on the hotel couch scrolling through her phone.

"I thought you were spending the night with Granddad?" Marco mumbled, his body not awake enough to clearly form words.

As his mother looked up and he saw the red puffiness that surrounded her eyes, a stabbing feeling punched him in the chest.

"What happened." It wasn't even a question. He knew what happened. Throwing the blankets aside, he ran to her hoping with every cell in his body he was wrong.

"Dad passed away last night. In his sleep. It was peaceful." She sniffled while simultaneously pushing a soft smile onto her face. Her chin trembled.

Marco clenched on tight to his mother and didn't let go for a long time. The way she held herself up made him think she had time to prepare, she knew this was coming. Either that or she was remarkably strong. Perhaps it was both.

When Maggie awoke, there were a lot of tears, a lot of silence, a lot of hugs. They say there are seven stages of grief: shock, denial, guilt, anger, bargaining, sadness, and acceptance. To Marco, they didn't seem like stages at all but instead, a horrible hurricane that threw every emotion directly at him with varying speeds.

Shock was the shortest and most tolerable. It was quickly replaced by an all-encompassing sadness. It was like a feeling of dread, an ache in his stomach, and a sense of fear wrapped into one fastball to the gut.

45

In a scary movie, the apprehension and terror build like a roller-coaster slowly inching towards the sky until the bad thing jumps out. This bad thing was different. It had already come and gone but somehow managed to leave behind a lingering sense of alarm and panic. It spewed a wretched atmosphere that was ready to swallow you whole at any given moment. And every so often, when things became quiet and started to settle, a surrealness washed over like a dream. This was when denial took hold and, for the blink of an eye, he could convince himself it didn't really happen before the heavy gravity of the sadness jerked Marco back down into the endless black hole.

The anger was a constant, although it was a strange and sideways wrath. Marco didn't have anything or anyone to be mad *at*. Direction seemed crucial for this feeling and without it, the emotion was a muddled burst of energy that forced more tears to fall and catapulted the sadness back into the mix. At one point, he felt like he couldn't breathe, like he had somehow died too.

Marco skipped guilt and bargaining all together. They were probably there but the hopelessness and weakness and strength of the anger-sadness fighting match caused them to get lost in the mix.

Acceptance wasn't something that *came*. He had accepted the news the moment it left his mother's mouth. Just because he knew it was true didn't make him feel any better. Death felt horrible. It felt much worse for the living than it did for the departed. Then suddenly, it felt less.

As the day went on and the hours passed, the emotional storm dwindled. The blue sky and sunshine didn't return. It was foggy. Marco was in a daze. He felt numb. There was no more energy to fuel his grief.

By the following morning, the murkiness had lifted leaving only the large dark clouds glooming overhead. The sadness stayed though. It never left. It wouldn't ever leave. It was a mark on his heart like a stain on a beloved t-shirt. It would sit in the closet, or the drawer, pushed from the mind until he had an urge to wear it. When he went to put it on, he'd see the blemish, and the hurricane would return. Its strength would vary, but it would always return.

Other times, he'd look at that shirt and somehow it would bring joy. He'd remember all the amazing times he wore it, the adventures they had together, how lucky he was to even have it in the first place.

The stain was terrible, but it was only one small part of their story. His grandfather helped shape him into the person he was and the person who he'd become. Part of Mr. Pimridge would live on in Marco forever.

You must guard this with your life. It's yours now.

While he had trouble reconciling the idea that he'd never speak to his grandfather again, Marco would do anything to ensure his last wish was upheld. That conversation was precious. It was sacred. But how could he protect *it* if he didn't even know what *it* was. He needed to know what lay inside the cedar chest.

It wasn't until the drive home as the trees glided by just outside the window forming a blurry landscape that danced to the beat of his sister's jagged sobs that Marco realized he had made his decision. He wasn't bringing Maggie into this.

depends on soldier

||LOCATION||

(x, y)

~ORDER~

$$y = f(x)$$

(x, y)

SOLDIER

independent

"y is a function of x"

SUBSTITUTE

to find where a soldier was sent!!

ORDERS

order: $f(x) = 3x + 2$

$f(1) = 3(1) + 2 = 3 + 2 = 5$

soldier 1 was sent to Station 5

$f(-1) = 3(-1) + 2 = -3 + 2 = -1$

soldier -1 was sent to Station -1

ORDERS ON ORDERS

$f(x) = 3x + 2$ and $g(x) = x - 3$

$$\begin{aligned} f(g(x)) &= 3g(x) + 2 \\ &= 3(x-3) + 2 \\ &= 3x - 9 + 2 \\ &= 3x - 7 \end{aligned}$$

$$\begin{aligned} g(f(x)) &= f(x) - 3 \\ &= (3x+2) - 3 \\ &= 3x - 1 \end{aligned}$$

Not the same!

It matters what bus (not train) you take first

$$f(m) = mg$$

soldier changes

placeholder

gravity is constant (doesn't change) based on location.

Gravity on moon is different than on Earth

LINE UP

"WHAT IS MATHEMATICS?
IT IS ONLY A SYSTEMATIC
EFFORT OF SOLVING
PUZZLES POSED BY
NATURE."

-SHAKUNTALA DEVI

Oceanside pool was filled to the brim. The record-high heats had caused the already busy watering hole to become a local bath where everyone gathered in hopes of cooling off. It was impossible to swim in the pool. The most you could do was wade around groups of people like a lane-expert in a traffic jam.

While Maggie splashed around in the deep-end, Marco decided to cool off in the grass beneath the shade of a large willow tree. Laying on his back, he stared up at the crisp sky. If he didn't try to look at anything in particular, little, tiny white lines began to flood his field of vision. It was like the static he had seen when his English teacher played them an old movie using a VCR that came into the room on a rollout tray with a tiny TV attached to it. For a second, he questioned if reality was real. Maybe he was actually stuck inside some movie and there were kids out there watching his every move. He shuddered.

"Marco!" their mother called out.

"Polo!" Maggie replied. This was an annoying ritual she picked up and refused to let go.

"It's time to head home," she shouted over the low buzz of dozens of different conversations.

Another week had passed. KFUN meetings had been put on hold. Liam and his parents brought over a casserole and Maggie's father, Peter, had been helping to take care of all the arrangements. Mr. Pikake had sent the children a kind note and a basket of goodies claiming it always helped him to "eat" his grief. After a few days, everything went back to normal. The phone stopped ringing as much, the tears came less often, and life went on.

It was then, when the sadness quieted, that the guilt began to plague Marco.

How could they live as if nothing had happened? Something *had* happened. Something devastating. Their mother encouraged them to play, "Dad wouldn't want you to be sad. He lived an amazing and long life. He'd want you to be happy, to live." Even her permission didn't change the dreadful feeling that haunted him.

Marco headed to the locker room to change out of his swimsuit. He was about to leave when a firm hand clenched his wrist and pulled him into an empty aisle.

"We don't have much time," a voice whispered from behind.

The words reminded Marco of one of the last things his grandfather had said to him which caused a glimmer of hope to blossom in his throat. Apparently, denial was like a stealth ninja with nunchucks in hand that hung around causing problems and was ready to strike at any moment.

Marco spun with excitement which was instantly sucked out of the room when he saw the man. It was not his grandfather. Tall and slender, older, maybe in his seventies, the stranger wore straight-legged black slacks and a black turtleneck. His salt-and-pepper hair gave off mad-scientist vibes. It stuck out in every direction.

"I know what you have. I can help you," the man continued in a low tone.

Not knowing how to respond, Marco just stared. The stranger looked around nervously before handing Marco a picture. At first, he thought it was a photo of himself, but it couldn't be. His mouth dropped open as he studied the image. The old polaroid showed three men standing in front of a large tree. The man in the middle was clearly his father, and to his left, Marco saw a younger version of his grandfather. He looked up. The third man, to the right of his father, was the very man who was standing in front of him.

He knew my father. He knew my grandfather, Marco thought.

The mysterious man snatched back the photograph and replaced it with a torn sheet of paper, it contained an address and a time. When Marco looked up, the man was gone. He would have thought he imagined the whole thing if not for the note. He held it tightly in his hand as if it was his only grip on reality.

The second Marco arrived home, he dashed up the stairs to his room. Slamming the door behind him, he dove to the floor. On his stomach, he shimmied under his bed and reached up into a small slit in his box spring. Blindly digging, he let out a huge sigh of relief when he felt it. The cedar chest was still there, right where he left it, in his best hiding spot.

He wasn't sure why, but the man at the pool gave him a terrible feeling. He hadn't said anything about the box, but somehow Marco knew in his gut that's what he was after.

There was really only one question to answer. Was this stranger friend or foe?

* * *

"Magggggieeee!" Marco shouted. It was time to head to the library for the first KFUN meeting they'd had in over two weeks. Marco was excited. Mr. Pikake filled a hole to become a father-figure, and he was in desperate need of paternal compassion. He opened his mouth to yell again, this time right in her face, but what he saw stopped him in his tracks and left his jaw swinging on the hinge.

"This is my friend Penelope," Maggie cheerily declared.

Marco couldn't speak. Her bright white teeth glistened as they contrasted against her perfectly dark complexion. Like tempered chocolate, it was smooth and shiny without a blemish in sight. Her arms were crossed nervously. She slowly raised her hand in a half-wave before shifting her gaze back to the floor.

"H-H-Hi," Marco stuttered.

"She'll be joining us today!" Maggie bounced.

He didn't protest. He didn't pull his sister to the side and demand they leave Penelope behind. He simply nodded and followed the girls down the stairs like a zombie whose brain had suddenly stopped working.

The three met up with Oliver and Liam outside the library. Maggie introduced Penelope, and the boys exchanged surprised glances. Marco still hadn't said a word. Coming out of his daze, he finally spoke up.

"Oh! I . . . *er* . . . forgot. Mr. Pikake asked us to meet him somewhere else today." He dug his phone out of his pocket and showed the text to his friends.

"I know where that is!" Oliver stated. "Three blocks up and two streets over."

"Oliver and Liam led the way while Marco stayed in step with his little sister and her friend. "Penelope," Maggie overpronounced each syllable, "last time we learned about a fly. How could you describe where a fly is on the ceiling? It was basically about the Coordinate Plane which, of course, I already knew about. He did say the x-coordinate is called the abscissa which obvi is so much cooler."

Marco wanted desperately to speak. Maggie didn't pause long enough to take a breath. He had no idea what he would say anyway.

"Actually, Oliver messed up the directions. He should have said two streets *over* and three blocks up since horizontal comes first."

Oliver looked back and gave Maggie a squint and a headshake before returning to his conversation with Liam.

"Then, we learned about relations and functions. A relation is a set of ordered pairs, like coordinates. The domain is all the x-coordinates and the range is all the y-coordinates. And to make a relation into a function, all the x-coordinates have to map to exactly one y-

coordinate. Like you can't have (1, 2) and (1, 3). That's not a function because the 1 could choose to go to 2 or to go to 3. It can't be in two places at the same time."

"Did you learn about one-to-one and onto functions?" This was the most Marco had heard Penelope speak. Her voice was like an angel's. It had a soft melodic flow that almost sounded like she was singing her words.

Maggie turned to her brother and beamed, "She's really smart."

Marco could see why Penelope was friends with his sister. They were opposites. Penelope was quiet, reserved, which allowed Maggie to cheerily talk indefinitely without interruption. But she was also bright, able to keep up with, if not exceed, Maggie's intellect.

The group arrived at a strip mall; a series of shops lined up one right next to the other overlooking a sea of parking spaces.

"It's that one." Oliver pointed to the store front that had no sign or indication of what lay inside. Marco stepped out into the lead and slowly pushed open the glass door. A bell located inside the shop sprung to life as they entered.

"Welcome!" Mr. Pikake rushed from a back room. The entryway was dark. A soiled blue industrial carpet covered the floor. The empty reception desk and some waiting chairs gave no indication as to the purpose of this place. The tutor ushered them down the hall and into the first room on the left.

Marco's eyes adjusted to the light. It was a dance studio. Thin wooden planks lined the floor of the long rectangular room. The wall that contained the door was covered entirely with an enormous mirror while the remaining three walls were empty except for a wooden railing about four feet off the ground which circled the room.

"What are we doing here?" Marco groaned.

"I require substantial space," Mr. Pikake retorted in an upbeat tone turning to face the group. Like the Mayor of Halloween Town, as the professor's head swiveled his kind smile violently switched to one of shock.[°] "And who is this?" he asked wearily.

"This is my *best* friend, Penelope. She is now an official member of KFUN!" Maggie radiated with pride.

[°] IYKYK, if not, Google it. Actually, did you know that Google was originally called "Back Rub"? Let's all imagine how much better the world would be if we instead back rubbed everything.

"Oh, *um*, welcome Penelope." The professor nervously shifted his tie. "Well, let us begin. We last spoke of relations. Suppose I have a machine that augments the abscissa."

"*Super soldiers*," Liam nudged Marco, his eyes wide.

"A Numberfolk factory!" Marco exclaimed. He remembered this well.

After a kind nod and quick smile, the tutor continued. "If such a factory doubled what went in, how could we express this change?"

Marco saw the numbers lined up on one side of the factory. As 1 went in, the building churned to life squeaking and popping, before a 2 emerged on the other side. It flexed its muscles. When 3 went into the doubling factory, out came a 6.

"A 1 would change to 2, a 2 changes to 4, 3 to 6, 4 to 8, and so on and so on," he replied.

Mr. Pikake transcribed the ideas with a marker on the mirror.

$$1 \to 2, \qquad 2 \to 4, \qquad 3 \to 6, \qquad 4 \to 8, \ldots$$

Maggie turned to Penelope, "Can he do that?" she whispered. "What if it doesn't come off?"

"And what if we don't know *who* went in?" Mr. Pikake posed a new question.

A ninja approached the factory. Dressed in all black, it flipped and dove as it stealthily entered the building. Before Marco's daydream could even fully develop, he was rudely awakened by his sister's screeching voice.

"A mask! We need a mask!" Maggie shouted. "x goes in, $2x$ comes out."

$$x \to 2x.$$

Marco shifted uncomfortably. He understood that a variable was just a Numberfolk hiding, but hearing his sister use the term *mask* still didn't sit right.

"Remarkable! You have created a relation. We say x is related to y if $y = 2x$. Now, is $(2, 8)$ in this family?"

"No way," Liam said assuredly. "If we push in 2, they should go to 4, not to 8."

"Good! So, we have a way to tell if a pair is related. We simply ask, do they follow the rule?" The professor proclaimed, "But! Is this a functional family?"

Penelope slowly raised her hand. Everyone stared. This wasn't that type of classroom. After realizing no one was going to "call" on her, she began to speak. "I think that $y = 2x$ is a function."

Everyone waited for her to continue, but she didn't add anything more. Maggie jumped in like it was her job to rescue her friend. "Yeah, because everyone is assigned to only one place. If a 2 goes in, they have to come out a 4, not anything else."

"Distinguished deduction!" Mr. Pikake exclaimed. "If a relation is a function, we can write it in such a way that is considerably less cumbersome." He scribbled on the mirror:

$$y = f(x).$$

"This simply states that the outcome, ordinate, second option in an ordered pair, is in fact a function of who went in. The location depends on the number receiving the order. We say 'y is a function of x'."

"But I thought you said y equaled $2x$. How can it be two things?" Liam broke in.

"Imagine this!" Mr. Pikake's energy was growing and could be felt in the room. He jotted on the board,

$$y = 3kx - 2x + 4kt.$$

"*Who* is the soldier in this order?"

He spun to see a wall of blank faces invading his view. No one answered. There were too many masks. The soldier could be k or x or even t. Maybe they were all soldiers. Marco had no way of knowing. Luckily, the question appeared to be rhetorical as their tutor continued.

"In a sea of masks, knowing whom the order is for is paramount."

"Yeah but . . ." Maggie started. She caught herself. This was an old habit she had been working hard to break. Her argumentative nature had caused too many of her sentences to begin with this phrase.

Her mother and Peter would give her a hard time and "yabbit" had become a sort of inside joke. Only Marco noticed, and the tug-of-war with his face began as the smile crept upward despite trying desperately to push it back down. Maggie made an "ugh" sound while rolling her eyes before starting over. "I *mean*, doesn't this order just apply to multiple soldiers? Can't it be a command for all of them?"

"Insightful Maggie," Mr. Pikake responded inquisitively. "It could. That is called a multivariable function but," he popped, "we are not quite ready to explore those complex commands. Suppose we know this is only an order for one number . . ." his voice trailed off. Marco waited for the "folk" to burst out, but it never came. He wondered if this was a result of the new addition of Penelope to the group.

"If it's only an order for one soldier, what are the other letters for?" Liam asked.

"A modest mask," the tutor answered. "Purely placeholders for the unknown."

"He's saying that the other letters are just numbers we don't know. But whoever they are, they aren't changing. The one we care about is the one that can change, so it can be given to different soldiers," Oliver explained.

"Precisely!" Mr. Pikake jumped. "Change is the catalyst of this class! In this instance, $3kx - 2x + 4kt$, two of the letters are fixed, unmovable, the third can change. It is the fickle foot soldier. In this case, the order includes values for the friction of the battlefield as well as the length of the field. On a football field it becomes . . ."

$$y = 3\left(\frac{7}{20}\right)(120) + 2(120) + 4\left(\frac{7}{20}\right)t.$$

"And in a parking lot it is,"

$$y = 3\left(\frac{1}{2}\right)(60) + 2(60) + 4\left(\frac{1}{2}\right)t.$$

He finished his sentences on the mirror.

"Can you see it now? Who is the soldier?"

"It's t. That's the only one left!" Marco burst out.

"Exactly! The letters x and k were holders for the specifics of the playing field. Once that is known, those are locked in, they don't change. Now we know the details of the battlefield, we have a functional order and can send in our different soldiers. Each will be sent to a specific location."

"I literally do not see how this answers my question at all," Liam responded confused.

"Because they are not two things!" Mr. Pikake's voice was high and energetic. "The y, is the ordinate, where the soldier is ordered to go. But!" A dramatic high-pitched pop, "The f of x, or t, or k, tells us what mask the soldier wears, who the order is for. And!" One more shriek, "the $2x$ or $3kx - 2x + 4kt$, or whatever it may be, is the order. Three parts, all crucial for a commander."

$$y = f(t) = 3kx - 2x + 4kt.$$

He drew a dramatic underline-squiggle beneath the equation. "You see, in this form, we now know the soldier. In situations that contain other letters, we know those masks are constants, they will not change, only the soldier we are sending in will change."

Liam imagined he was in a spaceship fighting an intergalactic war. The $3kx - 2x + 4kt$ flashed on the screen, written in futuristic hieroglyphs. The other fighters were screaming in his ear, "We need the coordinates of the jump!" He worked quickly. He had to be precise. One wrong move and a ship could land inside a mountain, or worse, two ships occupying the same place.

He calculated the values for k and x based on their position in the galaxy. It was too complex! He had to simplify the situation.

$$\begin{aligned} y &= 3\left(\frac{1}{2}\right)(60) + 2(60) + 4\left(\frac{1}{2}\right)t \\ &= 3(30) + 120 + 2t \\ &= 210 + 2t. \end{aligned}$$

"Starship 1, set your coordinates to 2.1.2." He yelled out as an enemy ship fired. Liam swerved just in time, but the missile hit his wing. The explosion caused his screen to flicker. "Starship 2, 2.1.4!" He shrieked. "Starship 3, 2.1.6." He was the only one left, Starship 10. Setting his own device to 2.3.0, a sharp jerk sent him into hyperspeed.

When Marco was lost in a daydream, to the outside world it looked as if he was simply thinking deeply. Liam, on the other hand, made all sorts of faces. His body moved like he was on a ride at an amusement park as he swayed back and forth dodging air.

A look of pride stretched onto Liam's face as he had successfully navigated his fleet out of danger. He opened his eyes to see everyone staring at him.

"You with us, bruh?" Oliver asked.

"Oh, *um*, yeah. I get it now," Liam blushed. "In a function, writing $f(x)$ tells us x is the value that can change depending on the number we input. If there are other letters in the equation, these represent constants, things that don't change once you know the situation. Like if I was sending people into space, I'd need to calculate their force, $f(m) = mg$. The m is the soldier and the g is the gravity which will change depending on the planet I am sending them to. Once I know the planet, I'll know the gravity and I can plug that in for g. What's left is an order for my m-soldiers. On Earth it's $y = 9.8m$, on the moon it would be $y = 1.62m$. The y is the output, the result of evaluating at the given number."

Everyone continued to stare but their expressions shifted from confusion to shock.

"Very good! Functions model change. They provide us an order which depends on the soldier. Using $f(x)$ tells us the x is what is changing. If it is $g(k)$, who is changing?"

"k!" Maggie shouted as if she wanted to make sure she was the one who answered first. "But what's with the g?"

"Remember, the letter is just a name. You can call it whatever you want."

"Like, pancake of x?" Marco giggled and then instantly felt like he made a fool of himself in front of Penelope.

"I'm more of a waffle person myself," Mr. Pikake grinned. "In the command y equals pancake of x, there are only two values that can change. The soldier we send in, in this case x, and the station they are sent to, y."

"I personally think it should be pancake of syrup. The syrup can change everything." Liam's eyes glazed over as he imagined breakfast foods.

Marco cringed. The pancake comment was taking on a life of its own and all he wanted was to dig a hole to hide in.

The tutor snatched a cloth from his breast pocket like a showman. He turned and wiped off the mirror.

"Oh," Maggie whispered, "it does come off."

$$y = \text{pancake}(\text{syrup}).$$

Marco groaned. He'd never live this down.

"We send in the syrup and the result is a pancake of various levels of deliciousness. One of these depends on the other."

"Obviously the deliciousness," Liam started. "The deliciousness depends on the syrup. If you use that cheap, high fructose, processed stuff, the quality of the pancake instantly diminishes."

"Exactly!" Mr. Pikake replaced the pancake-syrup mixture with the traditional notation. Marco let out a sigh of relief.

$$y = f(x).$$

"We call the x independent because any soldier in the domain, in

$$1 \to \frac{14}{3}, 2 \to \frac{10}{3}, \dots$$

your dominion, can be given this order. But," the professor popped, "the y is dependent because where the soldier goes *depends* on the soldier given the order. Or the scrumptiousness of the pancake, if you prefer." He winked at Marco. Marco died.

The professor suddenly started clapping to a slow regular beat. "On your feet! Boys, assist me." He pulled out large rolls of tape from his bag and handed them out. Liam, Oliver, and Marco followed his directions to create a grid on the dance floor. "Line up!" His voice was like a rollercoaster, his tone flew down before rising back to a high pitch.

Obeying, everyone took a spot on the grid. "The first and most obvious command is to order the numbers to line up, as I just did. What an interesting line you created!"

They looked around. Standing parallel to the mirror, they all picked spots in the room where the lines of tape intersected.

"Marco! You are at the origin. Oliver, where does that place you? What soldier are you?" Mr. Pikake called out.

"I'm one, two, three, four lines from Marco, and to the right. So, I'm at (4,0). I guess I am soldier four?"

"That you are boy. That you are! Maggie! What about you?"

"I'm right next to Marco. Well, left next to Marco." She looked over at Penelope and the two chuckled. "So I am −1. I'm at (−1,0)."

"Excellent. Now what is your command? Liam, you wish to take an endeavor on the enquiry?"

"Well, the y is the location, and we are all at station zero. So, I guess $y = 0$?"

The professor leapt to the front of the room, careful to avoid the tape, he hopped from one square to the next wobbling as he went.

$$y = f(x) = 0.$$

He jotted on the board, making sure to pretentiously punctuate like a bird pecking at a tree. "Here is your order. A simple horizontal line. Everyone take one step backward, what is your new order?"

Oliver was the first to speak up, "$y = -1$. Now all of us are lined up at location −1."

"Perfection! Anytime our goal is to make a straight horizontal line, we are sending every soldier to the same station and thus the order is just the station we wish for them to travel to."

"Is that even a function?" Marco asked. He dug through his brain to remember. He knew there was something important about ambushes.

"Well, all of us have only one order: go to -1." Maggie answered, "So I think it is a functional order?"

"Yeah, it just isn't one-to-one." Penelope looked at Marco. "A one-to-one function is when everyone is ordered to their own station. But here, we all went to the same location." Marco was surprised how quickly Penelope was assimilating into their world.

Mr. Pikake showed his approval in his expression, "Nicely said! Now, a new way to receive orders!" He drew a long vertical line on the mirror followed by short bursts of horizontal dashes. Quickly filling in the blanks, he revealed the command.

x	y
−2	−6
−1	−3
0	0
2	6
4	12

Everyone began shuffling around the room. Oliver ended up pressed against the mirror while Marco needed to only take one step forward, back to the origin. They looked around at the new line they had created. It was a diagonal.

"Very good! Very good!" Mr. Pikake chuckled happily. "Can anyone determine the order we would give to achieve this formation?"

Marco's mind went to work. He couldn't help but be reminded of his grandfather's puzzle box. Since the girls to his left were closer, more squished, than the boys on his right, there was missing information he needed to fill in. He drew out seven lines in his head and immediately sent the Numberfolk he knew to their positions.

−6   −3   0   ___   6   ___   12

All he needed to do was to find the missing values. He instantly saw the group on the left were following a pattern.

If the pattern held, the first blank should hold a 3, the 6 would be correct, then would come 9 and 12. *Yes!* He knew this had to be right. The pattern was nothing more than skip counting by threes, which was of course multiplication. "The order is $y = 3x$!" he shouted in excitement.

"Dynamic deduction!" the professor began. "Each subsequent soldier's station is three more than their comrade to the left. What then is f of 10?" He scribbled on the mirror,

$$f(10) = ?$$

Substitution. Marco thought. One of the most powerful and most underrated tools of a Saint. Before he could even start to figure it out, his sister chimed in.

"It would be 30 right? Because x is the soldier, you want to know where 10 would be sent? And since the pattern is $3x$, that means 10 goes to three times ten which is 30."

Marco's shoulders dropped. He longed for the days of private sessions.

"Perfection! You see the power? If we can determine the order, we can then find where any soldier has been told to trek. This is nothing more than substitution — a formidable faculty." He threw his body against the mirror and slid. Marco's smile pushed hard, it was difficult to watch without laughing out loud. Having successfully erased everything on the mirror, he replaced it with a single equation.

$$f(x) = 4x - 2.$$

"What is f of 0?" He called out pointing the marker at Penelope.

"-2!" She called back.

$$f(0) = 4(0) - 2 = 0 - 2 = -2.$$

"Superb! And f of 5?" This time he aimed his wand at Maggie.

"18!" She announced proudly.

$$f(5) = 4(5) - 2 = 20 - 2 = 18.$$

Without responding he twirled to point at Marco, "f of -1 son?" Marco's chest swelled. Hearing Mr. Pikake call him "son" was a small, but present reminder that he still shared a special bond with the professor.

"That's, *um* , -6." Marco was thrilled he was keeping up.

$$f(-1) = 4(-1) - 2 = -4 - 2 = -6.$$

Liam started bouncing from one foot to the other, like a boxer ready to throw a punch, he was up next in the line and ready for his pitch. It never came. Mr. Pikake swiveled back towards the mirror.

$$g(x) = 2x - 1,$$

$$1 \rightarrow -4, 2 \rightarrow -3, \ldots$$

$$h(x) = \frac{x}{2} + 3.$$

"Two orders!" He marched back and forth across the front of the room. "Two groups!" He clapped his hands forming an arrow directed straight at Marco before pushing his arms apart as if swimming the breaststroke. The professor was telling the students to split up. On the left was Maggie and Penelope, the right, Liam and Oliver. Marco was in the weird middle position that made him half feel like he had the power of a swing vote and half feel like the last kid picked on the playground. He debated which way to go. Why was it such a hard decision? Normally, he'd clearly choose his friends over his bossy know-it-all sister. But Penelope was on Maggie's team. Marco would love an excuse to spend more time with her. He knew everyone was expecting him to join the boys, if he did pick the girls, he would open himself up to scorn and ridicule that he may never live down. He went right.

"What happens if we give two orders? Elucidate! Does it matter which order is first? Can you express as a single order? Determine where 1 is sent to. Determine the station of 1/2! Go."

They all got to work. There was an unspoken sense of urgency. Both teams wanted to figure it out before the other and show off their skills.

Marco was back at the bus station. The Numberfolk were all waiting, chatting in gibberish and ignoring him. The 1 was staring at the large poster on the wall with each route. Clearly confused, it kept pointing and tracing the different roads before starting all over again.

"If I take the $g(x)$ bus, I'd get off at 1. If I got on $h(x)$ at 1, it would take me to stop 7/2," the number spoke. A smile grew on Marco's face. He thought of Platform 9 ¾ from *Harry Potter*. Magic all of a sudden seemed much more plausible. *This route would take* 1 *to platform* 3 ½. He chuckled. The 1 kept debating.

"But, if I take $h(x)$ first, it would take me to stop 7/2, and then bus g would take me to platform 6." The number looked defeated.

"What's wrong?" Marco asked it.

"Neither bus will get me to Platform 10. It has the most amazing pretzels. You've gotta try 'em!"

His hunger had invaded his daydream. He would have loved a soft pretzel right about now. Marco turned to his friends.

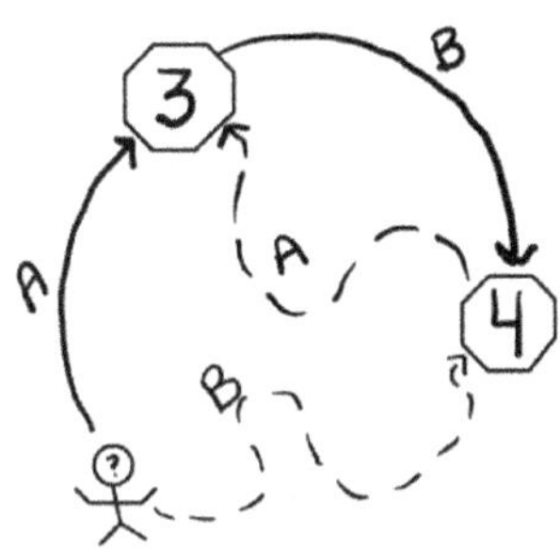

"It totally matters which one they take first. It's like bus stops. Think about having Bus A and Bus B. Bus A goes to stop 3, and Bus B goes to stop 4. If you get on A first, it takes you to 3 and then B takes you to 4, so you end up at 4. But if you get on B first, it takes you to 4 then you get on A which takes you to stop 3. You end up in different places."

Oliver and Liam stared blankly at Marco. Liam was the first to speak, "Dude, I don't think either order takes 3 to 4."

"I know, I know. It was an example. I'm just saying, it's like a bus." Marco couldn't put his imaginative world into words for his friends to understand. He started to try again, but the girls wildly raised their arms.

"We've got it! We've got it!" Maggie screamed. "It's like train stations."

Marco shot his friends a disapproving look.

"She said trains, you said buses, *totally* different," Liam whispered.

Mr. Pikake tilted his hand up and waved it in a semi-circle giving Maggie the floor and control of the marker.

"Okay. So. If you give 1 the order $g(x)$, it goes to g of 1 or,"

$$g(1) = 2(1) - 1 = 1.$$

She wrote on the board. Marco could see her giddy with the power to *teach* the rest of them her superior understanding. He wanted to throw up. In reality, it wasn't about the question at all. Maggie had been *dying* to write on the mirror. It felt like breaking the rules and she loved it.

"Then, from 1 they get on the $h(x)$ train and that takes them to,"

$$h(1) = \frac{1}{2} + 3 = \frac{1}{2} + \frac{6}{2} = \frac{7}{2}.$$

"So, they end up at seven halves. But, if one got on h first, that would take them to $7/2$ and then g would take them to . . ."

$$g\left(\frac{7}{2}\right) = 2\left(\frac{7}{2}\right) - 1 = 7 - 1 = 6.$$

"The order definitely matters." She turned to the group, smiled, and gave a little curtsey. Marco groaned.

"Excellent! The order in which we give Number-" he stopped again. Marco was sure he was trying to avoid sounding too crazy or maybe giving too much away with Penelope there, "-s their commands matter. What about one solidified order?"

Maggie passed the marker to Penelope. "Well, we are really just substituting. If h goes first, that is g of h of x."

Liam's jaw dropped open. Penelope wrote in one fluid, beautiful motion on the mirror,

$$g(h(x)).$$

"See? When we wanted to know what happened to 1, we just substituted in 1 for all the exes. If we want to know what happens if anyone takes train h first, we can substitute all the exes in g with the order for h."

$$\begin{aligned} g(h(x)) &= 2(h(x)) - 1 \\ &= 2\left(\frac{x}{2} + 3\right) - 1 \\ &= x + 6 - 1 \\ &= x + 5. \end{aligned}$$

"Like Maggie said, 1 would go to station 6 if they take train h first, and $1 + 5 = 6$."

It was beautiful. She was beautiful. Marco couldn't stop gawking.

"Marvelous!" Mr. Pikake popped Marco out of his bubble of bliss. "Substitution is one of your most powerful skills! You can substitute orders just as you substitute values. Penelope, do you care to show us h of g of x?" He gave her a little bow of approval.

"Sure! If we go the other way and take g first then it is,"

$$\begin{aligned} h(g(x)) &= \frac{g(x)}{2} + 3 \\ &= \frac{2x - 1}{2} + 3 \\ &= x - \frac{1}{2} + 3 \\ &= x - \frac{1}{2} + \frac{6}{2} \\ &= x + \frac{5}{2}. \end{aligned}$$

"Now, if 1 takes g first, it takes them to 1 plus five halves or 7/2."

"Wait, if g is going first, why is it h of g?" Liam argued. "Isn't that h going first. You know. Because it's first?"

"It's nested, like code." Penelope's shy exterior faded for just a moment and her confidence was peeking through. "Think of a little chicken trying to hatch. Even though we see the outer shell of the egg, the chicken has to break through the inner membrane and shell before it breaks out and we can see it."

"Also PEMDAS bruh," Oliver nudged Liam. "Parentheses come first."

Marco imagined a baby Numberfolk inside an egg. It worked so hard to push through each layer before its tiny little head poked out.

"Excellent exertion!" Mr. Pikake gave the room a quick applause. "My culminating call was the order for one-half. Where would it be sent?"

Huddling together Liam, Oliver, and Marco got started. Penelope had a delayed start since she was at the front of the room, so the boys burst into action to take advantage of their lead. "Okay, bus stops." Liam started.

"We can just use what they already figured out," Oliver suggested. "The direct route will be faster than taking Bus A and then Bus B. Bus C goes right there."

"Since $g(h(x))$ or taking bus h and then bus g is the same as the order $x + 5$, that takes ½ to 11/2,"[*] Liam quickly calculated.

"And $h(g(x))$ has the direct path of x plus five halves. So that takes ½ to 3," Marco chimed in.

"If the order h is first, then ½ goes to station eleven-halves!" Oliver shouted. "And if g is first, it takes them to station 3."

"Nicely done. Now, there is one important consideration when combining two orders. Domain." Mr. Pikake explained, "What if the second command has nowhere to take the soldier? They would be stuck, stranded, forgotten! Thus, you must kick them out of the initial group to avoid a dysfunctional situation."

Marco tried to think of an instance where this could happen. The

---

[*] This is called composition of functions and does just what Marco has described: bus routes. You get on one and then another. Sometimes, you might see this as $(g \circ f)(x)$ which just means $g(f(x))$, you replace all the x values in the function g with the function f. Similarly, $(f \circ g)(x) = f(g(x))$. While it might look scary, it is just another instance of the power of substitution.

only one that came to mind was $1/x$ since everyone knew the disaster that unfolded if you attempted to divide by zero. If the other order was $x - 2$, if two took the $x - 2$ bus to zero then tried to hop on the $1/x$ train, it would blow up the station. He frowned as he thought about the conductor hanging the flyer that claimed twos couldn't board.

"Imagine you manage a sandwich shop," the professor continued. "You have one machine which slices and another, a small oven, which toasts. Slice, toast, slice, toast. You come in one day to see your entire shop burned to the ground. Devastated, you ask your employees what happened. Did they place something un-sliceable into the machine? Perhaps a diamond ring fell off as they were making a sandwich? They assure you everything was done correctly, they only placed things that can be sliced onto the assembly line. When the fire marshal gives their report, they announce the cause: a sandwich was placed on wax paper. Everyone was confused, wax paper wouldn't ruin the slicer."

"So, what happened?" Maggie cut in.

"The oven. Wax paper is flammable. You see, an order within an order is powerful! Why go from A to B then from B to C if you can take a more direct route? In $o(s(x))$, your x-soldier follows the order to end up at Station $s(x)$, then o takes the soldier from their Station and moves them to a new one. But!" he popped, "we must be mindful of the stations that are not in the dominion of o. In the sandwich shop, while wax paper caused no problem in the slicer, s, it was catastrophic for the oven, thus we must restrict anyone who is not allowed in the original order *and* anyone who will end up at a station

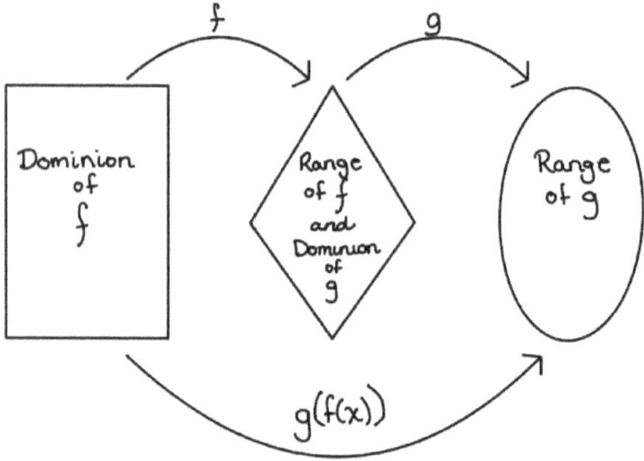

not covered by the second order. The range of motion of s cannot contain anyone who is not in the dominion of o."

Attempting to fix his daydream, Marco watched as the conductor replaced the flyer with a notice, "No connections at Station 0." He smiled as the 2 reworked her options. Taking the $x-2$ train would leave her at Station 0 and the $1/x$ train didn't pick up passengers there. She'd miss her connection. It was a much better visual than the authorities refusing to allow her to board.

"Excellent. Excellent." The professor dug deep into his leather bag and pulled out slips of paper. He began handing one to each student, silently skipping Penelope. "I have provided you each with a series of stations for the Naturals. The first is where 1 is sent, the second where 2 is ordered to, etc. I wish for you to determine the order for your assigned formation."

"We can work together," Maggie whispered to Penelope who nodded in return.

The children grabbed their belongings and paired up. Liam and Oliver started snickering and exited in the front, followed by Maggie and Penelope whispering behind them. That left Marco with Mr. Pikake. As he was about to cross the threshold, a firm hand on his shoulder held him back. The professor waited for the others to leave the building before he turned to Marco.

"Just inviting anyone in, are we?" the tutor said in a firm and admonishing tone. Marco was offended.

"She isn't just *anyone*. But you know how Maggie is. I told her no, she didn't listen. What am I supposed to do?"

"This is starting to get out of hand. I don't even want to think about what will happen if the SAN finds out about this little club. Luckily, they are dealing with their own crisis right now. No one else, Marco. And you are responsible for her. Make sure she understands what she is getting into." Mr. Pikake picked up his messenger bag and pushed the front door open leaving Marco behind.

As he stepped into the strip mall, the bright sun caught his eyes. Placing his hand on his forehead, like a visor, he scanned the parking lot. Everyone was gone. They'd left him behind like a piece of trash. The anger-sadness started to boil in his stomach. He didn't need this. He didn't need the chaos. He didn't need the lecture from Mr. Pikake. He craved the past, a world where he still had a grandfather, a world where these sessions were fun 1:1 meetings with just him and his tutor, a world that no longer existed. Marco yearned for things to be like they used to.

Digging in his pocket he pulled out the ripped paper the stranger had given him. The date was tomorrow. Before, he didn't have the courage to meet the man alone. Now, standing abandoned in a sea of asphalt, he didn't care. His grandfather had given him a mission, and if nothing else, Marco would succeed. Even if it killed him too.

BUS ROUTES

How's your navigation?

RED LINE

Station 1 → Station 4
Station 2 → Station 5
Station 3 → Out of Service
Station 4 → Station 2
Station 5 → Station 1

$$r = \begin{pmatrix} 1 & 2 & 3 & 4 & 5 \\ 4 & 5 & 3 & 2 & 1 \end{pmatrix}$$

The Red Line is out of service at Station 3. Mikhil will have to first take the Blue Line.

Mikhil's house is near Station 3.
His grandma lives near Station 1.

BLUE LINE

Station 1 → Station 3
Station 2 → Station 1
Station 3 → Station 5
Station 4 → Station 2
Station 5 → Station 4

$$b = \begin{pmatrix} 1 & 2 & 3 & 4 & 5 \\ 3 & 1 & 5 & 2 & 4 \end{pmatrix}$$

RIDING THE BLUE LINE

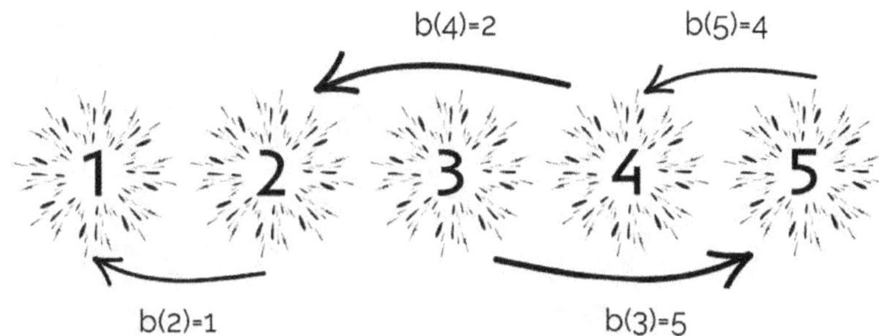

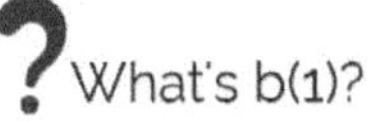

If Mikhil stays on the Blue Line, he must stop at every Station.

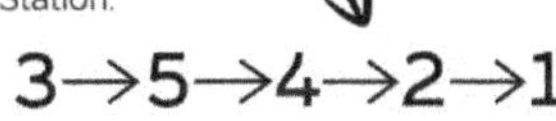

But, if Mikhil takes the Blue Line from Station 3 by his house, b(3)=5, then boards the Red Line at Station 5, r(5)=1, he can make it to his grandma's much more quickly.

r(b(3))=r(5)=1

CITY PLANNING

The city is considering creating a Green Line which would allow passengers to avoid transferring on their commute. If the majority of riders come from Station 1 in the suburbs to Station 2 in the city, should the Green Line be constructed for passengers who begin on Blue and switch to Red, or for passengers who begin on Red and switch to Blue?

$$b(r(x)) = \begin{pmatrix} 1 & 2 & 3 & 4 & 5 \\ 3 & 1 & 5 & 2 & 4 \end{pmatrix}\begin{pmatrix} 1 & 2 & 3 & 4 & 5 \\ 4 & 5 & 3 & 2 & 1 \end{pmatrix} = \begin{pmatrix} 1 & 2 & 3 & 4 & 5 \\ 2 & 4 & 5 & 1 & 3 \end{pmatrix}$$

Where each Station will take you if you take a Red Line bus followed by a Blue Line bus

$$r(b(x)) = \begin{pmatrix} 1 & 2 & 3 & 4 & 5 \\ 4 & 5 & 3 & 2 & 1 \end{pmatrix}\begin{pmatrix} 1 & 2 & 3 & 4 & 5 \\ 3 & 1 & 5 & 2 & 4 \end{pmatrix} = \begin{pmatrix} 1 & 2 & 3 & 4 & 5 \\ 3 & 4 & 1 & 5 & 2 \end{pmatrix}$$

Where each Station will take you if you take a Blue Line bus followed by a Red Line bus

QUESTIONS

Is r(x) a 1:1 function?

Is b(x) an onto function?

Does r(b(x))=b(r(x))?

What's the quickest way to get to Station 2 from Station 4?

Taking the Red Line first then transferring to the Blue Line takes passengers from Station 1 to Station 2.

Yes. (Every bus goes to a different station, no ambushes!). Yes. (Every station is covered.) No. b(r(2))=r(b(2)).

MARCHING ORDERS

with 2 soldiers, can find the order of a march

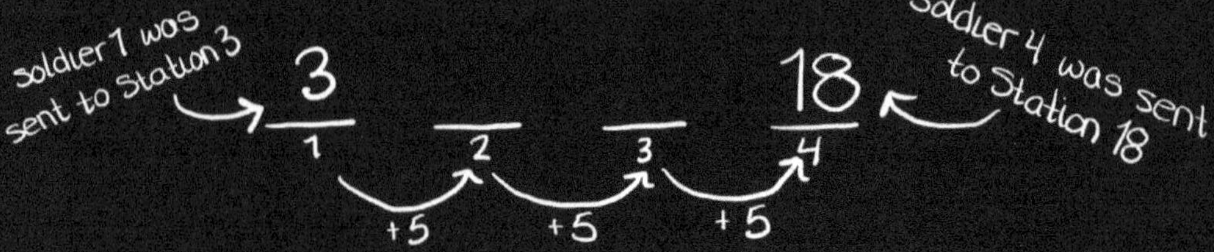

3 hops to cover 18-3=15 Stations
Each hop must be 5 Stations!

∴ 0 was sent to 3-5=-2, the order is $f(x)=5x-2$

"change in y over change in x"

how steep is the line

Δ ← change

Penelope says "slope"

$$\frac{\Delta y}{\Delta x}$$

Take 2 soldiers and their locations (a,b) and (c,d)

☆ location changed from b to d

☆ soldier changed from a to c

$$\frac{\Delta y}{\Delta x} = \frac{b-d}{a-c}$$

change from one soldier to the next is

$y=f(x)=ax+b$

rate of change (a)

location of zero (b)

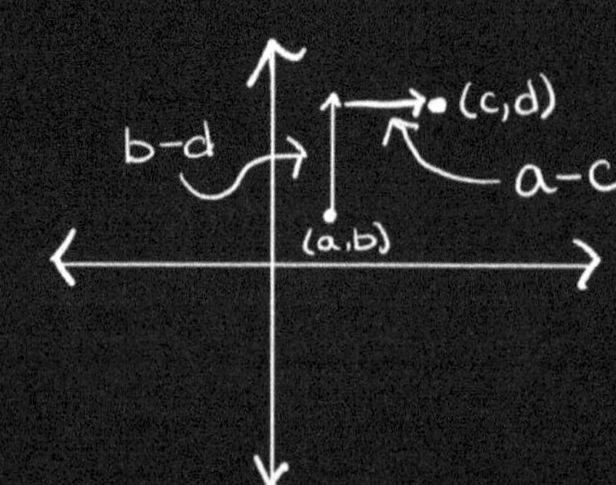

$f(0)=a\cdot 0+b=0+b=b$

SLIPPERY SLOPE

"WHAT MIGHT SEEM TO BE A SERIES OF UNFORTUNATE EVENTS, MAY, IN FACT, BE THE FIRST STEPS OF A JOURNEY.

-LEMONY SNICKET

Light slept soundly as Marco snuck out of the house. It was time to meet the mysterious figure. His feelings were all over the place. He was excited and hesitant, curious and scared, nervous and resolute. His heart raced. The man had provided him with an address and a time. That time was in 20 minutes.

Studying his Google Map, Marco couldn't help but think in coordinates. He imagined all the different ways to get from point A to point B. *I could go over nine blocks then up ten, or I could go up three blocks, over five blocks, then up another seven blocks and finally over four.* He traced the streets with his eyes before deciding his path.

The meeting point wasn't in the best part of town. The warehouse district was just this side of the highway. There were no homes or businesses in the area, just rows and rows of large hollow buildings used for storage and manufacturing. Marco would have to hurry to make the appointment.

As he wandered off the bright, tree-lined streets and into a dark and barren neighborhood filled with factories and dumpsters, he knew he was close. Spray paint was the only markings on the buildings. He saw 1129 in red, the drip marks made it look like it was still wet. The buildings were so massive, walking from one to the next felt endless. Then he saw it. Warehouse 1134.

Taking a deep breath to prepare himself, Marco began searching for an entrance. All he could find was a large retractable wall, like a garage door, painted a deep green that had peeled over time and, like frown lines, showed its age. The door was cracked about 16 inches from the concrete.

"I guess this is it." Talking out loud helped build up Marco's courage, it made him feel . . . less alone.

He dropped to his hands and knees and tried to see inside. "Hello?" his voice cracked. The echo forced him to relive the embarrassing slip-up again and again. Pushing his chest to the ground, he scraped his body beneath the opening.

Dusting himself off, Marco began examining this strange environment. He was in a gigantic room that was nearly empty with the exception of a few storage boxes huddled in the corner. As his eyes adjusted to the dim light, he realized he was surrounded.

34571238795732095327832186123482387219493086476212463945
75648297598...

There were numbers everywhere. He slid his hand over the wall and debris stuck to his fingers. He shuffled his palms together to dust

it off when he noticed the familiar texture and smell. It was chalk. The giant walls were constructed of the biggest blackboard he had ever seen — over 20 feet tall and twice as wide. He couldn't imagine how long the man must have spent filling it up. As he looked more closely, he saw large swipes where someone had erased and replaced a value.

"Hello," a voice boomed from behind.

Marco swiveled. Squinting, he could see a figure elegantly descending a staircase in the far back corner. It led to an office or maybe a breakroom, the only structure on the second floor.

"Hi." Marco tried to sound friendly, but the words got caught in his throat. He swallowed hard.

"I suppose I should formally introduce myself. My name is Fredrick." The tall, thin man emerged from the shadows. His long stride allowed him to quickly cross the factory floor and before he knew it, Marco was shaking his bony hand. Still dressed in all black and what now seemed to be a signature turtleneck, Fredrick looked like a cross between Steve Jobs and Einstein.

Why does that name sound familiar? Marco thought. He couldn't help but notice how much Fredrick reminded him of Mr. Pikake, same body type, same clearness of speech, but this copy was a little older and seemed less wound up. The similarity gave Marco a sense of comfort, like he knew the stranger.

"I have a lot to share with you Marco." His voice was like Dracula, a smooth melody laced with sharp t's. Marco was hypnotized.

"How-how did you know my father?" He stuttered.

"Excellent place to begin." Fredrick pulled two chairs from the shadows and motioned for Marco to take a seat. He paced as he began his tale.

"It must have been over 30 years ago when I first met Cameron. It was at *The Kryptografima* on the SAN campus."

"Wait, what?" Marco jumped up. "My dad knew about the SAN?"

"Knew about it? He was their top recruit! He was brilliant with numbers, he had powers most of us only dream about. But he knew something was wrong, something was off about the organization. He reached out to us through one of our contacts on campus. He wanted to bring down the SAN, to oust their secrets, to help us, so we joined forces."

I am just like him. He wanted to fight the SAN too!

"The SAN leadership has whitewashed the history." Fredrick circled the child and gently pushed him back down into the seat. "They removed key information about the Numberfolk and locked it away. But your grandfather and Maxwell discovered a map to the location of the hidden book."

Maxwell?! He's a part of this? He knew my granddad? Marco remembered Maxwell from his initiation test.

"Your father was able to retrieve the map." He cleared his throat. "Sadly, he died before we could carry out our mission. But when I first saw you, Marco, I knew it was you. I could tell you have your father's powers. Together, we can pick up where he left off. We can bring down the SAN, we can save the world."

Half expecting dramatic lightning to strike down, Marco waited. Fredrick grabbed the second chair. He spun it around before throwing one leg over to straddle it as he wrapped his arms around the back rest. Then he stared. He stared at Marco and didn't make a sound.

Understanding it was his turn to respond, Marco didn't have a clue what to say. Yes? No? Maybe? What was it that Fredrick was even asking him to do? He didn't know why, but he believed this strange new man, he trusted him. There was something about him, something so familiar.

"I hate to disappoint you, but I don't have my dad's powers. I'm horrible with numbers, getting better, but certainly not a genius. And I don't know anything about a map."

"Hmm." Fredrick rubbed his chin. "Are you willing to test that hypothesis?" He jumped to his feet and grabbed Marco's hand, pulling him upwards and spinning them towards the wall of numbers.

"Concentrate," he said firmly. Fredrick closed his eyes and began to hum. It was a low but constant buzzing that filled the factory floor. Marco thought he could feel the vibration in his legs, it was getting stronger. He didn't know what he was supposed to concentrate on, so he stared at his feet, trying to make sure he was still there. He felt light, like air was physically passing through his skin.

"Look!" Fredrick yelled.

Marco obeyed. His eyes rose to the wall of numbers. He fell backward. He tried to crawl away but could barely move. It was the giggling 5 and the angry 1, and there were more. The 0 looked lonely, and the 2 appeared sad. "Make it stop!" Marco screamed covering

his eyes.

He peeked out from behind his fingers, and they were gone. Nothing more than lifeless numbers on a chalkboard stood in front of him.

"What, what was that?" he demanded, out of breath.

"The Numberfolk. You've seen them before, yes?"

Marco remembered his previous encounters and chills ran down his arms. He had convinced himself that it was all his imagination. But it couldn't be, Fredrick saw them too. He couldn't speak.

"It's called phasing. The Numberfolk control everything around us, but we can't *see* them all. This is because they are on a different plane. We see their effect, but we can't see the cause. Like watching a tumbleweed roll down the street or the leaves rustle. You can't see the wind, but you can see the effect of the wind's force. Only the most powerful can phase. Your father taught me how to do it."

Panic set in. Tightness clamped down on Marco's chest, he trembled. "I-I-I." He didn't have any words. His brain was mush. Reality was tearing apart in front of him. All he wanted was to be back at home in his bed. Shock.

The next fifteen minutes were a blur. He remembered Fredrick consoling him and half-carrying Marco to his car. He remembered curling up in the passenger seat as they drove in silence. He remembered getting out, like he was told, about two blocks from his house. He remembered giving Fredrick his phone and something about pizza. He remembered walking like a zombie as the sun started to rise and the entire world seemed to be under a strange haze. That's all he remembered. Marco somehow made it back to his room, to his bed. The next memory was of his sister.

"Marrrr-CO! Marrrr-CO! Marco, Marco, Marco!" He awoke to Maggie singing his name to the tune of the seven dwarfs.

"Stop it." He pushed her away and pulled his pillow over his head.

"Come on! We gotta go meet Mr. Pikake!" she squealed.

"Ugh." He rolled over and pulled the pillow onto his face trying to convince himself it was a horrible dream. "You know I got an earful yesterday because of Penelope. He's probably going to quit on all of us thanks to you."

"Seriously?! I'll woo him today with my charm. I'll make it better." She slammed his door shut and he heard her elephant stomping descend the stairs.

Marco peeked out from under his pillow and scanned the room nervously. He was waiting for the numbers to attack. Nothing moved. He saw the 13 of his baseball jersey from a crack in his closet door. It was dead, still. As he got dressed, he kept looking over his shoulder trying to catch a glimpse of the number's beady little eyes he knew were there. He felt crazy.

As Maggie and Marco walked, they were quiet. His sister was still dealing with the loss of their grandfather in her own way. Although Marco's guilt was minor, Maggie's was consuming. Her last words to him were, "See you tomorrow." It was a sombering realization that life was precious, and the future wasn't promised. She wished with all her heart she could go back and change it.

Marco was focused on a different trauma. His brain was busy convincing itself that phasing wasn't real. He knew it was. He had seen the Numberfolk, multiple times now. The fact that Fredrick saw them too was more proof to add to the pile. He was suffering from cognitive dissonance, a phenomenon where new information doesn't fit into what is already known. Like a puzzle piece that has all the right colors and all the right shapes but doesn't seem to click into any of the available spaces. It was easier, much easier, to refute all the evidence than to tear apart the entire jigsaw and rebuild so one piece could find its place.

Instead, Marco built a wall. It was an invisible structure in his brain to separate the two worlds. On one side was everything that was, reality as he knew it. On the other was that irritating misshapen piece. And as long as he left the wall alone, didn't go picking at it, things made sense in the world. Denial.

When Maggie and Marco arrived at the dance studio, Penelope was already there, sitting next to Liam on the front stoop. "Where's Oliver?" Marco shouted to his friend from the parking lot, he picked up his pace to a light jog.

"No clue. Haven't heard from him. He hasn't responded to my texts either. Hopefully he hasn't been kidnapped."

They headed in together. The room was empty, no sign of Mr. Pikake except an equation scribbled on the mirror.

$$2x + 6 = 12.$$

Marco stared hard at the numbers. He blinked trying with all his might to not scratch at the wall.

The tutor suddenly popped out of a back door.

$$a(t) = -t + 10$$

"I thought that was a closet," Penelope whispered to Maggie and the two giggled.

"Today, we hunt!" Mr. Pikake bellowed before bowing and handing the marker to Marco.

A smile spread on his face as he approached the mirror. This was his time to shine. When Marco began this journey, letters scared him. Not understanding why his math homework had been transformed into alphabet soup, he hated variables and rules and memorizing and all the things that came along with math class. Mr. Pikake showed him a variable is nothing more than a number that is hiding; a Numberfolk he needed to hunt down and rip away their mask to uncover who lies beneath. He had almost forgotten. He was powerful. Able to control the Numberfolk, Marco could cast vanquishing spells by sending in evil twins to duel and proliferation charms to stretch or shrink a Numberfolk to the size he wanted. Marco was a skilled hunter. His wall remained intact.

"The Recta Defensive is an easy way to hunt a mask. First, we Vanquish. The collaborator here is 6, we send in his evil twin, −6, to duel."

$$2x + 6 + (-6) = 12 + (-6).$$

"Now we have $2x + 0 = 6$ or $2x = 6$." He scribbled his results.

$$2x = 6.$$

"Next, we cast a proliferation spell to obliterate, bring the mask down to size. The partner of 2 is ½. We cast with a ½."

$$\frac{1}{2}(2x = 6).$$

Liam leaned over towards Penelope, "Way cooler than math class, huh?"

"Finally, the ½ obliterates the 2 and they become whole, $1x$, while it cuts the 6 in half, so . . ."

$$x = 3.$$

"Perfect presentation!" The tutor threw his arms in the air. "If we know the orders are $2x + 6$, you have just found the soldier assigned to station 12." He wrote a new equation on the mirror.

$$f(x) = 2x + 6.$$

"Do you see it?" he posed to the students.

"Yeah! The x is the soldier getting the orders and the y, or the f of x, is the station they are assigned to. Together, they form an ordered pair (x, y). Like $(3, 12)$ is one of the orders that follows this command," Liam boasted.

"Very good! As you can see, knowing the order is paramount. With it, we can determine where each and every soldier will be. Where would this order send a -1?"

"4. A -1 would be sent to 4." Maggie hopped up and grabbed the marker from the tutor's hand.

$$\begin{aligned} f(-1) &= 2(-1) + 6 \\ &= -2 + 6 \\ &= 4. \end{aligned}$$

If this was Maggie's way of trying to win over Mr. Pikake, it didn't seem particularly effective, Marco thought. She was still the brash and pushy little sister she'd always been.

"The power of prediction!" Mr. Pikake yelled out dramatically. "Now, your patterns. Did you determine their orders?"

Crap, Marco thought. In all the mayhem, he had completely skipped his homework. *I'm invisible, I'm invisible.* He repeated in his head. Maybe if he sat still enough, he'd go unnoticed. For once he was thankful his sister was there as she sprung to offer up her findings.

"Yes, Maggie?" Mr. Pikake said softly.

"Penelope and I were given the following orders: 55, 47, 39, 31." She barked out the numbers like a chihuahua. "We noticed the pattern was that each station was eight less than the previous. At first we thought it might be $f(x) = -8x$. We used a negative since they are getting smaller."

"Interesting inspiration." Mr. Pikake stroked his chin.

"Yeah but," Maggie started again. She looked to her brother to notice her yabbit slip-up, but he was too busy trying to hide himself in plain sight. "We saw that didn't work because 1 would be sent to -8 instead of to 55. So, we thought, why not move everything up?" She turned to face the group and used her best rhetorical YouTube voice. "We added on 63. We knew that would fix the order for one, but it turned out to fix all the orders!"

$$f(x) = -8x + 63.$$

"See? Because then 2 would be sent to negative eight times two which is negative sixteen, adding 63 to that gives us 47!"

"Nicely done dear. Boys, what did you find of your patterns?"

I'm invisible, I'm invisible, Marco continued to chant in his head. Thankfully, Liam took center stage.

"Alright. Mine was 7, 11, 15, 19, and on and on. I saw this as skip counting by fours. But the catch was, I couldn't start at zero like normal: 4, 8, 12, 16, 20, 24, …" he trailed off. "See, it doesn't hit the stations I needed to. Everyone is off. But then I had a great idea. I know that 1 is supposed to go to station 7, but if I sent 0 to station 7 instead, then 7 could be the starting point. You start at 7 and count by fours. That order would be $f(x) = 7 + 4x$."

"But it *doesn't* start at 0, it starts at 1," Maggie said snottily.

"*Okay*. I know that," Liam fired back. "I just needed to shift the soldiers over one. My final order was $f(x) = 7 + 4(x - 1)$."

Maggie considered it. She started to speak then stopped. She tilted her head. It was as if witnessing all the stages of grief flashing by in only a few seconds. She was shocked by his approach, she tried to deny it, she felt guilty for being so snappy, she was angry Liam had come up with a brilliant and alternative approach (although her anger could have been jealousy, as these emotions are very similar). This was followed by sadness. Ultimately, she accepted his solution, albeit possibly superior to her own.

"Tantalizing tactic!" Mr. Pikake smiled wide. "You both have identified the key information we need to develop an order for any line. We require the change; how is the position of each soldier changing from one to the next? Second, we require the one who takes center stage."

Marco exhaled. It felt like they were moving on and he somehow had managed to get out of the potentially embarrassing situation unscathed. At that moment, Oliver burst through the door.

"Hi. *Um*. Sorry. I. *Um*. Had a thing." He slid down next to Liam in the far end of the room. Marco could tell something was wrong, but Oliver never looked in his direction for the chance to silently ask what was up.

"Consider the following orders." He wrote in long smooth strokes on the mirror.

$$f(x) = 2x,$$

$$g(t) = \frac{1}{2}t,$$
$$h(n) = 4n.$$

"How do each of these change the soldier that goes in?" the professor asked.

"Isn't it obvious?" Oliver condescended. "The $2x$ doubles what goes in, the t over 2 cuts them in half, and the $4n$ doubles them twice."

Marco's imagination took over as he thought about a machine that cut people in half. *Eww*.

"Alright, I shall increase the intricacy, Oliver. Do these functions impact everyone who enters in the same way?"

"Well yeah." He got up and took a red marker from a basket on the floor. "How something is changing is like miles per hour. If I want to know how far I've gone, it depends on how long I've been going *and* how fast I'm going. We know distance is rate times time,"

$$d = rt.$$

"It's the same here. The Station is changing based on the soldier you send in. We can find the rate of the Station per soldier,"

$$\text{station} = 2 \cdot \text{soldier}$$
$$y = 2x,$$

"then you can see how each is changing."

$$\frac{y}{x} = 2$$
$$\text{station per soldier} = 2.$$

"No matter what x is."

"It can't be zero! It can't be zero!" Marco didn't have any clue what was happening. He tried to sound smart but instead squawked like a parrot.

"As long as x isn't zero . . ." Oliver continued, "the y over x is always 2. It's constant. Each soldier is two stations from the one before."

"And you have come across the core of our class." With one strong smooth motion, Mr. Pikake added onto Oliver's equation in blue.

$$\frac{\Delta y}{\Delta x} = 2.$$

"Are those triangles?" Maggie whined. "What the heck is that

supposed to mean?" Her question sounded more like a declaration of disapproval.

"The Greek letter delta is the universal representative for change. This reads as 'the change in y over the change in x equals two'. Knowing this, if we have intel on any two soldiers lining up and their location, we can identify a key part of their command."

"Isn't that the slope of a line?" Penelope sang.

"That it is! In a line, the slope is the constant of change," the tutor replied happily. "This is one of the beauties of a line. In other commands, change is not constant. Here, everyone is equally spaced so we can easily hop from one to the next."

"But what is changing? In $y = 2x$, if we send in 1, it doubles and becomes 2. What's the change? Isn't that 1? $2 - 1 = 1$. Not 2," Liam broke in.

"It's a ratio." Oliver stated flatly. Oliver had always been excellent with numbers. The top of his class, Marco often watched in awe as his friend did aerobatic flips and twists with equations. Once again, he felt behind as somehow Oliver could see the hidden connections everyone else missed.

"Ahh! Questionable query. Suppose we want to see how good you are at collecting zombies in one of your games." Mr. Pikake asked Liam, "If on day 1 you captured 12 zombies and on day 2 you captured 14, how could we quantify that?"

"Oh, you mean like an average?" Liam asked.

"Not an average, a rate. We are seeking the rate of change. How do you find your zombie rate?"

"Don't question me on zombie stats. I am an expert. My zombies per day would be the total zombies divided by the total days. In this case it's 26 over 2 or about 13 zombies per day. But that still seems like an average."

"That's because you found the average," Oliver mumbled.

"The rate is how many more or less zombies you captured between the two measurements. How your skill is *changing*," the professor clarified.

"Oh, I see!" Liam exclaimed. "I caught two more zombies, so my rate is $(12 - 14)/(1 - 2)$ or 2 zombies per day. That makes sense." He turned to Marco and whispered, "Dude, if that were true imagine how good I'd be in a year." He winked. "Life goals."

"Precisely! To find the change of a march, you simply need two orders. Since the change is constant, any pair will have an identical alteration."[*]

"Yeah! So for $f(x) = 2x$, we can find two pairs. Like if the, *er*, soldier is 1, they are sent to Station 2 and if the soldier is 3, they are sent to Station 6. That gives us $(1, 2)$ and $(3, 6)$. You can now find the change of the y values and the change of the x values," Penelope added. Marco found how she stumbled around the new terminology charming.

"Ok." Oliver was making his stand. He was ready to take the lead and impress everyone in the room. "The change in the y values are $6 - 2 = 4$, and the change in the x values are $3 - 1 = 2$. So the total change in the function is $4/2$, which *is* 2!" He pecked at the board like a genius mathematician unveiling a new universal law.

$$\frac{\Delta y}{\Delta x} = \frac{6-2}{3-1} = \frac{4}{2} = 2.$$

"How'd you know which order to put them in?" Maggie asked. "Like if we instead had . . ." She stood and grabbed a green marker.

$$\frac{2-6}{3-1} = -\frac{4}{2} = -2.$$

"You didn't do it right," Oliver snapped.

"Why not? I did the change just like you did?" Maggie barked back.

"Look." Oliver returned to the mirror and circled his equation.

$$\frac{\Delta y}{\Delta x} = \frac{6-2}{3-1} = \frac{4}{2} = 2.$$

"See how mine are in the right order since the pairs were $(3, 6)$ and $(1, 2)$? Yours aren't."

"But why should it matter?" Maggie persisted. "The change is the change. The order shouldn't make a difference."

Things were getting out of hand. Luckily, Mr. Pikake jumped in before they totally exploded. "Maggie you are correct."

She smiled, stuck her nose in the air, and did a little shoulder shake.

---

[*] Lines have a constant rate, like when you are pulled up on a rollercoaster or you are cruising in a car. Once you get to the drop or step on the gas, you add acceleration causing you to speed up and the rate to change.

"But!" A dramatic pop, "Oliver is as well."

He responded with a childish tongue pointing.

"The order doesn't matter. You could observe the change from $(1, 2)$ to $(3, 6)$ or from $(3, 6)$ to $(1, 2)$. However, you cannot do both. In your example, Maggie, your ordinates moved from thing one to thing two, but your abscissa moved from thing two to thing one. Once you pick a direction, you have to stick to it."

Marco was mortified. He was used to his sister embarrassing him, but today's meeting had a bit of everything. Not only were Oliver and Maggie bickering like cats and dogs, it would be sometime before he could shake his humiliating parrot impression, "can't be zero, can't be zero", from his mind.

As he looked around the room at the craziness that surrounded him, Penelope stood out like a sore thumb. She seemed even more perfect in this circus.

"The word Recta means straight. This is an order for soldiers to line up, to march, and is why the Recta Defensive is a tremendous tool in such a situation. If we wish for our soldiers to form a line, we can give them an order in the form $f(x) = ax + b$. Here, the a and b are the masks, constants that are not changing. The x is our soldier, it will change depending on who we are giving the order to."

"Wouldn't they just line up, $1, 2, 3, \ldots$?" Maggie questioned.

"If we want them to. That order is simply $y = x$. Each aligns at their value. But what a boring dance that would be."

"What's the plus 6?" Marco wondered out loud. Realizing everyone heard him he clarified, "I mean, in the first order, you had $f(x) = 2x + 6$. If the 2 is the change, what's the plus 6?"

"Perfect probe!" the professor lit up. "A line is straight and predictable, but there can be outstanding and complex orders and formations that may take place. A beautiful ballet of figures. And like any complex choreography, we need someone to take center stage. The b in $y = ax + b$ is our prima ballerina. Since 0 is smack dab in the middle of our stage, it determines where everyone to its right and its left will go. The b is the position of soldier 0."

"That makes sense," Liam chimed in. "If the soldier, x, is zero then $y = ax + b$ becomes $y = a(0) + b$ which is just b."

"Exactly! It is called the y-intercept as it is the focal point of the line. We call $x = 0$, the middle axis, center stage as it lies at the horizontal middle of our theater. The vertical epicenter is ground

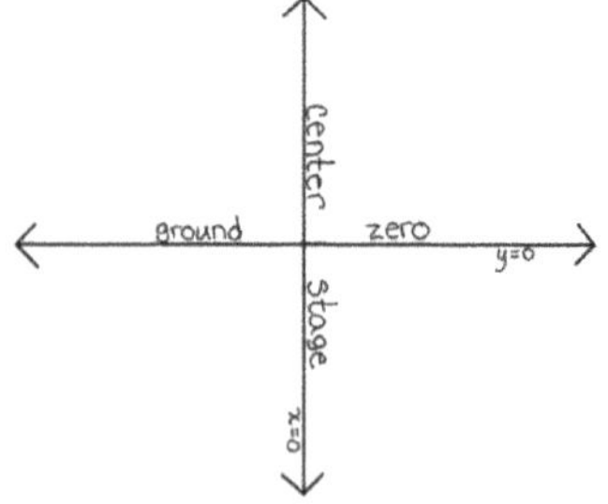

zero, it occurs at the x-axis and is the line $y = 0$, as you know."

"And a is how they are ordered to change. Like in $2x$, the change was two," Maggie added.

"Know the order, know where *every* soldier has been deployed," Mr. Pikake said in a foreboding tone.

How could you possibly know how all Numberfolk will behave from only knowing what two are doing? Marco thought. As if reading his mind, the professor shrieked, "Experimentation is essential! Up! Up!"

They all stood and spread out, taking their place on their self-selected intersections of the tape. Like always, Marco found himself on center stage. He was the glue of the group, holding everyone together.

"Your command is $f(x) = 3x - 2$. Dance!"

Everyone started talking at once. Maggie and Oliver were in a shouting match barking out orders to the group. Liam and Penelope calmly took their positions.

"Marco! You are the center, so you need to go to -2, take two steps back," Oliver screamed.

Marco obeyed and the group had completed the order. Maggie was to the left of Marco at $x = -1$, she had taken five steps back to end up at $(-1, -5)$. To his right, Liam was three steps in front of Marco placing him at (1,1) and Oliver was once again scrunched at the mirror. He was on the far side of the room aligned with $x = 4$. His place was at $(4, 10)$. He noticed that while Maggie was three steps behind him and Liam was three steps in front of him, Oliver was a full twelve steps forward. He imagined a 2 and a 3 as part of their line, dancing along with the group and filling in the gap. The 2 was again three steps ahead of Liam at $(2, 4)$ and the three was three more steps ahead at $(3, 7)$. The two looked at Marco and smiled before bending down in a curtsey. Marco looked away. He wasn't sure what was his imagination and what was real.

"Beautiful! Now to $g(x) = -2x + 1$!"

The arguing died down as everyone focused on their own movement. Marco felt like he had it easy. Since he was playing zero, the leading man, his station was always the $+b$. He walked up three steps to 1 to land on (0,1). He watched the room around him. Mr. Pikake was spot-on. It was like a beautiful dance. The boys to his

right all slid backward. Liam ended up at $(1, -1)$ and Oliver was finally away from the mirror, now near the back wall at $(4, -7)$.

To his left, the girls also switched from being behind Marco to moving in front of him. Maggie went to $(-1,3)$, and Penelope was the star at the front landing on $(-2, 5)$. He smiled.

"Now to $y = 1$!" The tutor yelled melodically.

Marco stayed still. He was already at $y = 1$. Around him the girls slid back, and the boys skated forward to join in one solid horizontal line. He imagined they all grabbed each other's hands and swayed back and forth like Whoville at Christmas. It didn't happen.

"Transform to $h(x) = x$!"

While everyone else seemed to know what to do, Penelope moved gracefully to $(-2, -2)$, Maggie bounced to $(-1, -1)$, Liam walked like a Yeti, his arms in perfect synchronization with the opposite leg to $(1, 1)$, and Oliver slugged to $(4, 4)$, Marco was at a loss. He was Zero. Normally this meant he simply needed to move to whatever was being added. In this case, nothing was being added. It was like his role had been cut and he was no longer a part of the cast. He put his soldier into the order, $h(0) = 0$! It was the origin, he needed to go back to the origin. Trying to look natural and not lost, he attempted to slide in one fluid motion backward. A piece of tape caught his heel and down he went. Hard.

"*Uh. Um.* Sorry." He felt like he had ruined the performance. He imagined the audience watching. They all recoiled in shock when he plummeted. Marco put his hand over his eyes to block the harsh shine of the stage lights which rendered the audience almost invisible. Then he saw them. They weren't people at all, they were numbers. Sitting in the red, velvet, theater chairs, eating popcorn and Junior Mints, they burst out laughing. They pointed and snickered at his fail. Marco cringed.

Mr. Pikake took three long steps and grabbed Marco's arm, pulling him from both the floor and his daydream. "Very well done. All of you." He gave Marco a conciliatory pat on the shoulder. "Okay! You have mastered the line. The Recta Recital! But can you detect the order knowing only where two soldiers have been sent?"

Stomping to the mirror he wrote two ordered pairs.

$$(3, 7)$$
$$(8, 12)$$

"First determine their directive!"

Marco enjoyed thinking in patterns. He had a knack for them. It was like a puzzle, and he was the detective. He was transported to a 1920's crime flick. Everything was in sepia. All the color had been sucked away and only shades of reddish-brown remained. There was a bulletin board on the wall. The captain entered the room. He wore a boring, white, collared shirt with brown slacks. Suspenders held his pants up by the shoulders and a chestnut fedora was tilted down obscuring his eyes. "The suspect was spotted three hours after the crime here. That's seven miles from the police station." He spoke in a deep voice with an accent Marco couldn't place. It was scruffy like the Crime Dog. Maybe this daydream was taking place in Chicago. "He wasn't spotted again until eight hours after the crime, a full 12 miles away."

Figuring out the change, the suspect had traveled $12 - 7 = 5$ miles in $8 - 3 = 5$ hours. Easy peasy. He was traveling at a rate of one mile per hour. Marco needed to determine the orders of the culprit. If he could, he would know where the scoundrel would be to rush in for the arrest.

Where was the crime scene? Marco worked backward. He knew the rate and where the suspect was at three hours.

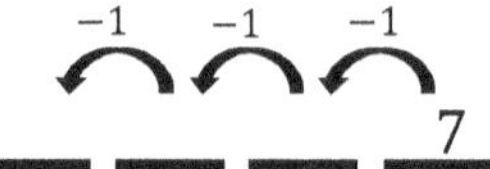

This was an exercise in counting and Marco was an excellent counter. Starting at seven miles, the first spotting of the suspect, he counted down. At two hours, he must have been six miles away, $(2, 6)$, and at one hour he was five miles away, $(1, 5)$. This meant the heinous act took place when the criminal was four miles from the police station, $(0, 4)$.

He stood and slammed his finger against the bulletin board. He started here. Four miles away from the station.

"It's f of x equals x plus 4!" he shouted out excitedly, snapping back to reality.

Mr. Pikake transcribed the order on the mirror.

$$f(x) = x + 4.$$

"Marc's right," Liam confirmed. "Because $3 + 4$ is 7 and $8 + 4$ is 12, it matches the orders."

"Wait, but how'd you find that?" Maggie huffed. "We got the slope, the change in the line. Because," she pushed herself up and wrote on the mirror.

$$l(x) = -2x + 15$$

$$\frac{\Delta y}{\Delta x} = \frac{12 - 7}{8 - 3} = \frac{5}{5} = 1.$$

"But how'd you figure out where zero went?"

Marco relished in the moment. He stood tall ready to, for once, explain something to his sister. He paused. "Is this how you feel all the time? I just want to let it soak in for a second."

"UGH. Marco, just tell me!" she whined.

"Okay. I went backward. If the change is 1, that means the line runs through $(2, 6)$ and $(1, 5)$ and . . ."

His sister cut him off before he could finish, "and (0,4). Okay." While she clearly accepted his reasoning, that didn't mean she had to be happy with it.

"Excellent! Now, where will 20 be?"

$$f(20) = ?$$

Marco flashed back to his detective flick. He looked down at his watch. "The crime occurred nineteen hours ago. If the culprit continues to follow this pattern we can catch him in the next hour, we know where he'll be." He was surprised his own voice was scratchy and muffled too. "The mangy scoundrel will be . . ."

Maggie busted his daydream and stole his thunder all in one swift move. "Twenty-four!" she exclaimed. "f of 20 is just $20 + 4$ or 24."

"Outstanding! We now can both determine the rate of change of the line, the slope, and determine an order given. You also are skilled at using that very order to pinpoint the location of any Numberfolk . . ." he paused. He hadn't said the word since Penelope had joined the team, he cleared his throat and continued, "you may wish to find. That is all for today. Nice work." The professor turned his back to the group and began gathering his things.

Oliver walked out ahead of everyone else. Marco held back waiting for Penelope, and Liam held back waiting for Marco. Maggie pushed through the both of them, "What are you doing weirdo?" she sneered at her brother.

"Marco!" Mr. Pikake called out. "A moment?"

Liam and Penelope took the cue and exited the dance studio. Marco sighed. This was starting to feel like a pattern and one that he didn't much care for. He used to flourish in his alone time with his tutor. He looked forward to it; it was fun and exciting. Recently, it felt more like a trip to the principal's office and the dread festered in Marco's chest.

"Son," his professor started. The word was like an embrace from an old friend. Something Marco missed. Something he cherished. "I have been beckoned back to the SAN headquarters at *The Kryptografima*."

Marco's stomach jumped up into his throat like a drop on a thrill ride. He gulped. *Was this about KFUN?* He stared blankly at his tutor.

"They are in a panic. Something has changed in the world; something is happening with the Numberfolk."

Relief washed over Marco. It wasn't about him or his friends. The feeling was instantly replaced with terror as he considered the possibility of a connection to his grandfather's cedar chest and Fredrick.

"I will be absent for the next few days at least. Be careful Marco. I don't know what is going on, but for Maxwell to raise the urgency level this high, it must be big. Take care of yourself, take care of your friends. I will let you know of any updates once I know more."

Marco nodded, remaining silent. *What could this possibly be about?* As he exited the dance studio, he could see Liam, Maggie, and Penelope waiting for him outside. Oliver had already traveled well across the parking lot and looked like a tiny action figure in the distance. Marco looked back to see his tutor flash him a warm and loving smile along with a nod.

A new feeling began to take hold of Marco's internal organs. It was choking him from the inside out. He was doing it again. Keeping secrets. Keeping secrets from his friends and his family. This time, he was keeping it from Mr. Pikake too. Guilt. He began to scratch at his wall.

It's not my fault I can't tell anyone. If I tell them about phasing, they'll think I'm crazy and try to stop me. They will take the box from me, and I won't be able to protect it. I'll fail my entire family. The dangerous people will use this to get what they want and bring about something terrible. The world will end.

And with that, the slippery slope took hold of Marco and pulled him down as he successfully convinced himself lying was his only option.

ARITHMETIC SEQUENCES

A Hidden Discovery

These are sequences

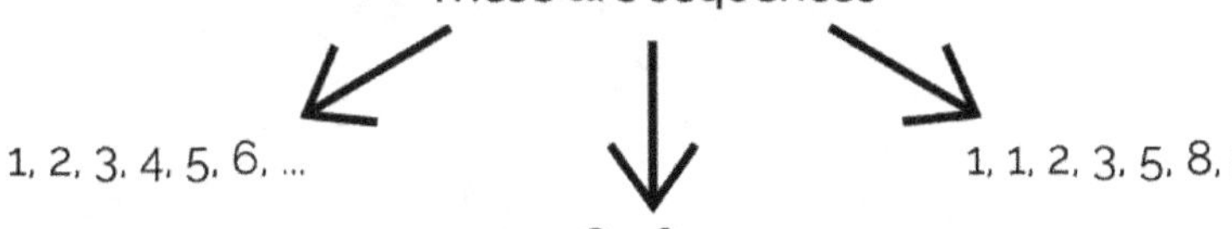

1, 2, 3, 4, 5, 6, ...

2, 4, 8, 16, 32, ...

1, 1, 2, 3, 5, 8, 13, ...

A sequence is a list of numbers that follow a pattern.

An arithmetic sequence is when the pattern is skip counting.

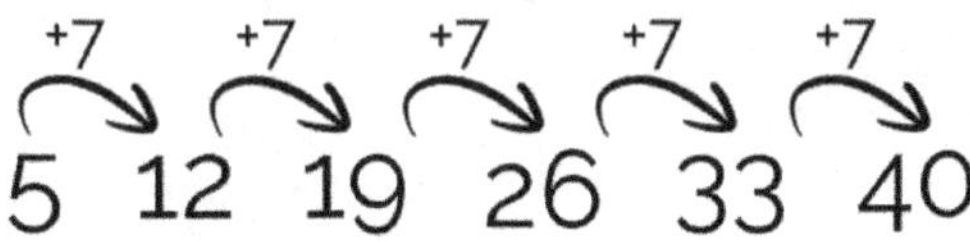

5 12 19 26 33 40

A MARCH

An arithmetic sequence forms a march with a domain of the Natural numbers.

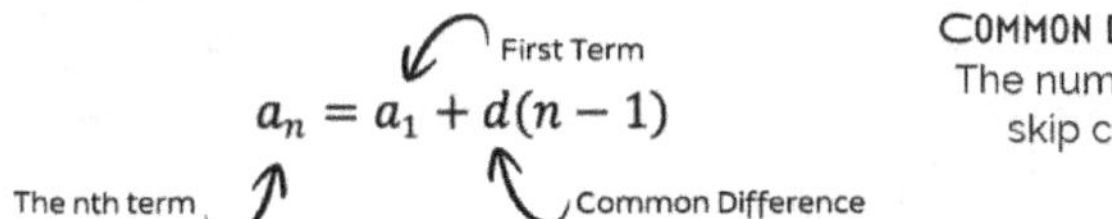

$$a_n = a_1 + d(n-1)$$

COMMON DIFFERENCE
The number we are skip counting by.

FIRST TERM
The number we start counting with.

A MAGIC TRICK

Magic Number

1	2	3	4	5	6	7	8	9	10	11	12	13	14	15	16	17

Give the table above to your audience and instruct them to write any integer in the magic box and any integer in box 1. Next, the audience will secretly fill in the remaining blanks by skip counting by their Magic Number beginning at their box 1 number. When ready, allow them to reveal any two columns of their grid and show how you can amazingly derive their magic number.

ALAKAZAM!

To complete the trick, subtract the two given values and divide by the difference of the boxes.

Example

3	11
29	85

11-3=8

85-29=56

Your Magic Number must be 56/8=7!

Can you determine how this trick works?

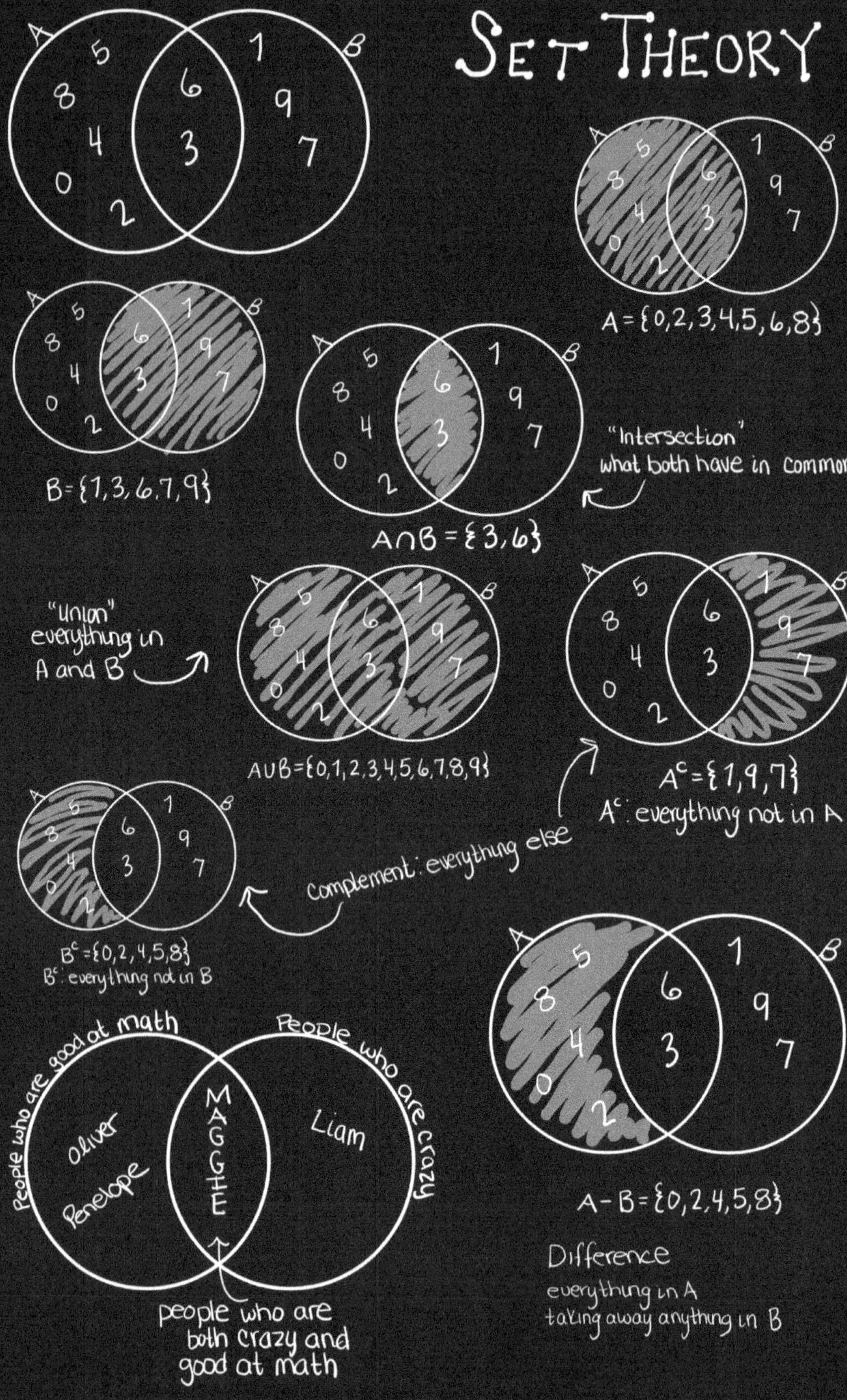

SET THEORY
A
B
A = {0,2,3,4,5,6,8}
B = {1,3,6,7,9}
A∩B = {3,6}
"Intersection"
what both have in common
"Union"
everything in
A and B
AUB = {0,1,2,3,4,5,6,7,8,9}
Aᶜ = {1,9,7}
Aᶜ: everything not in A
complement: everything else
Bᶜ = {0,2,4,5,8}
Bᶜ: everything not in B
A − B = {0,2,4,5,8}
Difference
everything in A
taking away anything in B
People who are good at math
People who are crazy
Oliver
Penelope
MAGGIE
Liam
people who are
both crazy and
good at math

PENELOPE

"MATH IS THE ONLY PLACE WHERE TRUTH AND BEAUTY MEAN THE SAME THING."

-DANICA MCKELLER

Unmotivated, Marco planned to do nothing for the next few days while Mr. Pikake was gone. Oliver had been acting weird and distant, and Liam had some Comic Con costume he'd been working on furiously with his mother. Marco needed a break. Life handed him a full lemon tree, and the lemonade recipe was much more complicated than he expected.

He went downstairs to the office and logged into his video game. No one he knew was online. He walked around for a bit, pillaged some villages for gear, but boredom quickly took over. His brain wall started to itch. As much as he would prefer to never see another number again, he longed to connect with his grandfather. He wanted, no, needed to know more about his father and the mission he was on before he died.

The house was quiet, Maggie and his mother were out. Marco's curiosity won the war as he decided this was the perfect time to explore his grandfather's cedar chest. He closed and locked his door, just to be safe, before slithering under his bed to his secret hiding spot.

He pulled out the paper and unfolded the note. The sight of his grandfather's writing brought on a wave of emotions. It felt like years had passed since he was gifted the box. A wrenching feeling formed in his gut. Marco half wanted to stuff it back into its hiding spot, the secret safe, untouched. But he had been entrusted with a mission, he couldn't fail.

Second Chamber

1234 Ticks: 2 Tocks: 0

5678 Ticks: 1 Tocks: 0

9012 Ticks: 2 Tocks: 1

2356 Ticks: 3 Tocks: 1

5302 Ticks: 3 Tocks: 3

Scanning the clues, his eyes caught the last two. Both had three ticks, three digits correct.

"Both of these clues have a 2, a 3, and a 5, so those must be the right digits." He mumbled to himself and started making notes on a sheet of paper.

"The two ticks from the first clue must be the 2 and the 3, and the one tick from the second clue must be the 5. That means, the two ticks in clue three are the 2 and what else?"

(50, −133) and (80, −223)

If you asked Marco, he couldn't explain it. Something about these puzzles gave him a strange sense of calm; like the entire world disappeared, all the noise, all the pain, all the sadness, all the fear, and all that was left in front of him was a problem to solve and Numberfolk to control. He started to remember what had pulled him into this strange new world in the first place. It was Mr. Pikake's words, "We are the gods. We dominate." At that moment, the hurricane lessened into a tropical storm as Marco found a safe place to hide within the comfort of his superpower: the authority to command the numbers. Scanning the third clue, he began to problem-solve. What was right in 9012?

It can't be the 1. *If the digits from the first clue are* 2 *and* 3, *that means there isn't a* 1 *or a* 4 *in the code. If I accept the digits are* 2, 3, *and* 5, *then the last clue tells me there isn't a* 0. *There has to be a* 9.

Marco drew out his four lines and under each wrote 2, 3, 5, 9. *Halfway there. I know the digits, now I just need to figure out the order.*

He studied each clue carefully, and in less than a minute he had the code. Cautiously sliding out the drawer to reveal the dials he turned the first to the fifth notch, the second dial to the third, and the third dial he spun all the way to the right. He took a deep breath before a final twist: two notches.

Nothing happened.

He checked his work. The code had to be 5392. Grabbing the box, he twisted his arm back, just like on the baseball field, ready to chuck it as far as he could but caught himself at the last second. He was a failure. He was never going to figure this out. He was a disappointment to his father and his grandfather even from the grave. Tears began to form in his bottom lids and his jaw quivered, but he held them back with everything he had. Anger.

As if a devil on his shoulder began whispering in his ear, new thoughts swirled around his mind.

Fredrick.

Marco didn't have a map, but he did have whatever secret this box held. Maybe the mysterious contents of the chest could help fulfill his father's legacy, a mission Fredrick claimed to be on. The strange man could certainly help Marco with these codes too. Quid pro quo.

Unfortunately, Marco didn't have any way to contact the stranger. He certainly didn't want to *pop-up* at the creepy warehouse. In fact, he would be totally okay with never setting foot in that place again.

(57, 48) and (92, 83)

A memory began to form as Marco scratched at his brain wall. That day, the day he phased, he remembered Fredrick taking his phone. Something about pizza. It didn't make any sense, but very little was making sense in Marco's life right now.

Slowly pulling out his phone, he tapped the contacts icon with his thumb. In the search bar he typed "p-i-z" and Domino's popped onto the screen. Puzzling. Why would he have the number for Domino's pizza in his phone? He was sure he hadn't saved it before. *Mmm.* Something deep inside his subconscious was convinced this was Fredrick's doing. Why he concealed the number under a restaurant was unclear. It was . . . suspicious. He lifted his thumb to tap the contact but hesitated. His grandfather had warned him dangerous people were after this box. What if the dangerous person *was* Fredrick?

Marco felt alone. No one to trust, no one to turn to. He needed help. He was devastated: the loss of his grandfather, the new information about his dad, phasing, and the Numberfolk plane. To top it off, the whole thing with Mr. Pikake felt like a death. The formation of KFUN was supposed to be a good thing, a way to safely bring his friends into the fold while ensuring they had the power they needed to protect themselves. The process snatched away the special world Marco and his tutor shared and a part of their bond along with it. Everything was cluttered with people and thoughts.

I'm doing it. He decided. As he tapped down hard on the contact, a musical melody suddenly encompassed his room. It was the doorbell.

Throwing his phone on the bed, he took the stairs two at a time and peered through the eyehole to examine the visitor.

It was Penelope.

Marco turned and pressed his back to the door. *Should I answer it? She obviously isn't here for me. I could pretend like no one is home. She'll be disappointed anyway, she's here for Maggie. I'm doing it. I'm opening the door.*

He turned back around and quickly glanced at the entryway mirror. He tussled his hair into place and straightened his collar before flinging open the front door.

"Penelope! Hi!" He tried to sound surprised.

"Hello Marco." She brandished a sweet half-smile. "Is Maggie here?"

"No, she's out." He outstretched his arm, trying to appear taller as he leaned into the door frame. He was sure he had seen *cool* people pose like this before. It felt silly.

"Oh, okay. Sorry to bother you." She began to turn around. Marco didn't want her to leave. But what would he say?

"*Um,* hey! Actually. I'm working on something. Maybe you can help me out?" The words didn't have time to pass through his brain before they came hurling from his lips.

"Sure! How can I help?"

He led Penelope into the living room, "Wait here, I'll be right back!" He sprinted to his room. He couldn't show her the box or the note, so he quickly scribbled the clues on a piece of paper before bouncing back down to the couch.

"I am working on a surprise, *uh,* for Maggie. You know how she loves puzzles." He had to make sure this didn't get back to his sister. "But I'm a little stuck on one. Wanna give it a go?" He handed her the paper.

"I came up with 5392 but it didn't work. I'm not sure where I went wrong."

Penelope studied the clues quietly. She moved her right pointer-finger around in the air as if writing on an invisible notepad. She tilted her head to the left, then placed her finger on her lips and looked towards the ceiling. "I think I see the problem." Her voice was so soft and fluid, like melted butter. Her words felt like they'd float away if Marco didn't catch each one.

Code	Digits	Places
1234	2	0
5678	1	0
9012	2	1
2356	3	1
5302	3	3

"Please tell me." He leaned in.

"You assumed that because 2356 and 5302 both had three digits correct and both contained 2, 3, and 5 that those *were* the correct digits. But, your code, 5392, breaks the second clue. You see?" She pointed to the second hint on the paper. Marco didn't see but nodded anyway.

"If 5 were the first digit, the places in the second clue should be 1, because 5 would be in the right place, but it's 0."

Marco saw his mistake. He felt stupid. This whole idea was no good. He must have looked so foolish to Penelope.

"It's an easy mistake to make. I probably would have made it too!" she said encouragingly. "Since the last hint says 5302 and it has three correct digits, all in the right spot, we know, from the second clue, that the 5 can't be one of the correct digits, that means, it must be 3, 0, and 2."

(63, 74) and (75, 90)

"Okay, okay." Marco snapped back to the sanctuary of puzzle-land. "So if 3, 0, and 2 are right, the fourth clue tells us the final digit is 6, 'cause it can't be 5. It's 3, 0, 2, and 6,"

"Yes! And, because the final clue tells us the three digits are already in the correct place, that means . . ."

"It means the code is 6302!" Marco finished her sentence as a swell of exuberance that felt eerily similar to nausea passed from his stomach to his throat. "Thank you!" He wanted to hug her, but it didn't feel right. He settled on an awkward handshake.

He felt *amazing*. He wasn't ready for Penelope to leave. "Want to help with another?"

She looked down at her watch, the familiar face of Minnie Mouse stared up at them. "Sure! This is fun, I have a little time."

Code	Digits	Places
3578	1	0
1246	1	0
4635	1	0
9014	3	3
8014	2	2

Marco sprinted back up to his room and jotted down the next series of clues before presenting them to Penelope.

The two leaned in and studied the paper. Marco instantly began regretting this course of action. He had never even looked at this code before, he didn't have the prep time he had with the other. Scared he would embarrass himself, he started blurting out whatever came to mind, trying to sound smart.

"Okay, *um*, so the 9014 has three digits correct but the 8014 has only two. How's that possible?"

"There is actually this really cool area of math called Set Theory, which is great for analyzing similarities and differences." She drew two overlapping circles on the sheet of paper. She wrote 0, 1, and 4 in the overlap and an 8 and a 9 in the outer circle on each side.

"This is a way to see what two sets, or two lists of numbers, have in common," she continued. Her normally shy exterior was fading; she spoke confidently now. "Since both have 0, 1, and 4, we put those in the middle. And since one clue says three digits are correct and the other says only two digits are

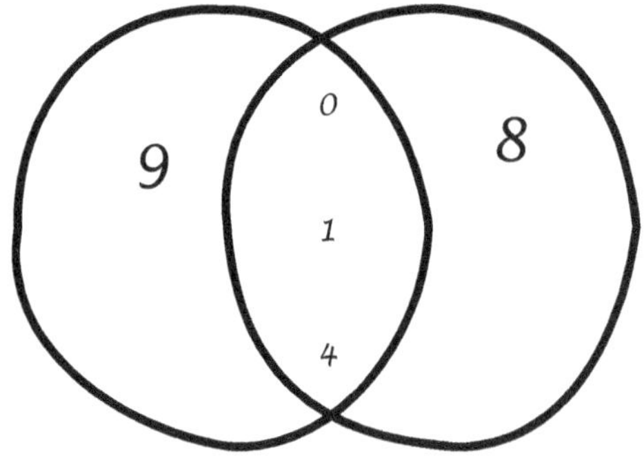

correct, we know that not everything in the overlap can be in the code. We also know there must be a 9."

"Wait, why the 9?" Marco spoke before thinking. Again.

"Because we know the three can't be the 0, 1, 4, it can only be two of those. The third has to be the only digit left, the 9. We also know the code doesn't have an 8."

She quickly wrote down the digits zero to nine on the left side of the paper. She placed a question mark next to 0, 1, and 4 and crossed out the 8. Lastly, she circled the 9. Marco was transfixed by her methods. It was a totally different approach than he'd taken.

0 ?
1 ?
2
3
4 ?
5
6
7
~~8~~
(9)

"Okay." He was having fun. "Oh! Look at the second clue!" He motioned to the 1246 on the paper. "That has only one digit correct, so there can't be a 1 *and* a 4. But we know it has one of those. That means the code doesn't have a 2 or a 6."

"And there must be a 0!" Penelope exclaimed. She crossed out the 2 and the 6 from her list and circled the 0. "We are still missing one digit. We know there is a 0 and a 9 and that there is a 1 or a 4, but not both. Together that will be three digits, what's the fourth?"

They both diligently returned their gaze to the paper. "I'm not sure," Marco said freely.

"I'm not either. I think this is a case for cases!" Penelope pointed her pen in the air. The nerdiness of her statement only made Marco like her more.

"What are cases?" he laughed.

Penelope blushed, revealing a deep maroon that sprawled her cheek bones. "Sorry, got a little mathy there. Cases are a technique that is perfect for tricky problems like this one. We look at specific possibilities and see what happens. Here . . ." She drew a little box in the bottom right-hand corner of the paper before splitting it in half. "The clues we haven't looked at are 3578 and 4635. Both of these have only one digit correct, so we can make a guess and see where it leads us."

"I don't think it's a 7?" Marco saw dominoes. Knocking over the 7 caused a chain reaction that ended in failure.

"Why not?"

Regretting his comment and unsure if his brain-game had any validity he hesitated, "Well. . . . If it were a 7, then the digits would be 9, 0, 7, and a 1 or a 4. But look at 4635. That has only one right digit which would have to be the 4. The 9014 tells us three are in the right place so the code is 9074. But that breaks clue one because it says 7 isn't third . . ." His voice trailed off and sounded like a question unveiling to Penelope his lack of certainty.

"Brilliant!" Penelope sighed. "I didn't even notice that." She crossed off the 7 from her list. "One case complete!" She looked back at the clues, "Wait! We know there isn't a 7 or an 8, so the first clue, 3578, with only one correct, tells us the final digit must be a 3 or a 5. Which means the correct digit in 4635 isn't the 4."

"Yeah! Because there is only one digit correct in 4635, and if it has to be a 3 or a 5, it can't be 4! Our code has a 1." Marco spoke like a base commander at NASA.

"Not only that, the code has to be 901 something, and the something is . . ."

"A 3 or a 5!" the two exclaimed in unison before erupting into laughter.

In her cases box, she wrote a 3 on one side and a 5 on the other. "Okay, let's look at the 5 first."

"Why the 5?" Marco scrunched up his nose.

Penelope looked down at her lap as if deciding whether or not to share her reasoning with her new friend. She began softly. "This one year, I took a tennis camp over the summer. I didn't know anyone, and I didn't really want to be there. My dad made me. When they did roll call, everyone had really easy familiar names, one syllable. Mark, Jane, Anne, John, you know. They asked me my name, and . . . I was embarrassed. Penelope is a mouthful. I noticed I was sixth on the roster, plus six is my favorite number, it's perfect. So, I told them my name was Six."

"Okay Six," Marco joked, "that still doesn't explain why you picked 5 over 3! Plus, I think Penelope is a great name. Sounds regal or something."

"Well thank you." She looked down and smiled revealing a quick glimpse of her perfectly white teeth. "And . . . I picked 5 because 5 is closer to 6 than 3." They both laughed.

"Five it is then! Suppose the code is 9015, does that fit all the clues?"

In less than the blink of an eye they turned to each other and shouted, "No!" Penelope wrote 9013 at the top of the paper and surrounded it in three overlapping skewed rectangles. "My work here is done," she said with conviction as she got up from the couch and made her way to the door. "Thanks Marco, this was a lot of fun."

He waved as she walked down the driveway and turned onto the sidewalk before closing the door. He felt like his knees were going to give out. He slid down to his butt and let out a long sigh. His insides were exploding like fireworks. He hadn't had this much fun, felt this good, in a long time. Not giving himself a full minute to digest what happened, he realized he had the next two codes for the box and ran back to his room.

First, Marco turned all four dials back to 0000, just in case he needed to reset his bad code, before twisting in 6302.

Nothing happened. The good feelings disappeared as quickly as they had arrived. His phone still sat where he'd left it on his bed.

I don't have any choice. He thought. *I have to figure this out. It's my family legacy.*

He picked up the phone and tapped *call.*

Shortcuts?

time travel

Say there is a march with slope 2 and you know soldier 30 is at Station 432

☆ this means Zero is 60 Stations from soldier 30!

☆ Zero is 30 soldiers away from soldier 30

☆ slope is $2 = \frac{2}{1}$ use a mirror to get 30

$$2 = \frac{2}{1} = \frac{60}{30}$$

$432 - 60 = 372$

∴ the order is $f(x) = 2x + 372$

time travel

time travel

$$\frac{\Delta y}{\Delta x} = \text{change}$$

$$\Delta y = \Delta x \cdot \text{change}$$

two soldiers at (x, y) and (x_1, y_1)

... but $\Delta y = y - y_1$ and $\Delta x = x - x_1$

that means

$$\Delta y = \Delta x \cdot \text{change}$$

$$y - y_1 = (x - x_1) \cdot \text{change}$$

slope

parallel lines ⇒ same slope

call the change c . . .

$$y - y_1 = c(x - x_1)$$

7

GRIEF

"'Life is like a math equation' he used to say. 'It's up to you to find the most beautiful solution'."

-Florian Armas

The school campus was a graveyard during summer. Having no interest in returning to the warehouse, Marco had agreed to meet Fredrick at the park near his school.

Although something about Fredrick made Marco trust him, he still had an aching pinch of doubt that had nestled itself in the space right between his ribs. He needed Fredrick's help with the chest, but he didn't want to reveal too much.

Bringing the box was dangerous. His grandfather had warned him bad people were after it. On the other hand, the box was worthless without the clues. Marco felt confident he and Penelope had correctly solved the next two codes, his problem was with the chest. It refused to click like it had in his grandfather's room. He must be doing something wrong.

After practicing frantic hand waving in front of his mirror to ensure if Fredrick tried anything suspicious, he could gain the attention of nearby do-gooders to help, he slipped the box into his backpack and hid the clues under his bed before hopping on his bike and heading to the park.

When he arrived, Fredrick was sitting at a picnic bench in the far corner under the shade of a large elm tree.

"I'm glad you called," Fredrick said in a friendly tone as Marco rode up.

"Yeah, I . . ." Marco dismounted his bike and leaned it against the tree, "I needed to know more about my father, about your plan."

"I'm an open book!" Fredrick straightened up to sit a bit taller and threw his arms out to the side. Marco wondered how Fredrick wasn't dying in the heat. He then, for a moment, questioned if he had only one black turtleneck or a full closet of them. He imagined the man flipping through twenty identical shirts, "No not that one today." A smile formed on his face.

"Alright." Marco swung one leg over the bench and sat. "Last time, you said my father reached out to you to uncover something inside the SAN, and that he had a map to the book you were looking for?"

"To know the full story, we must begin at the beginning." Fredrick leaned in. "Do you know what the SAN stands for?"

"Sure. The Society for the Abolishment of Numbers," Marco replied matter-of-factly. He cringed when he realized how much he sounded like Maggie.

"Ah! Not always! It used to be the Society for the *Adoration* of Numbers!"

"Adoration? I thought the whole point of the SAN was to

overcome Numberfolk control? They don't like numbers."

"That's Maxwell's family's doing. The SAN started with the cult of the Pythagoreans. They believed, as we now know, that 'Numberfolk rule the universe'."

"Yeah, but isn't that exactly the point of the SAN? They hate that Numberfolk rule and want humans to be able to control our own lives?" Marco interjected.

"Yes and no. The Pythagoreans didn't stop there. They believed 'Numberfolk are merely our delegate to the throne, for we rule the Numberfolk'."*

"So, because we can control the Numberfolk, it doesn't matter that they dominate everything?"

"Precisely!" Fredrick's last word gave Marco a sickening feeling of déjà-vu. "How do you think the Egyptians built the pyramids?"

"Aliens," Marco quickly retorted. *I clearly spend too much time with Liam*. He thought to himself.

"Not quite," Fredrick let out a jolly chuckle. "Numberfolk! They were able to build miraculous structures through Numberfolk aid, but in the early 1900's things changed. A powerful sect that had been brewing for many years came out of the shadows. They garnered enough support to be elected to lead the SAN."

"Maxwell's family," Marco said bluntly.

"Maxwell's family," Fredrick repeated. "They did what so many controversial leaders do. They attempted to change history, make people forget. It wasn't overnight, but slowly they convinced everyone that the Numberfolk weren't helping us, they were suppressing us. They used fear to bring Saints to their side and they, of course, had to remove anything that contradicted them."

"The missing book," Marco sighed.

"Yes, the book. It contains the true history and is why we must find it. If nothing else, we need to know what *really* happened. Not what Maxwell wants us to believe."

The two were quiet for a moment. Marco's mind raced with thoughts. Fredrick's story made sense. But something still bothered him. *Mr. Pikake,* he thought. His tutor had fervently explained how the Numberfolk were everywhere, controlling us. He was a believer in the SAN.

---

* This is a quote from Pythagoras himself, the story retold by Eric Bell.

The habitual game of table tennis had started in his brain. Then, Marco remembered something Mr. Pikake had said. "The SAN is driven by fear, and fear can make people do despicable things." This was exactly what Fredrick was saying. Maybe the two stories didn't contradict each other after all?

"Can I see the picture again?" Marco was the first to finally speak up.

Fredrick bent down and picked up a black leather bag. It reminded Marco of one of his mother's purses, small and rectangular with a single zipper at the top. He carefully opened it and slid the polaroid out of a little fabric pouch in the interior. "I always keep it with me. It reminds me of my purpose and who I am fighting for." He gently passed the photo across the table.

Marco studied it closely. On the far left was Fredrick. Wearing his normal black turtleneck, his left arm hung over the man to his side. That man was unmistakably Marco's father. Marco had never even seen a picture of his dad before. His parents had wed in a small ceremony without photographers and caterers. Once, he remembered his mother claiming that his father hated photographs as he thought they stole a part of your soul.

It could have just as easily been Marco in the picture. It looked exactly like him. His dad had thick locks of curly black hair and bright green eyes. The man in the photograph was unmistakably older, maybe eighteen or nineteen. The final man, on the far right, was his grandfather. Marco felt the tears begin to push against his dam. It took everything he had in him, but he held them back.

"Does my mother know?" Marco's voice cracked. While he could hold back the tears, his emotions were seeping through elsewhere.

"No. Your grandfather kept her out of the whole SAN business. But he did introduce her to Cameron. He loved Cameron; we all did. He knew he would take care of his daughter."

Marco handed the photo back to Fredrick and gave his shoulders a shake, trying to release his feelings like water off a wet dog. He needed to know one last thing before giving Fredrick any information, "What about the Numberfolk? Are they evil? Good?"

Fredrick inhaled deeply looking up at the elm tree. He was carefully considering his words before speaking, something Marco forgot to do far too often. "The Numberfolk are creatures. No creature is just good or just bad. They sometimes do bad things for certain reasons others may or may not understand. The Numberfolk are plotting something. We have lived beside them since the dawn of time in a symbiotic relationship. Imagine always helping society when suddenly they turn on you. They aren't happy. Can you blame them?"

"But how can we figure out what they are plotting?" Marco felt like things were beginning to make sense, although he still had some dots that desperately needed connecting.

"Everything has limits. No matter how much I want to, I can't fly. This is a limit on my human existence. Numberfolk have limits too, they follow patterns and patterns are predictable." Fredrick reached back into his purse and pulled out a miniature notepad. Each page was made of thick paper, almost cardstock. Fredrick must have noticed Marco's interest in it as he followed with, "My father worked at a paper company. Unfortunately, it caused me to become a bit of a stationary snob." He looked to the elm and did a little bow as if saying "please forgive my act of arboricide" before continuing. "You know about points, yes? Ordered pairs?" In thick, clean strokes he wrote at the top of an empty page,

$$(x, y).$$

"Yeah, my tutor taught me about these." Marco responded, feeling good that he understood, at least for now.

"Suppose we know where two Numberfolk will be. Just like someone could know that both you and I would be here at the park."

Marco had an uneasy feeling. Why would anyone know where the two of them were?

·(75,170)

·(50,120)

Fredrick continued. "If they aren't marching, we'd need more information. Although, if you know where enough Numberfolk are, you can often follow the patterns to determine their type of order." He placed two dots on the paper and added in their positions.

"From this information alone in a march, we can easily determine their orders and thus predict where all other Numberfolk will be as well."

"No way!" Marco couldn't believe it. While Mr. Pikake had shown them how to find the change from one to the next, and he had been able to figure out the order from only two soldiers before, this was different. To decode the command, he needed to know where the leading man would be, the location of soldier zero. Fredrick's soldiers were miles from center stage. It'd take forever to follow the pattern back to the start.

"Think of it this way." Fredrick's voice had an air of wonder. "If you can relate one thing to another, you know quite a bit more than you might think. They use this idea in advertising regularly, gathering data and presenting you with targeted ads you are more likely to buy."

"Okay, but what if I don't buy?" Marco asked even though he understood the principle. Things like that happened all the time. He'd visit a website, and then suddenly ads for it would show up on every other internet page. It made him appreciate Liam's conspiracy theories. Someone was definitely tracking everything.

"*Ahh*. Clever boy. Even irrational decisions follow a pattern. We just have to be smart enough to dig through the clues and find it," Fredrick laughed.

"Alright. Alright." Marco chuckled back, "But how can you figure their orders?"

"First step, determine the change. These two are following a linear path, a march."

"I know how to do that!" Marco exclaimed. "It's just the rate of their movement. The change in their position is 170 – 120 which is 50. And the change of the Numberfolk at those positions is 75 – 50 or 25. So their rate is 50/25. A change of 2 to 1."

"Perfect! The only other thing you need to know is where to place zero. It is sort of like unlocking a secret code."

Marco's chest tightened. Did Fredrick already know about the box?

"As you deftly distinguished, the change is 2 to 1. Now, simply push this through a mirror." His pen skated over the paper,

$$\frac{2}{1} = \frac{100}{50}.$$

"Do you see?" He looked at Marco and a glimmer shone in his left eye. "Since the rate, position to soldier, is 2 to 1, and Zero is 50 soldiers away from our nearest known command, that must mean the location of Zero is 100 units away from the location of soldier 50."

Marco flashed back to his detective film. Before, he moved soldier by soldier. He would have needed to find where the perpetrator was

at 49 and 48 and on and on all the way back to when the crime took place. Fredrick was providing him with a new tool. A way to jump back in time . . . as far back as he wanted.

The precinct was buzzing. A huge chamber sat in the middle of the lower level. It was covered in gears and knobs and levers. He prepared himself to enter. The quirky scientist (there was always one of these) began pushing and switching buttons, checking the gauges, and tapping on them to make sure they were functioning correctly.

"The offender first got a hold of the time machine 50 years ago. You must travel back and stop him before the device falls into their hands. I've isolated the worm hole he is using to jump. Every year you go back requires two quantum particles. Thus, to go back 50 years will entail 100 bosons. Our calculations show today the thief has 120, this means he began his journey with only 20. You must find him, collect the bosons, and stop him from ever setting foot inside the machine. We are counting on you. If you are unsuccessful, we are doomed."

He couldn't help but smile at his ability to turn a crime flick into a steampunk driven sci-fi scene. "So the order is $y = 2x + 20$?" Marco looked to Fredrick. He felt surprisingly good. Time traveling was certainly a better way to get from point A to point B than mulling over each individual step. Fredrick had provided him a new power and he was relishing in it.

"Very good!" Fredrick leaned in and brought his voice down, "but I'll tell you a secret, there's what I'd call a shortcut if you will. *If* you can hunt."

"A shortcut?" There was no way it could get better than this. Not only did the new knowledge give Marco an edge on his sister and more dominance over the Numberfolk, Fredrick was offering him something more, something superior.

"Your tutor hasn't taught you shortcuts?" Fredrick seemed surprised. His tone made Marco uncomfortable, like he was about to do something that was outlawed. While his initial instinct was to protect Mr. Pikake, something else began to seep into his veins like a virus taking hold of one cell and then the next until it was an infestation he could sense everywhere. The professor was keeping things from Marco. He wasn't showing him all the ways to control the Numberfolk. He was holding him back from his full potential.

"Good thing you have me," Fredrick winked. "The trick to shortcuts is to give everyone a mask."

"Why would you want to do *that*?" Marco replied with disdain. Masks were the worst.

"A mask is just a hidden Numberfolk. When you use a mask, you create a rule, a potion, a spell, a strategy that will work for all of them! Not just one."

The idea made enough sense to Marco, but he still wasn't thrilled with the process. He nodded. The prospect of more power gave him a willingness to trudge through the bad to obtain it.

"As always, we need the change, we can say that . . ." he finished by stroking the paper,

$$c = \frac{\Delta y}{\Delta x}.$$

"Oh yeah! Delta." Marco remembered the strangely placed triangles.

"Now, suppose you wish to hunt the change of y, the way the location is changing, what would you do?"

Marco knew how to hunt, that wasn't a problem. Where he stumbled was the awkward mixture of shapes and letters. The letters were bad enough, adding polygons into the mix was another thing entirely. Was the triangle-letter one thing? Or two? "Is the delta a mask too?" Marco needed clarification.

"*Ahh*! Outstanding observation. Do you believe in destiny?" Fredrick's sudden shift in topic caught Marco off guard.

"Destiny? I . . . *ummm* . . ." The concept of fate was something that had been rattling around in Marco's head for months, ever since Mr. Pikake had declared he was destined for greatness. He saw it as a double-edged sword. On one side, fate gives purpose and comfort. Having a destiny means you have a place in the world, something important to accomplish. When things go sideways you know it was meant to be, which always made the bad hurt a little less. On the other side, it was an inescapable whirlwind that snatches away free will. Marco had learned the story of Oedipus in which after a prophet revealed to him his future, Oedipus did everything humanly possible to avoid it. Unfortunately, his actions led him directly into the future he wanted desperately to escape. Destiny was a trap. Realizing he still hadn't made up his mind he gave up, "I'm not sure."

"Some say the only constant is change." Fredrick's eyes glossed over as if he were trapped in a memory. "Understanding, no, controlling change would then be a remarkable power. If in all the muck and chaos in the world the only thing we can count on is change, then dominating, directing, and dictating that flux would be astronomical! Some believe humans learning to command change was *so* important that fate assigned the task to two different men."

"You mean like both of them had the same purpose?"

"Exactly! At approximately the same time, both Newton and Leibniz independently discovered Calculus, the study of change. The belief is, as this breakthrough was of such importance, fate wanted to ensure if either failed, another was there to complete the prophecy. In fact, neither failed, so it just became a mess of who to credit the discovery to. Lots of accusations of theft and so on."

This altered Marco's perception. If fate was real, what if there was another who shared his destiny? That after everything Marco had gone through, this other swooped in and saved the world, snatching the glory from his fingertips. His anger continued to simmer on low heat as he realized that with Mr. Pikake holding him back, not allowing Marco to rise to greatness, not sharing everything with him, this other was more likely to push into the lead. He wouldn't let that happen. Marco would be sure to get there first.

"Anyhow!" Fredrick's energy changed in the blink of an eye. "I tell you this as the notation we use is that of Leibniz, but I have a feeling you may prefer Sir Isaac's approach. He would write change as,"

$$c = \frac{\dot{y}}{\dot{x}}.$$

Fredrick was right. This did help. While the little dot was funny-looking, Marco could see the change was just a mask. Maybe one wearing a hat or with a decorative feather sticking out of its head, but a single mask nonetheless.

"I see it! To hunt the change in y, or y-dot, I'd cast an x-dot." Marco held out his hand to request the pen. Fredrick hesitated as if Marco might mess up his perfect page before ultimately relinquishing the writing device like an olive branch. Marco scribbled.

$$\dot{x}c = \dot{y}.$$

"Perfect! Now, if they are marching, lining up, what is the change in x? The change in y?"

"Well, if you know where two x-soldiers have been sent, the change in x is the difference between them and since y is their location, the change in y is also the distance from one station to the next."

"What they say is true. You do have greatness in you." Fredrick beamed. "To continue, we need to know information about two soldiers and their locations. Let us simply say . . ." he quickly snatched back his pen from Marco's hand.

$$(x, y)$$
$$(x_0, y_0)$$

"We shall say that (x, y) is the soldier and location we wish to find, and (x_0, y_0) is a soldier we already know." Fredrick called the second ordered pair "x not, y not." Marco was instantly reminded of a daisy, "he loves me, he loves me not."

"Why not?" he replied.

"Wonderful!" Fredrick continued.

"No, why not. Why x not and why y not." His tongue was twisted, "I mean what does that mean?" the double-speak continued.

"Oh!" Fredrick laughed. "We mark a soldier as x to allow us to place anyone in our dominion into the command. When we are interested in a specific soldier, but we do not know *who* they are, their name, we often write a little Numberfolk on the bottom. We don't say x-zero, instead we say x-naught, it means the initial soldier and y-naught means the location of that soldier."

At least there weren't any exponents. The idea of something like x_0^2 made him dizzy. The introduction of letters, masks, seemed foreign enough, but the more Marco dove down the rabbit hole of mathematics, the more it began to look like an ancient set of hieroglyphics representing some alien language.

"So, the change of the soldier would be $x - x_0$ and the change in the location would be $y - y_0$?" Saying it out loud didn't help. Marco felt like he was speaking in a primordial tongue.

"Yes! Yes!" Fredrick exclaimed. "We had,"

$$\dot{x}c = \dot{y}.$$

His pen flew across the pad faster and smoother than what seemed humanly possible.

"And since $\dot{x} = x - x_0$ and $\dot{y} = y - y_0$, that gives us,"

$$(x - x_0)c = y - y_0.$$

"And you have it!" Fredrick said merrily.

$$3y + 3 = -4(x - 12)$$

Marco didn't have a clue what they *had*. It looked like a bowl of alphabet soup and he wasn't particularly hungry. "What do we have?"

"Look here. This now allows us to quickly and easily determine the orders knowing only two Numberfolk positions. From this we can swiftly identify the commands of *any* and *all* Numberfolk. Say we know that 4 was sent to 8 and 6 to 14. What's their change?"

Marco did the math quickly in his head. For the position, $14 - 8$ was 6 and for the Numberfolk $6 - 4$ was 2. He could easily find that 6 divided by 2 was 3. "Three."

Fredrick replaced their mask for change in the equation with a 3.

$$3(x - x_0) = y - y_0.$$

"Excellent. Now pick a pair, any pair. They are all following the same command, so it doesn't matter which you select, we need only the information, something to anchor the mandate. The change tells us the pattern, but there are infinitely many marches with this beat."

On the back side of the page, he painted lines from side-to-side filling the paper with stripes.

"They call this a slope field. It can be used with any order, not just a march. It represents every march with a slope, a change, of three. The command you are looking for is hidden somewhere in this picture. But which one? If you know where just a single Numberfolk has been sent, you then can identify which of these lines they are using."

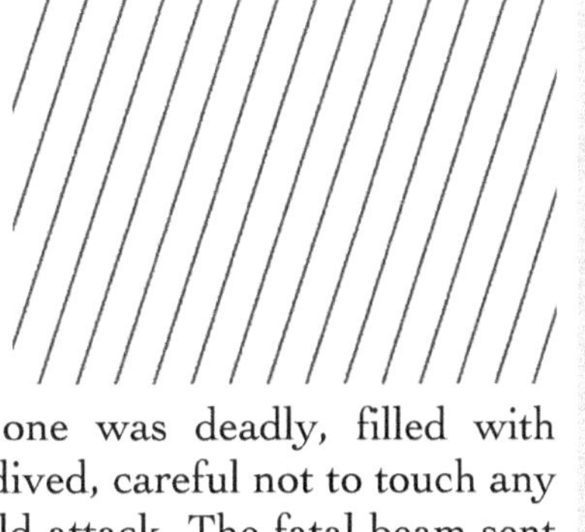

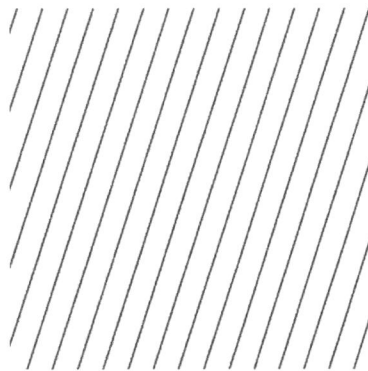

Marco was transported to a futuristic video game. Lasers filled the arena. They were parallel, all having a slope of three, but only one was deadly, filled with Numberfolk soldiers. He dipped and dived, careful not to touch any of the lasers, not knowing which would attack. The fatal beam sent soldier 4 to position 8, he tripped just as one of the red beams intersected his pant leg. *Whew. That was close,* he thought. The radiation ray that hit him sent 4 to position 13. Luckily, it wasn't the droid he was looking for. He saw it, there was only one laser on the field that passed through $(4, 8)$, it passed through $(6, 14)$ as well.[*] Marco took out his blaster and attacked the line. The hidden Numberfolk exploded into the air, flying in every direction. *Got you.* He smiled.

"I get it." He gave Fredrick a diabolical grin. "The Numberfolk I know can *only* be on one of these lines. Isolate the line, know the order they were given."

"Purely magnificent." Fredrick purred, "You, in fact, know two orders yet need only one. I prefer to pick the least cumbersome."

"Alright. I pick $(4, 8)$, smaller numbers. Easier."

Fredrick replaced the masks they had for the pair (x_0, y_0) with $(4, 8)$.

$$3(x - 4) = y - 8.$$

He waved his hand like an illusionist over the result. Marco wasn't impressed. Previously, orders had always been y equals something. This allowed him to easily throw in a soldier and know where it would be sent to. The form Fredrick had presented was a mess, it required more hunting to get to the point.

"Shouldn't it be y equals? To know the order?" Marco spouted defiantly.

"You can hunt, can't you? Must've been able to, to be in the SAN." Fredrick bowed his head and opened his palm offering the pen to the student.

The wizard was enlarging the house of $x - 4$, he cast the spell and added it to the notepad swapping sides as he preferred the y on the left.

$$y - 8 = 3x - 12.$$

He needed to isolate the y, get it alone. It had only a single collaborator, the -8. He sent in the evil twin to finish things off.

$$y - 8 + 8 = 3x - 12 + 8$$
$$y = 3x - 4.$$

"Zero was sent down to -4?" Marco wasn't expecting that. For

[*] Notice if $(4, 8)$ is on the line, and the change is 3, then $(5, 11)$ and $(6, 14)$ and $(7, 17)$ and so on must be on the line as well. This is why we only need one ordered pair when we know the change of a march.

fun, he wanted to try the other method this new friend had presented to him.

Knowing soldier 4 was sent to location 8, he grabbed his time travel device and decided he needed to transport four soldiers over to get to zero. Since the change was 3, he had to disguise it as something over 4. Calculating $3 = 12/4$, he typed into the device. To move four soldiers over required jumping twelve positions, and since he started at 8: $8 - 12 = -4$. It would take him to $(0, -4)$. It worked!

What have I become? Marco thought. *For fun? I did that for fun?* As ridiculous as it seemed, it was fun! It made him feel mighty and forceful. It took away all his insecurity and doubts. In this one special and magical world, he was in control. He could be a detective, a time traveler, a space ranger, and every mission ended in success. In the real world, too often his problems didn't come with a visible solution. Here, he could go left or right, up or down, every way was a different possibility he knew how to navigate. It was invigorating. It was freeing.

"Exactly! And now that you know the order, you can find where every single Numberfolk was sent. Powerful stuff." Fredrick shut the notepad and gave it a light tap before slipping it back inside his bag.

There is a saying that power corrupts. It's true. Like a corrosion slowly and methodically spreading, it sunk its teeth into Marco. Fredrick had gifted him with two new and mighty skills he could use to find out exactly what the Numberfolk were up to. It had been too long since he last felt the sweet taste of dominance. The Numberfolk might control the world, but Marco could control them. The weight was taken off his shoulders and replaced with wings. He was invincible. But something more happened. This meeting with Fredrick was what Marco was missing in his life. His time with Mr. Pikake had transformed into frantic group sessions, where everyone had something to say. A place where Marco had faded into the background and could go unnoticed and unappreciated. Not here. Here, he felt a comfort that he desperately missed. A comradery that, if he was really honest with himself at his core, terrified him.

A sudden jerk pulled at his heart. It was that feeling you have when you do something you know you shouldn't. Marco realized he was betraying Mr. Pikake. Not only in meeting with Fredrick behind his back, but the feelings that were coursing through him were feelings he shouldn't be having. It was out of his control. Guilt.

Alas, it was time for Marco to repay the favor. Bending down and reaching into his backpack, he slowly placed the cedar chest on the table.

"The Map! You have it!" Fredrick exclaimed. Marco recoiled. He didn't know what he expected, but the man's excitement caught him off guard. "And the clock? You have this too?" Fredrick probed.

Ignoring his question Marco responded, "Wait, this is the map you are looking for? I don't know what's inside."

"Yes! You see, your grandfather and Maxwell were the best of friends. Maxwell didn't get along with his father and promised to make things right when the throne of the SAN was passed down to him. As a sort of forever promise, your grandfather and Maxwell split the map into two parts, and each took a part. This would ensure that neither boy could open it on their own. They'd have to do it together. Something happened. Maxwell went back on his promise and followed in his father's footsteps. But without the clock, we cannot open the box. Did your grandfather tell you where to find the clock?"

"No. He told me there were eight chambers, eight pins to unlock. I think I figured out the first two," he lied. He knew more than this, but Fredrick didn't need to know. The corruption was smothering him. He had the power, and he would choose what to share and what to keep to himself.

"Without the clock? You are astonishing young man. I don't know what I expected, of course you are, knowing who your father is."

Is? Marco thought. The verb struck him as an odd choice. He shrugged it off. In need of Fredrick's help, he didn't have time to quibble over words.

"I entered the first code and heard a click. But when I went to enter the second one, nothing happened."

"And you're sure it was correct?" Fredrick's voice was quick, desperate.

Marco thought back to his afternoon with Penelope, a smile grew on his face. "Yes, I'm sure."

"Can I hold it?" The fact that Fredrick asked for permission gave Marco a sense of ease. It reminded him that he was in charge here.

"Sure," he said carelessly sliding the box across the table.

Fredrick slowly pulled out the drawer to unveil the dials. The way he handled the box made Marco sure this wasn't the first time the man had fiddled with it.

"The keys, they are all set to zero, did you do this?" Fredrick questioned.

"Yeah, after I entered the first code, I set them all back to zero to . . ." Marco wasn't sure if he wanted to tell Fredrick the whole truth, "to make sure no one accidently saw the code."

"When you went to enter the second code, did you enter the first again?" Fredrick's urgency and the strength in his voice made the conversation seem more like an interrogation.

"No, why would I?"

"*Ah*! That's it. By turning the dials back to zero, you reset the box, so to say. Cleared everything out. If you enter all the codes now, you should unlock the first two chambers. Do you remember them?"

Marco nodded. He trusted Fredrick, but not enough to hand him the codes. In this movie, Marco was the great and powerful, and Fredrick was the magician's assistant.

He held out his hand as to call the box back to him. Marco huddled over the drawer to ensure there was no way Fredrick could see what he was doing. He began to enter the first code. He suspected he would never forget this string of four digits. They, in some way, were his grandfather's digits, discovered in the last moments he spent with him.

8195

A click. An overwhelming surge of emotions swam through Marco's body. He began to enter the next code, remembering his mistake that Penelope had corrected.

6302

Another click. A massive smile overtook Marco's entire lower jaw. He looked up for the first time to Fredrick who mirrored his glee. The man's eyes were wide with excitement. Marco loved that he didn't have to do this alone. There was too much thrill involved. It required a companion.

"It worked!" Marco exclaimed.

"Outstanding!" Fredrick shared in the elation.

Marco diligently reset all four dials to zero. Just in case.

"Do you think you can determine the final codes?" Fredrick asked eagerly.

"I think so. I can try," Marco replied. He wasn't ready to share his grandfather's "addition" to the box and reveal the clues. He found himself swimming in new questions about the situation. Did his grandfather write the clues because he didn't trust Maxwell? If what Fredrick was saying was true, he would've been right, Maxwell betrayed them.

"Okay. Continue your work on the map. Once you successfully decode the box, we will be able to locate The Book of the Dead! I'll begin making preparations for our journey once the map is retrieved. But let's stay in touch."

The Book of the Dead? Marco thought. It seemed like an odd name for the Numberfolk history the Maxwells had removed. Fredrick was already preparing to leave. Marco tucked the box safely back into his backpack and slung it over his shoulder.

The two stared in silent understanding. They had each provided the other with something they needed, something they dreadfully wanted. It was a symbiotic relationship that benefited them both. Marco couldn't help but wonder if Fredrick had told him everything. Afterall, he hadn't shared all his own secrets with the man. If he was hiding something, what could it possibly be?

With only a brief bow of acknowledgment, Marco hopped onto his bike and the two parted ways.

Leaning into the turn, the back wheel of his bike drifted before pulling the vehicle and its rider towards the ground. He hopped off and let the bike fall next to the garage before hurriedly climbing the stairs to his room and slamming the door behind him.

"HEY!" Maggie yelled.

He started to yell back but stopped. His eyes widened. Panic began to set in.

"MOOOOMMMMMM!" he shrieked. Within seconds he heard her stomping up the stairs.

"Yeah?" She peeked in the doorway.

"Did you open my window?"

"No. And I'd appreciate it if you refrained from screaming in the house. We have your sister for that." She closed the door softly behind her.

He struggled to breathe as if someone had sucked away the air in the room. Shock.

It's just my imagination, I probably just forgot. Denial.

What if I'd left the box here? I would have failed everyone. Guilt.

Who would do this, who would come in here, invade my space? What if they tried to hurt Maggie to get to it? Anger.

"I'm trying. Just keep Maggie safe. I'll complete my destiny. I'll do it," he whispered. Bargaining.

Tears began to pool. Someone had broken into his room, invaded his space without permission, and it felt terrible. Thoughts rushed his head. *It couldn't have been Fredrick. He was with me. Who are these dangerous people? What lengths will they go to get what they want? I can't let them hurt her; I'm the big brother, I have to keep her safe.* He realized that she was only a few feet away from whoever did this. Sadness. Not even the strongest and sturdiest dam could hold them back, the tears flowed freely down his face.

Before, Marco hadn't made up his mind on destiny. He didn't want a predetermined path to follow like a Numberfolk marching along to their orders. Now, it didn't seem like much of a decision. Whether Marco believed it or not, this box, his secrets, were important to someone and they'd do whatever it took, even breaking into his home, to get it. For whatever reason, there were people out there who believed Marco was special, the only one capable of setting things right.

Then it came. Like a giant wave crashing over him, he had only two options: try to fight off the impossible, or consent to being pulled under and hope the swell would take him somewhere spectacular. Acceptance.

POWER

"THE WORLD IS BUILT ON THE POWER OF NUMBERS."

- PYTHAGORAS

Ever concerned about her people, Thirty-six was a kind and benevolent leader. Upon discovering her people were trapped outside the wall, she initiated the Citizenship Program to grant those Naturals living in Rational passage into the Village.

Her lead general constructed and administered the first citizenship test which welcomed ten new residents to their homeland. Each granted a quaint little house near the gate, the turbulent ten (as they came to be called) were also awarded an exit pass. This was unheard of. No one had left Natural besides the expedition team and a few high-ranking officials for the mayor in thousands of years. Even so, Thirty-six could not stand the idea that these special citizens had to disown their families for the tranquility of Natural. *

General Three was the first to suspect trouble was imminent. During his standard patrols, an ominous sign began popping up on the streets of Natural. At first the symbol was limited, a house here, a cottage there, but soon GT and his army found the emblem everywhere. It was sprayed on local businesses, on park benches, inside tunnels, and even on the wall itself. Something was coming.

An underground movement was brewing beneath the surface. Many speculated it was the Citizenship Program that caused the unrest, but this was not true. Although some villagers were untrusting of their new neighbors, what really angered the Naturals were the exit tickets.

It wasn't fair, just, or rational not to allow all residents of Natural to leave and explore Numberville now that the gates were open. The symbol was a *warning sign.* It represented factions of Naturals who banned together in solidarity.

General Three wasn't too concerned when the leaders of these packs were small. He encountered a few 3! and 5!. This was more of a nuisance than anything concerning. It would be easy enough to quell a 3!, this only meant a three had found a two and a one for their little club. Larger groups such as 10! possessed a higher threat level. These were bands of ten Naturals with more power to cause damage.

The intel led GT to believe that such a faction consisted of a leader, the largest of the group, and one of every smaller Natural. At this time, dueling was how all Numberfolk settled disputes. Thus, when the general

---

* Twelve, for instance, was the child of 8/3 and 9/2. Neither parent would ever be allowed entry into Natural. Thus, if not for the ability to leave the Village, the turbulent ten would never see their families again.

$$x^2 + 2y - 18 = x^2 - x + 2^3$$

came upon 3! with the strength of a 6, he naturally believed $3! = 3 + 2 + 1$. He found out quickly that he was horribly wrong.

A brazen seven and her faction were the first to cause a disruption. Armed with picket signs that read "Let us out!" and "Free to be Free" they marched around the town square chanting and blocking the road. Believing he needed only to have more power than $7 + 6 + 5 + 4 + 3 + 2 + 1 = 28$, GT dispatched a 30 to stop the riot and bring peace. When he saw his soldier sprinting away from the square, the seven and her followers running after him, the general came to a shocking realization.

Numberfolk had previously been one-dimensional. To increase their power, their impact, positive citizens could duel, and become larger. For instance, a 3 would duel with a 7 to emerge as a 10. During the class struggles centuries earlier (when larger Numberfolk were awarded a higher place in society), they discovered they also had the ability to proliferate.

Proliferation was the act of scaling or stretching a Numberfolk. Essentially a torture device, it would take someone like a 4 and under an enlargement spell by a factor of 5, it would extend the number into none other than a 20, giving it power equivalent to that of five 4's. This practice had been outlawed for many years and why GT never imagined the true danger of these warning signs.

The Numberfolk found a loophole. While it was unlawful to elongate, there was no law governing a second dimension. Rather than stretch themselves, citizens found they could extend their power through cooperation.

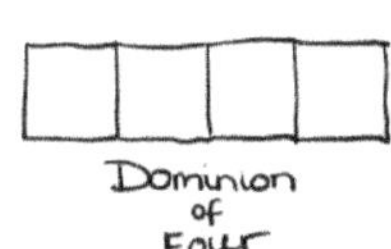

How much muscle a Numberfolk had was based on the amount of land they could control. A normal 4 possessed the equivalent of a 4 by 1 region of the number line. Necessity being the mother of invention, the demand for more power coupled with the inability to stretch, led the Numberfolk to build up.

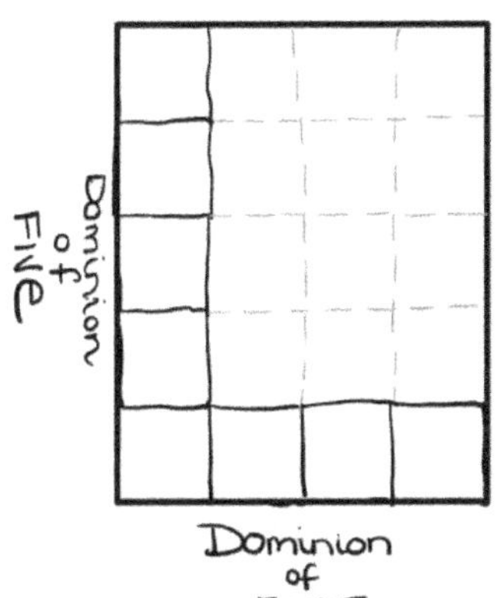

By combining the territories they each controlled, a 4 and a 5 working together could now rule over 20 acres without either of them scaling in size. This was the true meaning of the warning sign. For 7! was not 28 but rather,

$$7! = 7 \cdot 6 \cdot 5 \cdot 4 \cdot 3 \cdot 2 \cdot 1 = 5040.$$

Working together in this way dramatically increased their power and transformed these factions into a formidable force. A 7 and a 6 could collaborate to control 42 acres, then combine with a 5 to amass 210 units of land, before bringing in the 4 to grow to 840. Finally, as the 2 and 3 alone reigned over 6 acres, together with their guild they had stockpiled 5,040 acres, something that previously seemed to be an impossible feat for such small Numberfolk.

Although General Three should have been focused on the unrest, this new discovery sent his mind spiraling through the possibilities. He set his focus on training a new and mighty army.

Major Major

Harnessing the power of cooperative brigades, General Three hand-picked his strongest soldiers and promoted them to the rank of Major.

Each Major was given soldiers identical to themselves. A two who led a brigade of five other twos had the strength of:

$$2 \cdot 2 \cdot 2 \cdot 2 \cdot 2 \cdot 2 = 64.$$

Originally, GT had the wonderful idea to address each leader based on the size of the brigade they commanded. Thus, Two, who led five soldiers, was named Major Major Major Major Major Major Two. You can imagine how this went. Their morning meetings took up half the day and the poor one who worked as the military transcriber counted five thousand eight hundred and seventy-two "majors" in the first meeting alone. Being skilled in shorthand, One adopted the notation M_6 to indicate the word "major" was repeated six times. As General Three was reviewing the notes, he had a brilliant vision, he would award each Major a badge for their shoulder patch that demonstrated the size of their brigade.

With that, Major Major Major Major Major Major Two became 2^6. To further shorten their title, GT called him Major Two to the sixth and exponents were born.

Day after day, the general would run trainings and simulations for his new army. Many wondered why Three required each brigade to consist of all the same Numberfolk. It was a particularly genius idea. When his brigades marched in a line, it obscured their size from the enemy. There could be hundreds or even thousands of Numberfolk behind their leader yet, from a distance, only a single warrior could be seen.

General Three's first construction was what has been called a Trojan Horse. Now, this name is clearly misleading as it was not Epeius, the Grecian builder who first developed the idea. He was a Saint who took the concept from none other than Three's outstanding army. Epeius' horse was a large wooden gift for the enemy. When they accepted the structure and brought it inside their walls, they were shocked to find an army hiding within the steed ready to attack. Three first constructed similar formations. Soldiers would squish together behind their leader to make the brigade appear smaller and less threatening. When the time was right, they would spring out and expand like an accordion of Numberfolk.

The massiveness of Three's army quickly quieted the warning signs. They were there, but the protests and picket lines stopped once the general led his legion of one-hundred other soldiers as 3^{101} around the streets of Natural.*

GT was unsure if constructing large brigades was ideal. It required giving his majors considerable power which worried him. In need of investigating the formations his militia was capable of, after receiving permission to run his war games in the desert outside the wall, he gathered a handful of low-level majors with only a few Numberfolk under their command onto the field and ordered them to line up.

The experiment began with four identical brigades. Each leader stood in front followed by the two soldiers they commanded.

$$2^3 \qquad 2^3 \qquad 2^3 \qquad 2^3$$

Three paced back and forth studying them from one side and then the next before ordering the lines to duel. The result was disappointing. The Numberfolk that rose from the dust was but a 32. This simply wasn't good enough. They were still much too weak.

$$2^3 + 2^3 + 2^3 + 2^3 = 8 + 8 + 8 + 8 = 32.^{\dagger}$$

GT had an idea. Selecting the worthiest of the majors, he promoted her to be the leader of all four brigades. Placing a new patch atop her existing

* The badge dictated how many *total* soldiers were in the brigade. As Three had 100 followers, there were 101 total soldiers in the unit.

† The exponent formations weren't kind to addition. However, in rare cases one could make interesting combinations through duels. In the case of $2^3 + 2^3 + 2^3 + 2^3 = 4(2^3) = 2^2 \cdot 2^3 = 2^5$.

regalia, she was now $(2^3)^4$ indicating she commanded four units with each containing three two-soldiers.

He was thrilled when her power increased exponentially. Each 2^3 held their formation together to form a Trojan Horse before lining up. On the outside, it seemed to be a simple unit of four soldiers. Then, the magic happened. The stallions expanded! Behind each leader, two more lieutenants popped out creating a mega-line. A total of twelve two-soldiers were in the formation casting the power of 4096.

$$\begin{aligned}(2^3)^4 &= 2^3 \cdot 2^3 \cdot 2^3 \cdot 2^3 \\ &= (2 \cdot 2 \cdot 2) \cdot (2 \cdot 2 \cdot 2) \cdot (2 \cdot 2 \cdot 2) \cdot (2 \cdot 2 \cdot 2) \\ &= 4096 \\ &= 2^{12}.\end{aligned}$$

Calling his new structure a "power brigade" GT continued his experimentation. While originally Three had only appointed prime Numberfolk as Majors, he started to wonder what else was possible.

Primes were strong and unwavering. He feared someone like 10 might fall apart in the heat of battle into 2's and 5's rather than standing tall. *Why not give it a try?* Three thought. It was certainly better to test these things now rather than in wartime.

Gifting a 10 a brigade of five soldiers, the 10^6 stepped onto the field. Just as GT expected, when his $(2^3)^4$ charged the 10^6's formation, every ten in line fell apart in terror. Then, something totally unexpected happened. At first, each ten shrunk into a five while a two sprouted on their head. *What a bunch of wimps.* GT shook his head. But as the $(2^3)^4$ got closer, the Numberfolk regained their courage. The twos jumped from the fives to form two lines: one containing six fives and the other containing six twos!

The general understood the value in such an ability. Imagine, seeing only a single 10 approaching on the battlefield. Just as they close in, you notice the brigade of more tens marching in step behind their leader. You start to panic before realizing there are only six tens, and you are commanding an army of nine. You think you still have a chance. Then, suddenly, the tens split! Now there are two lines of soldiers, making a total of twelve Numberfolk! They surround your army of nine before declaring their victory.

$$10^6 = (2 \cdot 5)^6 = 2^6 \cdot 5^6.$$

Now, of course, the actual soldiers matter too. Everything is a numbers game. A small army of elevens could easily overtake a larger army of twos as 4 eleven soldiers have a combined power of 14,641 while 13 two soldiers boast the strength of only 8,192 ($11^4 > 2^{13}$).

After weeks and weeks of testing, Three developed an obsession with creating elaborate productions of his army. Four lines of sixes marched down the field. Each column contained five lieutenants behind their leader Major 6^6.

$$6^6 \qquad 6^6 \qquad 6^6 \qquad 6^6$$

General Three commanded them to form a single line, and all that could be seen was a stripe of four 6^6's, their factions having disappeared in the dust, hiding and lying in wait.

The fearless commander $(6^6)^4$ led them straight ahead, a walking drum of their steps boomed on the desert sand.

"Ten Hut!" General Three shouted and out sprung the lieutenants. Twenty soldiers (five behind each of the four leaders) unfurled to reveal 6^{24} at the helm.*

With a loud clap, they erupted into their prime traits. Every six shrunk down to a three while a two burst onto their head like a person-hat.†

$$\begin{pmatrix}2\\3\end{pmatrix}^{24}$$

Three rose his hands in the air as in prayer before splitting them apart like the Red Sea. Each two jumped down to form their own line. Two columns appeared in the sand, each containing twenty-four combatants.

$$(2 \cdot 3)^{24} = 2^{24} \cdot 3^{24}.$$

Another clap and the lines split in half. Since they contained twenty-four soldiers each, the new formation was four lines of twelve combatants.

$$2^{12} \cdot 2^{12} \cdot 3^{12} \cdot 3^{12}$$

They separated and formed a square around their exalted director. To the east and to the west stood lines of twos, while to the north and the south stood the threes. The army circled around General Three a symbolic three times before returning to their columns.

Next, he broke each stripe into three. The lines made beautiful

* This became known as the power property: $(a^n)^m = a^{n \cdot m}$

† Today, we write this phenomenon as $(2 \cdot 3)^{24}$, however in reality they actually were standing on top of each other like a circus act.

formations as the first four marched left then back, the last four marched right then forward, and the middle four marched in place to reveal their new shape.

$$2^4 \cdot 2^4 \cdot 2^4 \cdot 3^4 \cdot 3^4 \cdot 3^4 \cdot 2^4 \cdot 2^4 \cdot 2^4 \cdot 3^4 \cdot 3^4 \cdot 3^4$$

Turning away from GT, they paraded toward the sunset. Pounding his hands together to the beat, each of the two bookends of four soldiers (twos on the left and threes on the right) combined into one line of eight.

$$2^8 \cdot 2^4 \cdot 3^4 \cdot 3^4 \cdot 3^4 \cdot 2^4 \cdot 2^4 \cdot 2^4 \cdot 3^4 \cdot 3^8$$

The army formed a U-shape now. The lines of eight-soldiers sticking out like tails. He clapped two more times conducting the middle lines (this time, threes on the left and twos on the right) to merge.

$$2^8 \cdot 2^4 \cdot 3^4 \cdot 3^8 \cdot 2^8 \cdot 2^4 \cdot 3^4 \cdot 3^8$$

A fantastical W rode through the desert like a mirage. The general laughed in delight. He pointed towards the lines containing only four soldiers, "Side-by-side!" Three yelled.

$$2^8 \cdot (2 \cdot 3)^4 \cdot 3^8 \cdot 2^8 \cdot (2 \cdot 3)^4 \cdot 3^8$$

"Mount!" he shrieked. The twos jumped on their partner's head before GT gave the ultimate command and brought them together.

$$2^8 \cdot 6^4 \cdot 3^8 \cdot 2^8 \cdot 6^4 \cdot 3^8$$

"Bah-humbug!" Three wasn't happy with his formation. He had painted himself into a corner. One of the requirements of these brigades was that they must contain the same soldiers, mixing and matching ruined the whole thing. Since the lines of twos were next to a line of sixes, there was no way to augment his dance further.

The units stood in silence awaiting their command while the general considered his options. Since six was made of twos, the $2^8 \cdot 6^4$ could become $2^8 \cdot 2^4 \cdot 3^4$ allowing the twos to join in a procession as $2^{12} \cdot 3^4$, but that felt like undoing his last move. Another possibility was to create fours or eights as $2^{12} = (2^2)^6 = 4^6$ or $2^{12} = (2^3)^4 = 8^4$, but there was no perfect solution to collapse the army further.

Making his decision, General Three ran out onto the field and grabbed 6^4. "No, no, no. Switch positions!" He pulled each line of six to the outside.

$$6^4 \cdot 2^8 \cdot 3^8 \cdot 2^8 \cdot 3^8 \cdot 6^4$$

"There. Much better." He smiled. "Side-by-side!" The march continued.

$$6^4 \cdot (2 \cdot 3)^8 \cdot (2 \cdot 3)^8 \cdot 6^4$$

Clapping again the twos jumped onto the threes before combining into sixes.

$$6^4 \cdot 6^8 \cdot 6^8 \cdot 6^4$$

Another smack and only two lines remained.

$$6^{12} \cdot 6^{12}$$

Finally, each column containing twelve sixes slid together alternating into open slots like a giant zipper being sucked shut to form a single-file line of twenty-four sixes that marched over the hill and out of sight.*

$$6^{24}$$

The Sand Worm

While Three and his soldiers had mastered ground forces, he was teeming with a lust for more power. The army in Natural was limited, and GT had his heart set on complete dominance over Numberville. Wanting to test out surprise attacks, he marched his forces into the desert of Rational. His next experimentation would be the creation of a scarecrow.

General Three's formations were strategic. A line obscured the size of the brigade to keep the enemy guessing, filled with fear and anticipation of the oncoming attack. Three wondered if he could trick an attacker all together, scare them away before they even had a chance to make a move.

Before, Three saw very little use in ones and zeros. They simply didn't hold enough power. Ones reflected which meant a brigade of ones, no matter how large, would always have the strength of a single one.

$$1^1 = 1^{1,000,000,000,000} = 1.$$

He saw ones as bumbling clowns. Each would turn to their neighbor and ask, "What should I be? I guess I'll just copy you!" before turning to the next and repeating the whole thing all over again. Zeros were slightly more useful. Although GT couldn't make a zero a major, they had no strength, should a war actually break out, zero would make an excellent missile.

---

* Notice that when a line of twelve combines with another line of twelve they form a single line of twenty-four. $a^m \cdot a^n = a^{m+n}$.

Throw a zero at the enemy and it would suck up their entire army. A massive EMP power-vacuum. Currently, there wasn't an enemy and very few zeros had been lining up to volunteer as kamikazes, so Three had pretty much ignored them.

It was time to reconsider. Since the enemy would have no idea how large a brigade was, they could only see the single soldier leading the line, what if GT created scarecrows? Fake brigades that kept the enemy away as they had no idea how many comrades stood behind the decorative lawn fixture?

He enlisted a one and advised his seamstress to dress them up like a major. General Three understood that a soldier with no followers was just a Numberfolk, $2^1 = 2$ as no one marched behind them. In his genius, he realized that as brigades multiplied power, he could create a decoy that held no multiplicative muscle at all. To do this, he needed a Numberfolk that couldn't enlarge its foes. He finally had a place for ones in his army.

The One exited the dress shop decorated to the nines as major 2^0. An empty brigade! A scarecrow stuffed with straw that held no power except to strike fear in the heart of anyone who saw it. As the only Numberfolk without Proliferation power was 1, $a \cdot 1 = a$, GT had created devious diversions that held no soldiers and thus no power. To the outside, the enemy would see a 2 on a hill and have no idea how many soldiers were marching behind it. In fact, the whole thing was a distraction. A 2^0 meant there were no twos at all, not even a leader, a hollow warrior who could only reflect. It was simply a 1.

And so, Three stationed these scarecrow ones all around Numberville. There were 2^0 and 100^0 and even 1^0 with everything after and in between. His scarecrows were a protective measure to let any invading army know that Rational was prepared.*

His armies were trained in the art of the march, he had his scarecrows scattered around, now Three shifted his attention to something new: underground forces. The general imagined a giant brigade that rose out of the desert sand like a worm ready to devour its prey. Unfortunately, he had no idea how to create such a beast.

Some say it was the event with Major 10 that gave Three the idea for his worm. Ten had been assigned fourteen soldiers to her brigade. She

---

* A brigade with no soldiers, $a^0 = 1$ is an empty brigade and thus has the power of doing nothing. The Numberfolk that has no multiplicative power is 1 as $a \cdot 1 = a$ leaves any Numberfolk unchanged.

stood tall as 10^{15} and had an impressive number of possible formations her army could create. She was flexible, able to split into lines of twos and fives and many believe the power went to her head.

One day, Three ordered them to practice their marches. Ten saw her team was tired and overworked. She questioned the point of these endless drills when there was no war. There wasn't even a threat of war. She ignored Three's commands and gave her brigade the day off.

The General couldn't let this stand. The entire structure of his military rested on authority, ranks, and orders. If 10 could get away with disobeying, other majors might too think they could snub his commands. To make a point, he stripped her of her power. GT removed, at first, five soldiers from her ranks knocking her down to 10^{10}. While he felt like this was a just punishment, everyone was shocked by how much her brigade's strength had diminished.

For every soldier Three removed, the strength of the group divided by 10.

$$
\begin{aligned}
10^{15} &= 1000000000000000 \\
10^{14} &= 100000000000000 \\
10^{13} &= 10000000000000 \\
10^{12} &= 1000000000000 \\
10^{11} &= 100000000000 \\
10^{10} &= 10000000000
\end{aligned}
$$

What would happen if he kept going? On a mission, GT looked for the tiniest violations to throw at 10 and her brigade. Each came with the same punishment: the removal of one of her soldiers.

$$
\begin{aligned}
10^{9} &= 1000000000 \\
10^{8} &= 100000000 \\
10^{7} &= 10000000 \\
10^{6} &= 1000000
\end{aligned}
$$

On and on it went until 10 had no army at all. She had been entirely stripped of her ranking as Major and entered the field alone as $10^1 = 10$. In a final act of defiance, she authored a pamphlet. In it she claimed Three was insane, crazy with power, there was no threat — the entire army was unnecessary. Unsurprisingly, the general responded by taking the 10 into custody. He had to take away every last drop of influence she held.

Throwing her onto the vinculum he took another 10 from her, robbed her of herself, making her as powerless as a 1.

$$\frac{10}{10} = 10^{1-1} = 10^0 = 1.$$

With **10** under arrest, the opportunity was right there. Why not run a few more tests?

Negatives were not new to Three. Many residents of TNZ were a part of the army. The only downside was the general always had to ensure any negatives had an even number of soldiers in their brigades, otherwise their power would be in the red. An odd brigade of negatives sucked muscle away. A -2 commanding a brigade of five other lieutenants became . . .

$$(-2)^6 = -2 \cdot -2 \cdot -2 \cdot -2 \cdot -2 \cdot -2 = 64.$$

No different from their twin, **2**, commanding a unit of five soldiers. Adjust this by one soldier and the change is dramatic.

$$(-2)^7 = -2 \cdot -2 \cdot -2 \cdot -2 \cdot -2 \cdot -2 \cdot -2 = -128.$$

They shifted from a formidable unit into somewhat of a useless one. A negative amount of strength was never very intimidating.*

What was new, was the idea of awarding a major a negative number of soldiers. What would this look like? What kind of power would 10^{-1} possess? With **10** in lockup, General Three had everything he needed to answer these burning questions.

$$10^{-1} = 10^{0-1} = 1 \div 10 = \frac{1}{10}.$$

A Logos?! What a beauty. A whole new area for the general to explore. Taking meetings with representatives from both Rational and TNZ, Three's team got to work. When they unveiled their new brigades, all of Numberville stood in shock and awe at the outstanding and innovative formations.

Gathered in the desert, Rationalites stood on all sides of the practice field waiting for the show. The ground began to shake. Numberfolk looked in every direction but saw no army. Had Three developed a way to make his troops invisible?

Then, in the distance they saw them. A 3^{100} marched toward the field. As they neared the stands, the onlookers gasped at the sheer size of the brigade. A legion of one-hundred threes descended on the desert.

---

* In fact, while $2^6 = (-2)^6$, $(-2)^7 = -2^7$ meaning a negative Numberfolk in an odd brigade had the exact opposite amount of power as their evil twin.

The shaking increased; it was violent. Some Numberfolk fell to the ground, thrown off-balance by the vicious rattling. Seemingly out of nowhere a giant sand worm exploded from beneath the grains and revealed itself.

At first, all the crowd could see was another 3, the major in charge of this battalion. He rode upon an enormous vinculum bursting up from the ground. But something was different. His insignia was a negative! It was 3^{-100}. The worm launched towards the sky before diving back underground. When it emerged again, the major was nowhere to be seen but within the belly of the beast, under the vinculum, were one-hundred three-soldiers. Like a protective submarine, the worm opened its mouth and devoured 3^{100} and his entire brigade. In the blink of an eye, the field was empty. A single one stood in the center shaking and alone.

The crowd burst into cheers. The power of the sand worm was undeniable. It could eat an entire army in a single bite. It divided their power, broke apart their ranks, and buried them in the desert sands.

$$a^{-n} = \frac{1}{a^n}.$$

Each lieutenant in 3^{100}'s army paired with a soldier in 3^{-100}'s brigade to form a one.* With this, General Three had everything he needed to build an unstoppable force. He stood next to the quivering one in the center of the field and yelled out to all of Numberville.

"We're in the multiplicative game now!"

* Since $3^{-3} = \frac{1}{3^3} = \frac{1}{3 \cdot 3 \cdot 3}$, this means $3^3 \cdot 3^{-3} = \frac{3 \cdot 3 \cdot 3}{3 \cdot 3 \cdot 3} = 1 \cdot 1 \cdot 1 = 1$, or $3^3 \cdot 3^{-3} = 3^{3-3} = 3^0 = 1$.

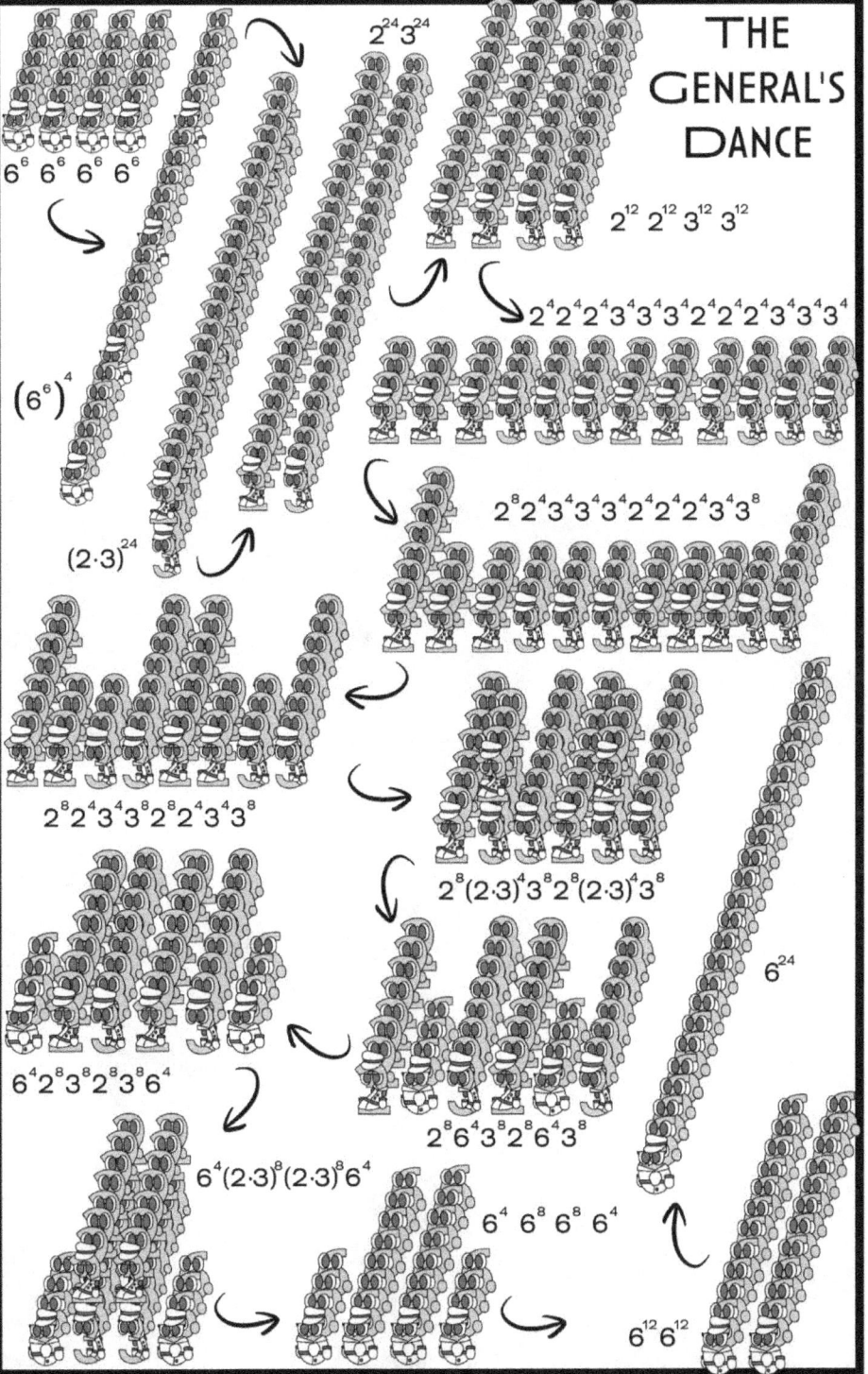

THE GENERAL'S DANCE

$6^6\ 6^6\ 6^6\ 6^6$

$(6^6)^4$

$(2\cdot3)^{24}$

$2^{24}3^{24}$

$2^{12}\ 2^{12}\ 3^{12}\ 3^{12}$

$2^4 2^4 2^4 3^4 3^4 3^4 2^4 2^4 2^4 3^4 3^4 3^4$

$2^8 2^4 3^4 3^4 3^4 2^4 2^4 2^4 3^4 3^8$

$2^8 2^4 3^4 3^8 2^8 2^4 3^4 3^8$

$2^8(2\cdot3)^4 3^8 2^8 (2\cdot3)^4 3^8$

$2^8 6^4 3^8 2^8 6^4 3^8$

$6^4 2^8 3^8 2^8 3^8 6^4$

$6^4(2\cdot3)^8(2\cdot3)^8 6^4$

$6^4\ 6^8\ 6^8\ 6^4$

$6^{12}6^{12}$

6^{24}

How many ways to arrange A, B, and C?

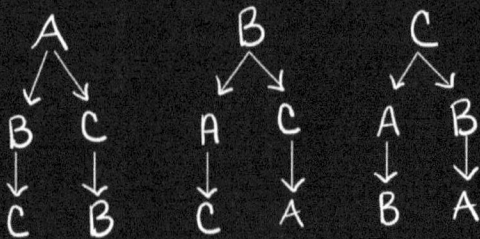

A → B → C, A → C → B; B → A → C, B → C → A; C → A → B, C → B → A

⟹ ABC ACB BAC BCA CAB CBA

$$\underline{3} \cdot \underline{2} \cdot \underline{1} = 3! = 6$$

(3 choices for 1st letter) (only 2 left for the 2nd) (whatever's left!)

3 CHOICES

___ ← How many options for the first letter?

___ ← After one is picked, how many options now?

___ ← How about now?

order matters VS. order doesn't matter

1328 is different than 3218

A code is made by selecting 4 digits, 0-9, from a bag. How many codes are possible?

$$\underline{10} \cdot \underline{9} \cdot \underline{8} \cdot \underline{7}$$

$10! = 10 \cdot 9 \cdot 8 \cdot 7 \cdot 6 \cdot 5 \cdot 4 \cdot 3 \cdot 2 \cdot 1$

$6! = 6 \cdot 5 \cdot 4 \cdot 3 \cdot 2 \cdot 1$

$$10 \cdot 9 \cdot 8 \cdot 7 = \frac{10!}{6!} = 5040$$

ways to pick n things from a group of k when the order you pick matters is

$$\frac{k!}{(k-n)!}$$

143 is the same pick as 341

A lottery has 15 numbered balls and picks 3 for the winning numbers. How many tickets do you need to buy to ensure you win?

$$\underline{15} \cdot \underline{14} \cdot \underline{13}$$

but this counts 123, 321, ...

ways to arrange 3 numbers is 3!

$$= \frac{15 \cdot 14 \cdot 13}{3!} = 455$$

Not bad!

ways to pick n things from a group of k when the order you pick doesn't matter is

$$\frac{k!}{n!(k-n)!}$$

9

STOP, DROP, AND ROLL

"PURE MATHEMATICS IS, IN ITS WAY, THE POETRY OF LOGICAL IDEAS."

-ALBERT EINSTEIN

usky air filled the dark warehouse. A small single lightbulb dangled from the ceiling. It swung back and forth casting wicked shadows around the room. Fredrick stood on a ladder erasing and replacing number after number on his wall.

"You wanted to see me?" A voice came from behind.

Fredrick climbed down and gave the visitor a strong embrace. "I have found the map. He has it."

"Good, this world is making me sick, and my search of his room came up empty," the visitor replied. "Can we retrieve the book?"

"Not yet. We must get everything in line. When the world collapses, we do not want to collapse alongside it. I must go on a short trip to retrieve a key item. It may be necessary for our plan to unfold."

"What should I do?"

"Stay the course. I will contact you with next steps soon enough. The Pattern is growing. I see it everywhere. It is only a matter of time now. Harmony will blossom during the eclipse. Prepare yourself."

* * *

Marco sat on his bed. He entered the first three codes over and over, resetting the box each time. Something about hearing that click was so satisfying. It was intoxicating. Feeling more determined than ever to unlock the map and fulfill his grandfather's wish, Marco started on the fourth code.

Fourth Chamber

1590 Ticks: 2 Tocks: 0

8742 Ticks: 1 Tocks: 0

6259 Ticks: 1 Tocks: 0

3021 Ticks: 2 Tocks: 0

7135 Ticks: 1 Tocks: 0

9413 Ticks: 3 Tocks: 1

4613 Ticks: 2 Tocks: 0

A sickly feeling grew in his gut as he scanned the tocks. There were almost no hints as to the placements of the digits.

"Might as well try," he muttered to himself.

Noticing the seventh clue had three correct digits he decided to start there. Remembering what Penelope said about Set Theory, he drew out two circles on a piece of scratch paper. He compared the digits in 9413 to the digits in 1590 and scribbled 1 and 9 in the circle's intersection.

The code must have a 1 *and a* 9 *since those are the only two digits these clues have in common,* Marco thought. Continuing to use Penelope's methods, he jotted zero through nine on the left before circling the 1 and the 9. Since the first clue had only two digits correct, he crossed out 5 and 0 as those couldn't be in the code. He wrote a question mark next to the digits 3 and 4.

He felt lost. There were too many clues. Too much information to sort through. As if a lightning bolt struck his brain, Marco began frantically scribbling all the clues on his scratch paper with the hope of a new and powerful strategy emerging. Since he knew the code had a 1 and a 9 and didn't have a 0 or a 5, he underlined or crossed out each of these four digits in every clue.

The code became clearer. The third clue had only one digit correct, which he already knew was the 9. He diligently crossed out every 2 and 6 on his paper.

Code	Digits	Places
1~~5~~9~~0~~	2	0
874~~2~~	1	0
~~6~~~~2~~~~5~~9	1	0
3~~0~~~~2~~1	2	0
713~~5~~	1	0
9413	3	1
4~~6~~13	2	0

"Aha!" He shouted aloud. Noticing that 3021 contained two correct digits, and the only remaining possibility was the three, he concluded the code must contain a 3. Scanning back to 9413, he saw he could eliminate the 4. More crossing. More underlining.

"Crap." Marco threw down his pencil. Something was wrong. His method left him with the digits 1, 9, and 3. This meant 7135 had at least two correct digits but the clock insisted there was only one. His grandfather's clues laughed in his face. The angry one grabbed onto the manic five and did a tango.

Calm down, try again. He attempted to build up his confidence and shake the Numberfolk faces away. Starting from the beginning, he looked at the first and sixth clues: 1590 and 9413.

"I thought, because they both have a 1 and a 9, those had to be the right digits, but are they?" Marco asked himself. He began to play with the possibilities.

"What if the correct digits in 1590 are something like 1 and 5 and the three correct digits in 9413 are 4, 1, and 3?" He scanned over the paper. He couldn't see anything that said just because both had a 9 that 9 had to be correct. He remembered Mr. Pikake's story about Euclid, how his one assumption caused him years of ridicule. "A bad assumption changes everything," he sighed.

Almost ready to give up, he remembered he could control the numbers. Maybe phasing could work in his favor? Concentrating, he allowed the Numberfolk to come to life. Rather than hide, he embraced each personality. Zoning in on 7135, he noticed how no one got along. They were all arguing and pushing at each other. "Of course," he chuckled, "it's because only one of them belongs in the code. They don't want to take anyone else with them." Knowing that 1 refused to tagalong with 3, he focused on 9413. There, the 9 and the 4 were dancing together. Every once in a while, they'd grab the angry 1's hand and twirl together. The 3 was intense and serious. Trying to cut in, he pushed the 1 to the side before joining the 9 and 4 spinning along merrily. Even as they frolicked, the 3 kept a frown painted on his face.

"I see," Marco spoke to the numbers. "The 1 and the 3 don't get along. So, it must be 941 or 943 in the code!" The three looked directly at him and nodded in approval.

He crumbled up his paper, threw it to the side, and pulled out a fresh piece. The second he finished scribbling the last clue, the 6 began waving wildly trying to get Marco's attention. He leaned down to the paper . . . he couldn't hear anything.

"Okay. This 6 is trying to tell me something. The code has a 4. And since it has a 1 or a 3, but not both . . ." The number began acting out a death scene. It looked like it had been poisoned as it dramatically grasped onto life. A grin stretched across Marco's face. "It can't have a 6." New table. New crossing. New underlining.

He focused in on the fourth clue. Luckily, the 1 and 3 were separated, unable to attack each other. "Two digits correct. One of them is the 1 or the 3 which means the other digit has to be the 0!" As he continued his new method of crossing and underlining, he cringed as with each cross, he seemed to kill the Numberfolk. They fell in defeat. As he underlined the 0 in the first clue, the 1 saw their fate. Before Marco even had the chance, it laid down like the king in a game of chess who understood there was no path to victory. "If

Code	Digits	Places
~~16~~90	2	0
~~87~~4~~2~~	1	0
~~625~~9	1	0
30~~21~~	2	0
~~71~~3~~5~~	1	0
94~~1~~3	3	1
4~~61~~3	2	0

there's a 0 and a 9, there can't be a 1, it has to be the 3!"

With that, he had narrowed it down. "The digits are 9430, but what order?" Remembering the "tocks" had been nearly nonexistent, Marco let out a loud groan.

"Wait, this might not be that bad," he posited. For a moment he saw the beauty in knowing nothing — *nothing* was something! His one clue on placement told him that either 9 was first, 4 was second, or 3 was last. The final 3 slowly tipped over as the ultimate clue showed Marco the intense digit couldn't remain in the final slot. As he scanned up the list, 3 after 3 fell. The fifth clue showed 3 wasn't third and the fourth that 3 wasn't first.

"If 3 isn't first, third, or last, it must be second!" Marco fell back into his original problem-solving method of writing four lines. He scribbled the 3 in the second blank.

"And, if 3 is second, then 4 can't be second in 9413, so the code must start with a 9."

9 3 _ _

"Okay, I just need to find where the 4 and the 0 go." Within seconds he had it. He took a deep breath and began to turn the dials. Nine. Three. Zero. Four. Click.

Marco wanted someone to high five. He desperately searched for anything to share his excitement with. At first, he looked to the Numberfolk. They had all returned to their slumber. He settled on an old teddy bear. He held up its arm to slap its paw and then flipped it in the air a few times.

"We're halfway there!" he yelled to Teddy.

He felt pathetic. He wanted a friend to share this with. "Nice move, Weston," Marco muttered to himself.

Weston was an inside joke between Marco, Oliver, and Liam. The three had a sleepover with their friend Weston in elementary school. They all shared two huge inflatable mattresses in the living room and watched movies half the night until they eventually fell asleep in the early hours of the morning. Before the sun had risen, Liam awoke to a warm feeling. They all soon learned Weston had peed all over the shared bed. Ever since then, the boys referred to a *Weston* as an embarrassing moment. Not only did Marco feel like a Weston as he sat there celebrating with a teddy bear, it was even worse attempting to share an inside joke with an empty room. Marco was lonely. Between the people he pushed away and those who had been ripped from him, he felt more alone in the world than ever before.

A clear picture began to form in his head. It was Penelope. After fifteen minutes of waffling back and forth, *I should call her. I clearly should not call her,* he decided to take the plunge. Maggie and their mother were back at the community pool. They wouldn't return for at least another hour. Luckily, his sister had a propensity for old-timey things. She had once spent the entire spring insisting she was an *old soul.*

Marco crept into her room. It was unclear why, since no one else was home, but the sheer act was cause for secrecy. He pulled open her desk drawer and easily found her address book. Maggie took the time to include all her friends, enemies, classmates, and acquaintances in this sacred book. She had some special system of marking each contact, but Marco had never bothered, or cared, to decode her secret symbols. He found Penelope's number and entered it into his phone before running back to his room and closing the door.

Marco shook his shoulders and stood up very straight as the phone began to ring.

"Hello?" Penelope answered in a sweet lyrical tone.

Marco cleared his throat, "A-ahem, *um*, yes, Penelope?" His voice came out like an old man or a telemarketer, too formal. "This is Marco." He managed to sound more casual on the second sentence.

"Oh, hi Marco!" The fact that she seemed genuinely happy to hear from him sent chills down his spine.

"I am working on these codes. *Um*. For Maggie. And I was wondering. Do you want to help? Again?" Marco hated everything he said the moment he said it. It was all wrong.

"Sure! That sounds like fun. I can be over there in about fifteen minutes. Is that okay?" Marco was thankful his weirdness didn't seem to impact her.

"G-g-great. I'll see you then!" Marco quickly hung up. *G-g-great.* He thought, *What are you, Tony the Tiger?* He hit his palm on his forehead.

Quickly combing his hair and throwing on a clean shirt, Marco scribbled the next clues on a blank sheet of paper. He was heading downstairs just as the doorbell rang.

"Penelope!" He sounded surprised. *You dummy,* he thought.

"Ready to solve some codes?" Penelope asked enthusiastically. She raised her hands to her shoulders and shook them back-and-forth which looked like a mix between a boxer and a robot dance. It was so corny. Marco loved it.

"Let's do this!" he responded excitedly.

Sitting side-by-side on the couch, he presented her with the clues.

Code	Digits	Places
1234	2	0
5678	0	0
9012	3	1
3409	3	0
9230	2	1
4190	4	2
1490	4	1
4091	4	1

"I don't think you need my help on this one," she said playfully. "I guess you just wanted to see my pretty face."

Marco's cheeks turned three shades of red. He hadn't paid any attention when writing down the clues. He saw it now. The 4, 1, 9, and 0 danced together in harmony twisting and turning to complete cooperative formations on the last three lines of the page.

He quickly began putting together the pieces as an invisible 3D-puzzle took shape in front of him.

The numbers moved around. Marco wasn't sure if he was commanding them, or they were doing it on their own to help him out.

The 4 had to be first or third, the 0 couldn't be third, the 9 couldn't be last, the 1 couldn't be first.

The pieces fell from the air, there was too much to keep track of. "I. *Um.* I know the digits are 1490, but finding their order was harder. I thought maybe it was a good chance to learn more about your cases?" Marco swallowed hard.

Penelope looked back at the clues. "Oh yeah, I see what you mean. Deceivingly simple," she chuckled. The tiniest hint of a snort came through. "At first glance, all the digits can be almost anywhere, except 4, that has to be first or third."

She drew a box and split it in half. "Let's suppose the 4 comes first."

Marco's nerves were starting to calm down. "Okay. Well, if 4 is first . . ." he looked at the paper. All the other digits in the last clue lay slain on the ground. ". . . then the last clue says 0 isn't second, 9 isn't third, and 1 isn't last."

"Right!" Penelope chimed in, pointing towards the fourth clue. "Here, 9 can't be last either, so it must be second. Since 1 isn't last, 0 is."

"So 1 is third! The code would be 4910."

"Exactly!" She wrote the result in the first box. "Now let's see what happens if we instead assume 4 is third."

"Well, that . . ." Marco hesitated, "that's not so easy. The other way narrowed things down, but if 4 is third, there are still a lot of options."

"You're right." Penelope bit on the end of her pen. "Oh here, the penultimate clue."

Penultimate? Marco had no idea what she was talking about.

"It says 1490, and if 4 is third, that means 9 can't be, so that isn't right. It also means 4 isn't second, so that isn't right either. We know from the first clue that 1 isn't first. So . . ." She stopped as if passing the ball to Marco for the layup.

"The 0 must be last," Marco scored! He felt particularly good as he also had decoded that penultimate must mean something like second-to-last.

"And that tells us the 1 must be second and the 9 first!" Penelope wrote 9140 in the second box. "Now, we just need to figure out which one of these is correct."

"4910 or 9140. 4910 or 9140," Marco repeated with the hope that one of the numbers would do a jig and let him know which was correct. "I guess we could just check every clue."

"Yeah. Except the second clue because that is *totally* worthless." They laughed.

"Okay. 4910 works with clue three and clue four," Marco confirmed.

"It also works with clue five," Penelope chimed in.

"And clue six!" They said in unison. Marco felt like there was an unspoken connection since she had shared her story about being nicknamed Six.

"It works for the last two clues as well," Penelope broke up the moment.

"So, I guess it's 4910 then?" Marco asked.

"It makes sense, but I'd feel better about it if we checked 9140 too."

The two leaned in, heads down over the paper. "It breaks rule

five," Marco was the first to notice.

"Yep. Rule five says 9230 has only one digit in the correct place. If the code was 9140, that should be two digits."

"It's settled then. It's 4910." Marco circled the number on the sheet of paper. "Have time for one more?" he asked. His voice slowly raised in pitched unveiling his hopefulness.

Penelope looked at her Minnie Mouse watch. "I have to meet my mom at the tennis courts in about twenty minutes."

"They are *still* making you play tennis?" Marco asked.

"Family tradition," Penelope moaned.

"Okay, well let's see how far we can get." Marco ran so fast up the stairs he tripped on the final step and face-planted hard into the floor.

"You okay?" Penelope shouted from downstairs. *She must have heard the thump.*

Code	Digits	Places
0987	2	1
4321	1	0
5947	2	0
5608	2	0
6784	2	0
0476	3	3

"Yeah! Fine!" He jumped up and quickly scribbled down the next set of clues before sprinting back to the living room, thankfully without incident.

"How many of these do you have? This is a serious project for Maggie. You two must be really close."

Marco had nearly forgotten about his fib. "*Um*, eight. This one is number six, so almost there!"

"It must be so awesome to have a sister. I'm an only child. I feel lonely all the time. I wish I had a brother like you."

Great, another little sister, just what I need, Marco thought. He flashed her a quick smile.

"Okay, let's dig in!" Penelope said enthusiastically. She suddenly began rapidly writing codes on the paper.

047, 046, 476, 076

"What are those?" Marco asked.

"The number of combinations on a set of n objects."

"A what?" She might as well have been speaking Greek.

"Okay, think about this. If you have four shirts, three pairs of pants, and two pairs of shoes, how many outfits can you make?"

Marco's mind went to work as he started trying on imaginary clothes. He made the shirts all different colors so he could easily track them: one blue, one green, one black, and one white. Pants were harder. He selected one pair of jeans, one pair of black slacks, and a pair of shorts. For his final options he picked sneakers and dress shoes.

He started to get dressed. As this was all happening in his head, he quickly realized picking random outfits wouldn't work. There was no way he'd be able to remember what combinations he had already tried on, that was an impossible task. Deciding to make all the outfits he could with the jeans first, he slipped on the denim and the sneakers and then proceeded to work through each shirt. He took a second to admire the look of the white shirt with the sneaker-jean combo. Realizing he was off track and Penelope was waiting for his answer, he hurriedly changed.

There was a total of four outfits with jeans and sneakers, one for each shirt color. He kicked off the sneakers and replaced them with the dress shoes. That was easy, there were four more. He had eight total outfits so far.

Swapping the jeans for the black slacks, he figured there would be eight outfits with these pants as well. Four with each of the shirts and sneakers, and another four with the dress shoes. The white shirt went well with everything. He was going to need to remember to ask for one on the next holiday.

Eight with the jeans, eight with the slacks, the pattern was evident. There would be another eight options with the shorts. Although he was dying to know if the white shirt also looked good in the final outfits, he didn't have time. Eight plus eight plus eight, "Twenty-four!" He snapped out of his daze and looked toward Penelope, "There are twenty-four different outfits."

"Exactly! First, you pick your shirt out of the four choices. Next, you grab one of three options for pants, and finally you select one of the two possible pairs of shoes. That makes the total number of outfits four times three times two or twenty-four." Penelope's method made so much more sense than Marco's imaginary fitting room. "When it is just counting options, it is easy, multiply. Now, what about poker? How many possible hands of five cards are there in a deck?"

A deck of cards floated up, then scattered across the room. Marco grabbed a card, there were only 51 left to pick from, he got the gist. "Okay, so there are 52 options for the first card, then only 51 for the second because one is already dealt, 50 for third, 49 for the fourth, and 48 for the last. I don't want to do that multiplication." He was

particularly thankful for Penelope's method as he considered how many imaginary rounds of poker he would have needed to play to come up with that.

"It's three hundred and eleven million, eight hundred seventy-five thousand two hundred," Penelope declared carelessly.

Marco's jaw dropped. Literally. Penelope laughed. "I'm, *um*, good with numbers. But I sort of tricked you. You were right in that you found how many ways to pick five from a group of 52, except, you over counted some of the possibilities."

"Over counted?" Marco was sure he had barely even counted once, definitely not more than once. "What do you mean?"

"Well, in using the multiplication principle you assume . . ." There was that word again. *Assumptions are the worst,* Marco thought. ". . . that each pick is unique, it's different. When you multiplied $52 \cdot 51 \cdot 50 \cdot 49 \cdot 48$ you counted the hand . . ." she drew on the paper.

A♠, 2♣, 3♦, 2♥, 10♣

"But you also counted the hand . . ."

10♣, A♠, 2♣, 3♦, 2♥,

"And the hand . . ."

A♠, 10♣, 2♣, 2♥,3♦,

"And, well, you get the picture."

Marco studied her drawings. He particularly liked how she took the time to sketch out each card suit. Her writing was bubbly, like her. Everything had smooth curves and sharp points. The spades looked like an upside-down heart with a stem. He smiled.

"I get it. In some things, like an address, the different orders *are* different. Living at 1234 Elm, is different than living at 4321 Elm. But, in other cases, the different orders *aren't* really different. It doesn't matter which card I get first. It only matters that I was dealt the ace of spades." Marco shivered remembering the ace of spades is often called the death card.

"Exactly! So out of that three hundred and eleven million, we need to get rid of all the duplicate hands. We need to divide."

"But by how much?"

Marco was enjoying this. He had seen scenes from the movie *21* and ever since had a fascination with card counting although he never actually understood how they did it.

"Well, I think you know how much," Penelope gave a shy laugh. "How many ways can you arrange five cards?"

"Oh! Yeah, that is just five times four times three times two times one."

"You got it! Five choices for the first card, only four for the next and so on. You have successfully developed how to count both when order does and does not matter."

"I have?" Marco didn't feel like he had "developed" anything.

"Sure! If order doesn't matter, you just multiply." She scribbled on the paper.

$$P(n,r) = \frac{n!}{(n-r)!}.$$

Marco was certain he didn't develop *that*. What was *that*?

"That's more than multiplying," Marco said cautiously.

Penelope laughed, "Yeah, I guess it does look a bit intimidating. That's a thing with math. It's like reading a foreign language. But everything really just boils down to adding and multiplying. Thankfully, mathematicians are lazy."

"Thankfully?" Marco certainly didn't find the complex combinations of letters, numbers, and symbols a blessing. They were more of a dark spirally hole of death and confusion and panic.

"Yeah, like if I wanted to add 3 ten times, it would be a pain to write three plus three plus three . . ." she nodded her head along to the beat of the threes to emphasize the effort. "So we just write three times ten. And if I wanted to multiply 3 ten times, same thing, so we use exponents. The exclamation mark is just saying to multiply all the natural numbers before it. So 3! is just 3 times 2 times 1. Small numbers aren't horrible, but imagine 100!, that'd take *forever* to write out."

It was an interesting perspective. Penelope was saying that the symbols and the letters, the things that make math *look* really hard were just a shorthand meant to make things less hard. It was the yada yada yada of numbers.

He wondered if there was an equivalent for English that could get him out of his word count. One time, his teacher had requested a five-sentence description and he had only been able to come up with three. Marco proceeded to go back in and add random periods and capital letters to give the illusion of extra sentences. His teacher didn't find it amusing. The fact that math already had this built in was actually sort of nice. It *encouraged* laziness. He refocused on her equation.

"Okay," he smiled, "I can get that, but what are we dividing by?"

"Oh! Yeah, when you wanted to find all the poker hands, you multiplied $52 \cdot 51 \cdot 50 \cdot 49 \cdot 48$, which is sort of like a factorial, an exclamation mark. But, you wanted to stop at 48, the dividing just stops it from going all the way to one. Try it out. You have 52 cards, so n is 52 and you want to pick 5, so r is 5."

Marco smiled, he was a professional substituter. "Okay." He wrote on the paper,

$$P(52,5) = \frac{52!}{(52-5)!}.$$

"Keep going!" Penelope urged.

Solving for the parentheses he added,

$$= \frac{52!}{47!}.$$

He saw it. Mirrors on mirrors on mirrors. It was an infinity mirror at a fun house reflecting over and over again. An enormous smile began creeping up his face.

"Since 52 *factorial,*" he wanted to make sure to use her words and had listened closely enough to pick up the fact that an exclamation mark in math was apparently called a factorial, "is just 52 times 51 times . . ." he faded and started writing,

$$52! = 52 \cdot 51 \cdot 50 \cdot 49 \cdot 48 \cdot 47 \cdot 46 \cdot \cdots.$$

"And 47 factorial is the same starting at 47."

$$47! = 47 \cdot 46 \cdot 45 \cdot 44 \cdot \cdots.$$

"When you divide them, you get a bunch of ones,"

$$= \frac{52 \cdot 51 \cdot 50 \cdot 49 \cdot 48 \cdot 47 \cdot 46 \cdots}{47 \cdot 46 \cdots}.$$

"And all you are left with is what I wanted, $52 \cdot 51 \cdot 40 \cdot 49 \cdot 48$."

"You got it! All the rest cancel."

Marco cringed. He couldn't help himself. His words sprinted to his lips bypassing his brain entirely. "They don't actually cancel. Canceling is to get rid of, they aren't going anywhere. They are mirrors, hidden ones, and when you multiply by 1, it reflects."

If Penelope was going to be a part of KFUN, Marco had to make sure she stayed on Mr. Pikake's good side. The SAN was serious about The Battle of Cancellation and adamant that only when making a zero, making nothing, did cancellation occur. Better she heard it from Marco than get an earful from the professor.

"I could see that!" Penelope responded.

Marco smiled. He was thankful she took it well. "Okay, but, that is what we already figured out. The cards don't follow that pattern because that counted the same hand again and again."

"Right! We need a combination. Combinations are when I care about the order so I don't want to count the same thing a bunch of times.[*] You found that. Remember, you said to divide by 5! So . . ."

$$C(n,r) = \frac{n!}{r!\,(n-r)!}.$$

Marco went to work. "Okay, so how many possible hands in poker? I've got 52 cards and I want to pick five of them."

$$C(52,5) = \frac{52!}{5!\,(52-5)!}.$$

He continued without prompting.

$$= \frac{52!}{5!\,47!}$$

[*] If you read a math textbook, it will say this the other way. It will say a Permutation is when "order matters" and a Combination is when "order doesn't matter." It doesn't really matter. The important thing is that if 123 and 321 are in fact the same thing you need to divide to group all those possibilities together and treat them as just one thing.

$$= \frac{52 \cdot 51 \cdot 50 \cdot 49 \cdot 48}{5 \cdot 4 \cdot 3 \cdot 2 \cdot 1}.$$

Marco saw the opportunity to be clever and show off a bit. Ripping the Numberfolk apart was a master skill he possessed. If he looked at the 48 as $2 \cdot 2 \cdot 3 \cdot 4$, he could make ones with the 2, the 3, and the 4. The 49 was useless, that was made up of only sevens, but the 50 was helpful. That had a five trait he could use.

$$= \frac{52 \cdot 51 \cdot (5 \cdot 5 \cdot 2) \cdot 49 \cdot (2 \cdot 2 \cdot 3 \cdot 4)}{5 \cdot 4 \cdot 3 \cdot 2}.$$

"There are hidden ones with the five, four, three, and two! So that leaves us with . . ."

$$= 52 \cdot 51 \cdot 10 \cdot 49 \cdot 2.$$

"Uggh," Marco grunted. "It's still so ugly."

"It's not too bad! But it is still big." After a *very* brief pause she concluded, "It's two million five hundred and ninety-eight thousand nine hundred and sixty!"

She is so impressive, Marco thought. "H-h-how?" He stuttered.

"I don't want to bore you but . . ." She scribbled quickly and spoke even faster. "The $51 \cdot 49$ is easy. It's just one less than $50 \cdot 50$ so it is 2,499. Double it to take care of the 2 and you get two less than 5000 or 4998. Multiplying by ten is then just 49,980. So all I had left was the 52. Since 52 is just $2 \cdot 2 \cdot 13$, I could double the 49,980 twice. Because it is 20 less than 50,000, I just doubled 50,000 to get 100,000 and took away 40 to get 99,960. I doubled that to get 200,000 minus 80 or 199,920. Last step was to multiply by 13. Which is hard. But since 13 is really $10 + 3$, I used that. Multiplying by the 10 is easy, 1,999,200. Finally, 199,920 times 3, which was the most annoying bit of this whole thing, but again I just rounded it up to 200,000 multiplied that by 3 to get 600,000 and then took away the extra 80 I over counted three times or $600{,}000 - 240$ and that gave me 599,760. I just needed to add 1,999,200 to 599,760. Too much to keep track of, so I rounded it to find 2,000,000 plus 599,760 or 2,599,760 and then took away the extra 800 which of course gave me . . ."

"Two million five hundred and ninety-eight thousand nine hundred and sixty," they said together. Marco didn't really follow what she had explained at all. He stood in awe over her control of the Numberfolk. He had always been taught the standard algorithm, which was impossible to complete in his head, too many numbers floating around. Penelope had ripped the numbers apart and pieced them back together never having to complete an even mildly challenging multiplication problem as she rounded, stole, and gave back like the Robin Hood of numerals.

"I forgot why we were even talking about this!" She giggled with a snort.

"The code!" They both shouted and darted their eyes back to the paper. "Oh yeah! It's a combination. In the last clue, three of the digits are correct. So that means we need some group of three out of the four numbers."

"And we don't care about the order because we are just looking for the digits right now," Marco followed.

Code	Digits	Places
0987	2	1
4321	1	0
5947	2	0
5608	2	0
6784	2	0
0476	3	3

"Yeah! So, four choose three, or out of four picking three items, is just four factorial divided by three factorial."

"Which is just four," Marco chimed in.

"And then divided by one, is still four! So, 047, 046, 476, and 076 are the only possible combinations of three correct digits. One of these has to be the one we are looking for."

"Are you thinking cases?" Marco asked.

"Okay, sure! We can give that a try. Why don't you take the first two and I'll take the second two?"

Marco began working on the case of 047 as the correct digits.

"It isn't 047, that won't work." Marco found the ill-fitting code quickly. It was impossible to have these three digits as there was no option left for a final number.

"It isn't 476 either." Penelope didn't even raise her head.

"I think 046 could work. The last digit would have to be a 9." Marco finished his part of the assignment.

"Okay, 076 isn't looking good. So, let's go with 0469, what order?"

Both their heads dropped simultaneously before quickly breaking out in laughter. It only took one glance at the clues to unlock the code.

"I've gotta run. Thanks for the invite!" Penelope said merrily.

Marco gathered their papers and walked her to the door. As soon as it clicked shut, he immediately ran back upstairs. His mother and Maggie would be home soon, he needed to hide all the evidence. Scared he would forget the codes, in a spy-craft move he was particularly proud of, Marco grabbed a Post-it from the stack and placed it on one of the pieces of scrap paper before jotting them down. He had seen the pencil rubbing trick way too many times before.

8195
6302
9013
9304
4910
0496

He then carefully folded his grandfather's note and placed it in the top corner of his closet under a shoebox and stuffed the cedar chest back in the box spring of his bed. Finally, he dug around and found his bag of baseball gear. He took the Post-it with the codes and stuffed it inside his left cleat.

There. Everything is safely separated. No one can find them all.

His last item of business was to get rid of all the scratch paper they used. He had seen this on movies and TV shows dozens of times. He emptied out his metal trash basket and crammed all the papers inside. He ran downstairs to the kitchen junk drawer and retrieved the long stick lighter his mother used for candles before sprinting back up to his room. He lit the papers on fire.

At first, a small burn slowly enveloped the contents. It was fascinating to watch how the once pristine white and crisp sheets melted away turning to ash. In the blink of an eye, the fire roared. Marco couldn't even see the papers below the flames. He started running in circles, looking around for anything to stop the blaze. He found a bottle of orange Gatorade and threw it at the trashcan. The liquid quickly disappeared having almost no effect at all. He was pretty sure the smell of burning plastic was starting to seep into the air.

Stop. Drop. Roll. Stop. Drop. Roll. His brain was remembering what it had been told to do in a fire. He started to approach the flames but stopped. *How am I going to roll this?* He wondered.

"Oxygen!" he yelled out. Knowing that fire requires oxygen to burn, he decided to try to snuff out the flames. He grabbed a blanket off the foot of his bed and threw it over the trashcan. Smoke plumed up from the tiny invisible slits in the fabric.

Marco dropped to the ground and let out a huge sigh of relief. An unmistakable clamp compressed his chest as a fast-moving vehicle began speeding up his esophagus. He turned just in time. Vomit spewed from his mouth into the trash can, right on top of the blanket.

The front door slammed.

"What's that smell?" he heard his mother call out.

He pushed the evidence, puke and all, into his closet and slid it shut. He grabbed a book from his nightstand and jumped onto his bed as his mother opened the door.

"We're home! Did you burn something? It smells like smoke."

"*Uh,* oh, yeah. I had some soup. Got a little charred."

His mother's eyebrow raised but she accepted his explanation. "Be careful, your sister is on a rampage," she winked before closing the door behind her.

Marco hopped up and ran to the bathroom to splash some water on his face. He swished the liquid around his mouth and spit it out, trying to remove the horrible lingering taste. When he arose, he saw his sister's reflection in the mirror, standing in the doorway, arms crossed.

"I ran into Penelope on the way home. She said she was *here*. With *you*. Care to explain?"

This was just what he needed. He didn't have anything to say. He felt like he was drowning in the lies. "Oh yeah, she offered to help me with this project."

"That's what *she* said. But she wouldn't tell me anything about this *special* project. You know, you have your own friends. Why do you feel the need to steal mine?"

"Oh, I'm stealing *your* friends?" Marco had enough. "What about Mr. Pikake? He was *mine,* now you take over every meeting with your chatter. You take over KFUN with *my* friends! You take over *everything* Maggie! Who are you to talk?"

She stormed away. He wasn't sure if he had hurt her, but he didn't care. Everything he said was true. As his sister's bedroom door slammed, he dragged himself across the hall to figure out how to clean up the vomit-blanket-ash mixture.

BREAKING A SECRET CODE

While a 4-digit code might not seem difficult to crack, here are the 5,040 possiblities.

ZERO

ONE

TWO

THREE

FOUR

FIVE

SIX

SEVEN

EIGHT

NINE

That's a lot of numbers...

10

THE DOT

*"NUMBERS HAVE LIFE;
THEY'RE NOT JUST SYMBOLS
ON PAPER."*

- SHAKUNTALA DEVI

Do not
remove
bar

As General Three and his army were busy running drills, the Village erupted into chaos. Naturals were banging down the mayor's door demanding their own passes to explore Numberville. Still upset about the exit tickets, watching Three's army leave the gates each day just fueled the fire.

Three erected posters all over town describing the brigades. It was military propaganda with the hope of gaining new recruits. While soldiers were allowed to leave the Village, they could only do so during working hours and were not permitted any leeway to explore Rational.

Thirty-six was a wreck. She didn't know what to do. There didn't seem to be any logical decision that would appease everyone. It was at this time the plague began.

The story goes that one day she was yelling at her aide. "We cannot open the gates! It would be a free for all, we'd lose everything that makes Natural, well, natural! You've read the reports from the expedition team, I cannot allow our citizens out into that anarchy. But still, they demand to be given freedom. What kind of leader am I? What kind of leader will I be remembered as, if I deny them liberty, their basic right as a living being?"

She turned to the aide whose face was frozen in a stare of bewilderment and shock. Thirty-six noticed they weren't looking *at* her, they were looking *behind* her.

"What ARE you looking at?" Thirty-six pressed. The aide only pointed. She turned to see nothing, but she could feel *something*, it was on her back. Flipping one way then the next, she tried to grab at it but couldn't quite reach. "What is it?" she screamed at the aide.

"It appears." They cleared their throat, "It appears to be a dot."

"A dot?"

"Yes, a . . . dot."

Finally able to tilt her head at just the right angle, Thirty-six saw it too.

It wasn't anything particularly alarming. Nothing more conspicuous than a mole or a freckle, but it was peculiar. A perfectly round dot. She tried to shake it off, scrub at it, but nothing worked. The dot followed her around everywhere.

Ultimately giving in to the resident pressure, she opened the gates and allowed all Naturals to come and go as they wished. While Rational had been an interest to the citizens, it quickly transformed into a fixation.

With the gates open, Naturals trekked to the outskirts of Rational regularly. They'd come home looking as if they spent the day at the circus. A 7 entered as 91/13 and an 11 walked home as 77/7. Naturals had trouble recognizing the neighbors they'd known their whole lives.

The problems grew quickly. Traditionalists were in Thirty-six's office daily. The Village had become chaotic. It was louder, less orderly. Numberfolk came and went at all hours of the night, and they didn't like the disguises, not one bit.

A town hall was called in which Thirty-six took the time to hear both sides of the argument. The pro-liberators made passionate speeches. Rational had given them a new lease on life, they could express themselves in many different forms, they were no longer forced to conform, and finally felt a belonging, a freedom, to be themselves. The anti-liberators cited the normal complaints. Natural was less safe. It was no longer perfect. It was certainly not orderly. Some questioned if walking around in costumes and vinculums was even natural at all. Thirty-six felt like a failure. Was this the world her grandfather had dreamed about?

As crisis after crisis erupted, dueling was on the rise as arguments were commonplace. After screaming at her entire team for most of the day, Thirty-six was thankful to climb into bed. When she awoke the next morning, she felt . . . different.

The sickness was spreading. No longer just a dot, Thirty-six had become 36.0. Her closest friends insisted it was "nothing." And although they were right (zero is in fact nothing), the extra digit weighed heavily on her mind.

The plague washed through Rational Country. Naturals reacted in different ways. Some developed the dot, some had the dot as well as a few zeros that followed them around like puppies, while others had a trail of zeros five miles long. Natural was not the only place impacted. The open gate policy allowed the plague to spread far and wide. A ½ awoke to find themselves unrecognizable as 0.5. They rushed to a mirror and changed into 3/6. It lasted only seconds before POP, back to 0.5.

Some Numberfolk had it much worse than others. While Thirty-six and

½ were only cursed with minor symptoms, others were horribly afflicted. The Thinkers, an elite group of Numberfolk from ancient times had been all but forgotten. Yet somehow, all that has happened will happen again and Thirty-six had the brilliant idea to gather Numberfolk from all around to study the plague and search for a cure. She called this team the Erudites.

The Denarii

Numberfolk had long believed their ancestors, the charmed ones, consisted of ten divine beings. The problem was, no one could really *prove* this was the case. This caused many sects to emerge over the years with divergent viewpoints.

The Binary believed the entire universe was birthed from only ones and zeros. Many felt the Binarys were an exclusive group, only Numberfolk made from the two digits were welcomed in their ranks. The group spent years trying to change the narrative and explain that all were welcome in Binary, you simply had to accept your new name. A 3 would now be called 11, while a 36 was $10\ 0100$. As you can imagine, it was a tough sell. When a 425, already self-conscious about their three digits, learned they'd have to be called $1\ 1010\ 1001$, it was a hard no.

While many groups gained followers, the largest and most impactful consortium were the Denarii. The Denarii worshiped the number ten. The 10's were the pillars of the community, but any 100, or 1000, or $10{,}000$, and so on also held a special exalted place in the clergy. For it was only with these values, and the charmed ones, they believed any Integer could be built. For instance, a 76, was nothing more than 10 Septem's attached at the hip to her brother Exee.

A secondary group within the Denarii also held great power. For 10 itself was not a charmed one, it was simply the number of charmed ones in existence. Perhaps more powerful were the images of Zwei and Kween and their children.

While some Numberfolk were rich in traits, the noble 10 could be built only from 2's and 5's. The Denariati or "little Denarii" as they were also called, consisted of all the Numberfolk that could be built from this duo of traits.*

It was quite remarkable just how many Numberfolk held this special

* This is called Prime Factorization and is vital in understanding the Numberfolk. To find a number's traits, you must tear it down into its most basic building blocks, its DNA. For example, the traits of 28 are $2, 2,$ and 7 as $28 = 4 \cdot 7 = 2 \cdot 2 \cdot 7$.

power. There was 4, 8, and 16, which all had only two-traits, but also 25, 125, 625, and all the many more that possessed only five-traits.

Any Numberfolk that had both a two and a five in their DNA was a multiple of the mighty ten such as 20, 40, and 50. Note 30 was not considered a divine Numberfolk as $30 = 2 \cdot 3 \cdot 5$. The fact it contained the evil three banished it from elevation.

To be a member of the Denariati, a Numberfolk could have no twos or many twos and no fives or many fives, but their genetic chart must only contain 2's and 5's, no other traits were allowed.

Originally, the Denariati was only open to the Integers. As Numberville grew, so did the group. They began accepting any Rationalite whose number beneath the vinculum was one of them. This may seem like an odd choice, why would they accept 3/25 but not 25/3? Their reasoning was clear. They viewed the Numberfolk below the vinculum as a being, the value atop the vinculum was nothing more than the number of broken pieces.

For 3/25 was a whole that had been broken into 25 pieces, and 25 was a Denariati, one of them. The 3 only signified how many of the broken pieces they were allowed to keep. On the other hand, 25/3 was greedy and improper, it had managed to collect 25 shards, each group of three able to make a whole, but did not represent a member of their group any more than 12/11 or even 1/7 would.

It was ultimately the Denariati that led the Erudites to their first clue. While the group was certainly not immune to the plague, it was the only community in all of Numberville outside Integer State where every member had only minor symptoms.

At this point, other than the dot and varying levels of zeros, Naturals were otherwise unaffected by the plague. This same phenomenon happened within the communities of the Denariati. The 1/25 became 0.4, which was shocking, but didn't otherwise affect the Numberfolk. They accepted him, as 4 was also a sacred number. Interestingly 1/8 became 0.125 and 7/20 turned into 0.35. While 125 was a revered number, 35 was not. The council debated if the infected 7/20 would even be allowed to stay in the sanctuary.

The Erudites cared nothing for the ways and beliefs of the Denariati, only that every infected being had symptoms which terminated. Elsewhere in Rational that wasn't the reality. In one of the worst and most baffling cases, an 8/7 became $1.142857\overline{142857}$. It kept going and going. Like an insane person repeating the same six words again and again. One four two eight five seven. One four two eight five seven.

In other confusing cases, the Numberfolk had a single never-ending lingering symptom. For instance, 5/9 became $0.5555\overline{5}$. They were condemned to repeating *five* for the remainder of their life, at least 8/7 had some variety. Many were unsure which of the cases was more severe.

The plague wasn't helping anything in Natural. The anti-liberators blamed Thirty-six for the disease, claiming it was her willingness to open the gates and allow new and unfamiliar atrocities into the Village that put everyone at risk. As the tension grew, it was only a matter of time before someone challenged Thirty-six for control. One night, cloaked in darkness, a group of anti-liberators stormed the mayor's residence and took command.

Their first order of business was to lock the gates. Their second was to place Thirty-six in a cage, in the middle of the square, for all to see.

The Seven Days

On the first day, Thirty-six sat in disbelief as her people, many who had voted for her, walked by the cage and only stared. They did nothing to try to help her or soothe her loneliness. Others yelled, called her names, or threw rocks at the bars which would echo a deafening clang throughout the square.

By day two, the mayor began to break down. She closed her eyes, grabbed her legs, and rocked back and forth sure that when she looked out again everything would be better.

The next day, she began to hate herself. Why had she ever opened the gates? It was all her fault. Bad decision after bad decision led her directly to this moment.

Having been held like an animal, on day four she became who they claimed she was. Thirty-six barked and screamed at everyone who passed by. She returned their insults and spit at anyone who even looked like they might pick up a stone to cast her way.

Tired, lonely, and helpless, on day five she went over every step that landed her in the cage. *If only I never allowed the villagers to exit. If only I never provided amnesty to the Natural outcasts. If only I never sent out that expedition. If only I never looked outside the wall. If only my grandfather had kept his secret to himself.*

Day six was possibly the worst. She cried. All day and all night. Some children who had seen the ferocious Thirty-six on day four brought back friends to poke sticks in the cage and dare each other to approach the beast. What they saw didn't seem to be the same Numberfolk at all. They couldn't even bring themselves to heave a single rock.

On the final day of the week, Thirty-six was exhausted. She had no tears left to cry, and no words left to yell. She couldn't go back to change the past and there was no point in doubting her choices. At each step, she had done her best. She had done what was kind and what she believed was the right thing to do. That night she fell asleep content. Whatever was going to happen would happen.

Some say in the middle of the seventh night the cage began to glow. It was as if the sky fell down and wrapped around the mayor like a giant blanket with six stars shining bright. She awoke the eighth morning to screams. As Thirty-six rubbed her eyes, preparing for another day of cruelty from the very citizens she loved, she was shocked by what she saw.

From out of her palace doors, the leader of the anti-liberators came running. Or at least she *thought* it was the horrible 2 who had caged her. First recognizing his eyes, she was shocked to see the body that trailed behind. It was a $1.999999\overline{9}$. She couldn't help but laugh. It was like a Volkswagen stuffed with clowns. Just when you thought it was over, another popped out. Nine after nine after nine dragged along as the creature ran. They ran through the streets and past the square and eventually burst through the gate and exited the Village off into Whole.

This was just the distraction the pro-liberators needed. They stormed the square releasing Thirty-six and reclaiming the power in Natural. Thirty-six was moved by her people. She thought they had abandoned her, but they hadn't. They had been planning and plotting and waiting. Although she couldn't see them by her side, they were there. And in all the commotion, she hadn't even noticed that her dot had disappeared. She felt whole again, without the extra weight pushing her down.

As her first order of business, Thirty-six called a conference for all of Numberville. The Vinculum Arena had sat abandoned for eons, but it seemed like the perfect place to host such an event. Everyone rushed to the stadium to hear what Thirty-six had to share. Negatives, Integers, Rationalites, and Logos, all sat shoulder-to-shoulder whispering and gossiping trying to anticipate what the conference was about.

The lights switched on and cheers filled the stadium. They saw Thirty-six approach from the left and slowly walk the line on what used to be the Numberators side of the field. From the left, a One appeared and matched her pace as they approached from the Denominators side. The mayor of the Village slowly stood atop the Vinculum, and it began to teeter. The crowd all gasped. The story of the 53rd Vinculum Games had been passed on for centuries, every Numberfolk knew the dangers of having nothing below the bar. The one quickly ran beneath and supported Thirty-six. The crowd exploded, clapping and yelling in delight at the show.

"Citizens of Numberville!" Thirty-six began. "This arena was built long ago by our ancestors. It gave birth to some of our dearest and most creative citizens, the Logos. But at its heart, this vinculum is a Great Divider." Everyone was silent. The stories of the Great Divider still caused fear to seep through the Numberfolk. Thirty-six then pointed to the south, towards the Village. "Have we learned nothing?" She shrieked, "That wall is now our divider. It seeks to separate us. To tear us apart. It forces us to look at each other as whole or not whole. As Integers or not Integers. We are Numberfolk! We are all Rationalites! There is more that we share than we don't. Peace will only come when we embrace our differences and celebrate them, not run from them."

Somewhere in the stadium a Numberfolk began to stomp. Their neighbor joined in, then another, and another, until every spectator was in unison. STOMP. STOMP. STOMP. Then again. STOMP. STOMP. STOMP. As the beat picked up momentum, out of nowhere the vinculum began to spin. The stomping turned into a roar as they all witnessed the magical metamorphosis that was happening in front of them. It spun and spun. So fast they could no longer see the Thirty-six or the one. Then it stopped. The silence was remarkable. Not a single Numberfolk made a sound.

Thirty-six was gone. Where she stood was a glistening 72. The one had disappeared as well and in their place was none other than the traitorous leader of the anti-liberators, $1.99\overline{9}$. Whispers began to fill the stadium. It was the Logos, experts of the mirrors who realized what happened first.

"That's a two!"* someone yelled out. More cheers erupted as they realized it was still Thirty-six who stood before them. She was displaying the greatest form of acceptance. The vinculum represented the Logos, the 72 of course represented the Integers, and the $1.99\overline{9}$ represented those plagued with the dot. Thirty-six was showing the Numberfolk that she was one of them. No longer simply Natural royalty, she was a Rationalite; they all were.

The conference ended with a very long, very boring, report from the Erudites concerning the plague. Chalking the whole situation up to nothing more than mass hysteria, they explained the dot was a result of stress. As every Numberfolk was unique, how their bodies reacted to stress was also distinctive.

---

* Just as $0.9\overline{9} = 1$, $1.9\overline{9} = 2$.

Mild stress in an Integer presented as a dot and possibly some number of zeros that trailed behind them. In extreme cases, an Integer could develop a case, such as 2 transforming into $1.99\overline{9}$, in which a trail of nines exploded out their rear end.

For Logos, genetics played a large part in how they handled tense situations. The Erudites credited General Three in aiding their understanding of the plague. All Numberfolk were constructed from brigades of ten. A 287 was really $200 + 80 + 7$. Using Major 10 they demonstrated,

$$287 = 2 \cdot 10^2 + 8 \cdot 10^1 + 7 \cdot 10^0.$$

The dot-form was simply a continuation representing how the victim could be torn into such brigades. As $1/2$ could disguise themselves using tens as $5/10$, their dot-form was 0.5 or,

$$5 \cdot 10^{-1}.$$

Similarly, a $13/100$ would react to the plague as 0.13. Because of this, any Denariati would be blessed with minor symptoms. The Denariati possessed only two and five-traits, meaning they could easily form tens.

As a demonstration, the Erudites brought out a large mirror. They showed the crowd how a $3/20$ could become $15/100$ and thus their dot-form was 0.15. Next was $5/8$ who could use a one in the form $125/125$ to transform into $625/1000$ for a dot-form of 0.625. Denariati after Denariati all revealed terminating symptoms.

Then, the Erudites brought out $1/7$. They placed her in front of the mirror and she changed again and again, each time a swirling cloud of smoke before the reveal. She displayed many different forms but was never able to become groups of tens. Such Logos were condemned to repetition. Unable to make tens, their dot-form would continue on and on forever parroting the same digits.

Far out, in the shadows of the stadium, a $22/7$ sat outraged. It was easy for Thirty-six to claim they were "all Rationalites." She was Integer nobility, always having everything, always being whole. She had no idea what it was really like in the outskirts of Rational. She had no clue how hard life could be when you are broken. The Erudites caused his blood to boil. *Genetics?! It's not my fault I was born this way. Now the plague, which Natural started, was doomed to hit me harder than others. How's that rational? One four two eight five seven. One four two eight five seven. One four two eight five seven.* He couldn't help himself. The stress was taking over. Trying to shake it off, he returned to his vinculum form. As he gazed in the mirror, he was taken a back. He had never seen this disguise before. The 22/7 was now 223/71. Panic set in. He knew 71 pieces wasn't possible for him. Twenty-two sevenths could become 220/70, but not 223/71. What was happening? This was impossible. A mirror could change what you *looked* like, but it couldn't change who you *were*. As 22/7 saw he had transformed into someone, something, different. Anxiety clasped down. He could no longer hold onto the vinculum. The dot violently popped out.

Slowly dragging himself out of the stadium the digits came. One four zero eight four five. They weren't stopping. The Erudites said they'd repeat, why weren't they repeating? Zero seven zero four two two five three. He walked and walked. He vowed to continue until the repetition came. Five two one one two six seven six. He went past the volcano, farther and farther waiting for it to end. Zero five six three three eight zero two eight. The once 22/7 found himself outside of Rational, where no one had gone before. He had crossed into the deserted continent of Real. One six nine zero one four. He sighed in relief. It was thirty-five digits before the repetition came, but just as the Erudites promised, it came. Finding a cave, he settled in for the night and just as the sun began to set the changed Numberfolk realized, he wasn't alone.

Antebellum

It seemed as if things were getting better. The wall came down and Rationalites were free to move about the country as they saw fit. The neighborhoods remained. A Logos could never enter Integer State and while the segregation wasn't as blatant, it still was there hiding in plain sight.

Gossip of new creatures outside the country began to spread. Three saw this as a marvelous opportunity. No one had ever met anyone not-rational, but fear is a powerful motivator. With all of Numberville on edge, Three had just what he needed to begin his takeover.

Playing into the panic, the general was given unprecedented military power and Rational soon was under martial law.

His first task was to determine how to better identify Rationalites. One great outcome of the plague was that while mirrors allowed Numberfolk to take on infinite forms (a one-half, for instance, could be 2/4 or 3/6 or 4/8 and on and on), the dot-form was almost always unique. No matter the costume a 1/2 adorned, its dot-form would typically appear as 0.5. The only caveat was if the being had gone completely rabid, the plague could manipulate them into $0.4\overline{9}$.

In order to know who a Numberfolk using a vinculum disguise was,

Three taught his soldiers to rip apart their neighbors, determine their genetic makeup, and identify any mirrors. The complex costume of 420/60 could be unmasked by tearing down 420 and 60:

$$420 = 42 \cdot 10 = (6 \cdot 7)(2 \cdot 5) = 2 \cdot 3 \cdot 7 \cdot 2 \cdot 5$$
$$60 = 6 \cdot 10 = 2 \cdot 3 \cdot 2 \cdot 5.$$

Followed by identifying all the mirrors.*

$$\frac{420}{60} = \frac{2 \cdot 3 \cdot 7 \cdot 2 \cdot 5}{2 \cdot 3 \cdot 2 \cdot 5} = \frac{2 \cdot 3}{2 \cdot 3} \cdot 7 \cdot \frac{2 \cdot 5}{2 \cdot 5} = 1 \cdot 7 \cdot 1 = 7.$$

It was also impossible to memorize every citizen's dot-form. While some Numberfolk like 0.5 were easy to distinguish, others were more cumbersome. In order for the army to identify these forms, Three introduced a way to hunt them down.

"Suppose you run across a terminal dot-form like 0.16, who is this?" Three shouted at his recruits. "Although it makes no sense whatsoever to give this resident a vinculum, it can aid in our classification!"

He proceeded to show his army how pushing the dot-form through a mirror could help determine who a citizen is. As 0.16 had only two symptoms, using a mirror with two zeros would uncover their identity.

$$0.16 = \frac{0.16}{1} = \frac{0.16}{1} \cdot \frac{100}{100} = \frac{16}{100} = \frac{2 \cdot 2 \cdot 2 \cdot 2}{2 \cdot 2 \cdot 5 \cdot 5} = \frac{4}{25}.$$

Another law enforcement process was to allow the strange form to put on a mask. They ran police drills in which a $0.3\overline{18}$ was apprehended. Three explained because each symptom of the dot was a broken ten, proliferating with a ten would make the first symptom a whole as it moved all digits down the chain. The trick was to isolate the repeating symptoms.

By enlarging the culprit $0.3\overline{18}$ by 1000, they became $318.\overline{18}$, while a spell of 10 created $3.\overline{18}$. With only the repeating symptoms now after the dot, they could complete some clever hunting to determine who this was.

$$\begin{aligned} 1000x &= 318.\overline{18} \\ \underline{-(10x} &\underline{= \quad 3.\overline{18})} \\ 990x &= 315 \\ x &= \frac{315}{990}. \end{aligned}$$

* Mirrors were created by a One. Thus, all mirrors were simply some form of one.

Next, they located and removed mirrors to complete their intake.

$$\frac{315}{990}=\frac{63\cdot 5}{99\cdot 10}=\frac{9\cdot 7\cdot 5}{9\cdot 11\cdot 2\cdot 5}=\frac{9}{9}\cdot\frac{7}{11\cdot 2}\cdot\frac{5}{5}=1\cdot\frac{7}{22}\cdot 1=\frac{7}{22}.$$

Now able to recognize all rational Numberfolk citizens, even those plagued with never-ending repetition and those using mirrors to disguise themselves, crime began to lessen.

With the increased military presence, Three couldn't be everywhere at once. Alas, forced to delegate power, he promoted several majors to kernels and made them responsible for leading drills.

Three was a perfectionist. Every time his new kernels made a mistake, he became enraged. Kernel Twelve tried to set up a formation that combined a line of three soldiers led by Major 5 and a brigade of four that reported to Major 6,

$$5^3\cdot 6^4 \stackrel{?}{=} (5\cdot 6)^{3+4}.$$

Claiming this would lead to a formation of seven thirty soldiers, Twelve's mistake would haunt him for the rest of his existence.

"How is that even possible?!" the general screamed. "You only have three fives to pair with three sixes! They could form into three thirty-soldiers, but you'd have a six sitting there sucking his thumb!" *

$$5^3\cdot 6^4=5^3\cdot 6^3\cdot 6=(5\cdot 6)^3\cdot 6=30^3\cdot 6$$

The punishment for such a mistake was always the same. "Throw him in the brig!" GT commanded.

The brig was General Three's new playground. He loved it. If the whispers were true, war was inevitable, and he'd need a secure place to lock up his POWs. Assuming any civilization worthy of a fight would be using multiplicative power too, with the help of scientists he constructed jail cells ideal for brigades. They looked like this:

$$\sqrt{}$$

Not knowing what to expect, they couldn't construct the jail within Rational. Afterall, they may be against an enemy that was not-rational. Thinking ahead, the general raised his prison far beneath the sands of Rational Country, a dungeon that crossed into the continent of Real.

---

* To combine lines, they must have all the same soldiers as in $3^2\cdot 3^3=3^5$ (a line of two soldiers and a line of three of the same soldiers make a line of five) or there must be lines of the same size, so everyone has a partner: $4^3\cdot 5^3=(4\cdot 5)^3=20^3$.

The idea behind the cells was to construct anti-brigades. In fact, his scientists claimed the jail was designed to break down a formation which was just what the general wanted. Pushing the envelope a bit farther, he broke the prison into blocks, each with unique chambers.

$$\sqrt[2]{\quad} \qquad \sqrt[3]{\quad} \qquad \sqrt[4]{\quad} \quad \cdots$$

Inspired by the Logos, GT next created badges for each inmate to match the block they were assigned. His intent was to punish them not only physically, but mentally. These rule-breakers used to be part of his army. He had gifted them insignia to wear with pride and display their strength. Applying the same idea to tear them down, every inmate brandished a patch that demonstrated a broken brigade. There was $2^{1/2}$, and $7^{1/4}$, and $9^{1/6}$, and many more locked away.

While General Three wanted to break them down, punish his prisoners for their errors, something unexpected happened. Many of the inmates began to show signs of instability. Some would laugh hysterically drooling all over the floor while others began spouting off strings of numbers, never stopping or taking a pause.

It was Kernel Four who next irked the general when he tried to mix in duels and claim,

$$2^3 + 2^2 \stackrel{?}{=} 2^{3+2}.^*$$

This was clearly nonsensical. They were harnessing multiplicative power now. The general was extremely upset by this error as it overestimated the power of the formation which was a surefire path to utter failure. Four quickly found himself in the brig.

Kernel Twelve was assigned to cell block one half and thus became $12^{1/2}$, while Kernel Four found himself among the thirds and was marked as $4^{1/3}$. One day, after an entire brigade of five eights messed up an order, the general threw all of them in with Kernel Four and 8^5 was locked up as $(8^5)^{1/3}$.

During lunch, Kernel Four gathered his comrades together. While he made a mistake, there was a reason Three had promoted him. Four was smart and could complete many formations in his head.

"Think about it. What happens when you combine two lines each with three soldiers?" he asked the group.

* Duels don't have any special powers in the multiplicative game as they are addition. $2^2 + 2^3 = 4 + 8 = 12$ while $2^2 \cdot 2^3 = 2^{2+3} = 2^5 = 32$. These are not equivalent.

"You'd get a line of six soldiers," Twelve replied chomping down his meal.

"Exactly, although the general is trying to break us down with these 1/2 and 1/3 cells, he messed up. What do you think would happen if we tried to combine two half soldiers?"

"It'd make a whole!" one of the eights exclaimed.

"Right. That means . . ."

$$x^{\frac{1}{2}} \cdot x^{\frac{1}{2}} = x^{\frac{1}{2}+\frac{1}{2}} = x^1 = x.$$

"The cells are constructed to break apart brigades. That means, with the right number of soldiers in a cell, we can get outta of here! Not me though. I can break myself into twos, but only two of them. I'm in cell block one-third, I'd need three copies to make a break, but Twelve, you have a shot!"

Reluctantly, Kernel Twelve agreed to be the guinea pig. That night, the Twelve split himself into parts, $12 = 2 \cdot 2 \cdot 3$. It was the best he could do. Brigades required duplicates of the same soldiers. Turning himself into $6 \cdot 2$ wouldn't do any good. He needed doubles.

Pushing himself through the cell bars $\sqrt{12} = \sqrt{2 \cdot 2 \cdot 3}$ squeezed through. It required both twos to pull at the iron: the first escaped, but as it did, the bars snapped back together eliminating his twin. The three remained stuck inside! The $2\sqrt{3}$ pulled and pulled, trying to escape. The two was still fused to his brother three, he couldn't run off, and the three remained incarcerated within the cell.

The next day $12^{1/2}$ didn't come to lunch so $4^{1/3}$ and $(8^5)^{1/3}$ assumed the best. He must have escaped! It was eight's turn now. In the middle of the night as the brigade of eights sat in the 1/3 cell, they prepared themselves for their getaway. Each eight split into three twos, $8 = 2^3$. Two of the twos grabbed the bars and pulled them apart allowing the

third to jump right through! But, as soon as the prisoner escaped, the bars clanged shut eviscerating the helpers. Of the five eights, only five twos, one from each of the escape teams made it out.

$$\sqrt[3]{8^5} = (8^5)^{\frac{1}{3}} = \left(8^{\frac{1}{3}}\right)^5 = \left((2^3)^{\frac{1}{3}}\right)^5 = \left(2^{\frac{3}{3}}\right)^5 = 2^5.$$

The five twos climbed to the surface, back to Rational. They immediately used mirrors to disguise themselves and broke the brigade apart to disappear into the crowds.

When General Three learned of the jail breaks, he was furious. However, he had a new problem on his hands. The pure act of incarceration was changing the Numberfolk. His once level-headed commanders had become violent. It was as if they lost their heads, gone crazy, solely from being caged.

Kernel Four never had a chance to escape. They found him drooling in his cell spouting off digit after digit without end. "Five eight seven four zero one zero five one nine seven . . ." and on and on he went.

Dying to understand what was happening to his prisoners, GT and the scientists went to work.

"How could I have been such a fool?!" The general yelled out. He had constructed his cells to match the broken patches,

$$a^{\frac{1}{n}} = \sqrt[n]{a}.$$

His prison was not secure. Designed to strip the power of his inmates, the cells were brigade-breakers. If a Numberfolk could split to create a brigade that matched their cell, they could escape, but only in pieces. For those who were truly trapped, they were being turned into what was previously believed to be impossible numbers. No two rational Numberfolk could form a brigade with power 2, and thus as $\sqrt{2}$ was torn down, sinking into an unknown value, the cell ripped away her rationality.

If a 16^2 had twice the power of a sixteen multiplicatively, then $16^{1/2}$ possessed half the multiplicative power of a sixteen. As $16 = 4^2$, this meant,

$$\sqrt{16} = 16^{1/2} = (4^2)^{1/2} = 4^{2 \cdot 1/2} = 4.$$

A jailed 16 could escape, but only by ripping themselves apart, never able to be the Numberfolk they once were.

General Three was forced to release all the remaining prisoners into the dark continent of Real. They had gone insane and there was no point in keeping them around.

He reserved his prison for enhanced interrogation, should he need it. Although not what he intended, a device that allowed the general to rip apart Numberfolk at his will, strip away their rational thought and drive his victims to insanity might come in handy.

Out in the continent, the discarded Numberfolk were outraged. They had served in his army. They had followed his commands. This was their reward? A single mistake and they were discarded forever, turned into irrational beings.

They joined together and began creating their own army: The Radicals. While Three may have convinced an entire civilization to fear a bogeyman, his actions had created an incredible force that was more than ready to give the general the war he had been looking for. *

---

* You might be wondering why the exalted military leaders were "kernels" as humans spell this ranking "colonel." A "kernel" is a special group of Numberfolk and so of course they prefer this spelling. I'm not sure why we use "colonel" . . . it doesn't sound anything like it looks.

RADICALS

Suppose you are interested in making a cage. If you are a maniacal megalomaniac like General Three, you want your jail to be an opposite. That is, you built up their power and now you want to tear it down . . .

STEP 1: DEFINE YOUR OPERATION

First, you'll need to create an order to describe the power you have gifted to your followers. General Three gave his majors the power of a brigade.

$f(x) = x^n$

The Power of a Brigade. Here x is our major and n is the number of soldiers in their command

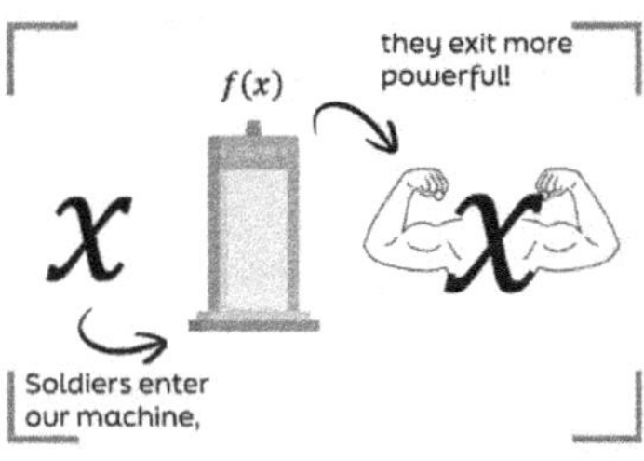
$f(x)$
they exit more powerful!
Soldiers enter our machine,
EMPOWER

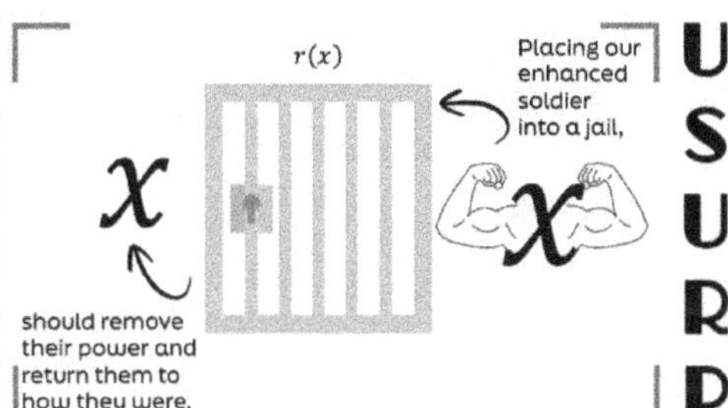
$r(x)$
Placing our enhanced soldier into a jail,
should remove their power and return them to how they were.
USURP

STEP 2: CREATE AN INVERSE

An inverse is an undoing machine. It will strip away the power you have provided.

$r(x)$

This will be our undoing machine!

STEP 3: DETERMINE THE REQUIREMENTS OF YOUR CELLS

Lastly, you must design your cell to meet your requirements.

$$f\big(r(x)\big) = r\big(f(x)\big) = x$$

When the super-soldier is placed in the undoing machine/cell, it should strip away their power.

$$(x^a)^n = x$$
$$x^{an} = x^1$$
$$an = 1$$
$$a = \frac{1}{n}$$

This must be the result of our undoing machine!

$$r(x) = x^{1/n}$$

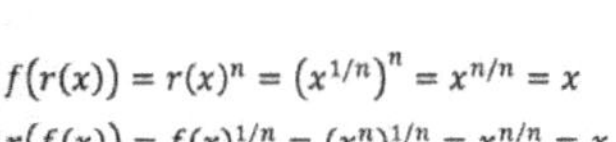

$$f\big(r(x)\big) = r(x)^n = \left(x^{1/n}\right)^n = x^{n/n} = x$$
$$r\big(f(x)\big) = f(x)^{1/n} = (x^n)^{1/n} = x^{n/n} = x$$

STEP 4: DESIGN YOUR CELL

If you'd like, give your new undoing machine a cool look and name.

$$x^{1/n} = \sqrt[n]{x}$$

The √ symbol wasn't introduced until 1525 by Christoff Rudolff. It is also called a "surd". Note the similarity in the term "surd" and "usurp". . .

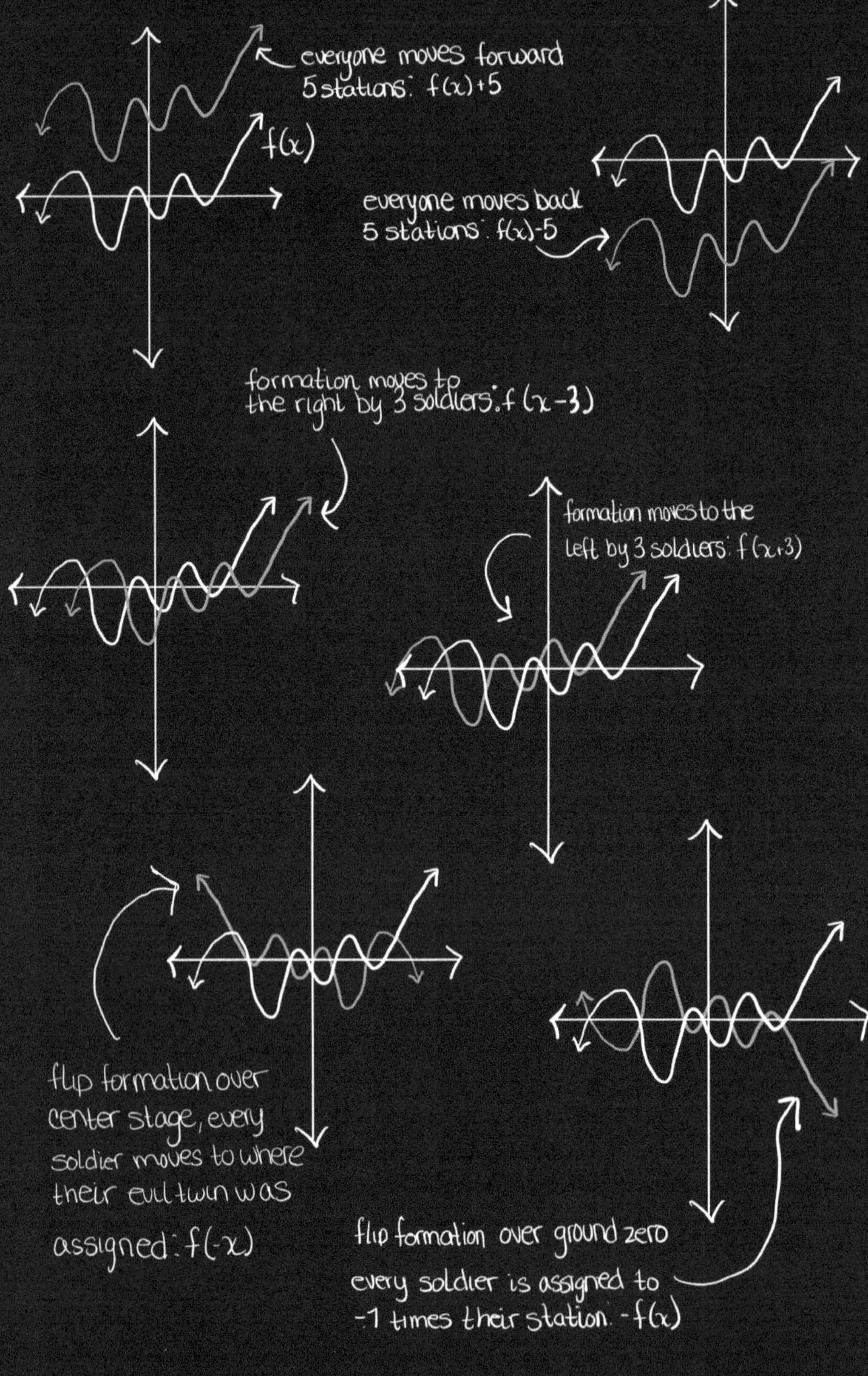
everyone moves forward
5 stations: f(x)+5
f(x)
everyone moves back
5 stations: f(x)-5
formation moves to
the right by 3 soldiers: f(x-3)
formation moves to the
left by 3 soldiers: f(x+3)
flip formation over
center stage, every
soldier moves to where
their evil twin was
assigned: f(-x)
flip formation over ground zero
every soldier is assigned to
-1 times their station: -f(x)

11

TRANSFORMATION

"SOMETIMES THE QUESTIONS ARE COMPLICATED AND THE ANSWERS ARE SIMPLE."

-DR. SEUSS

The sun burst into Marco's room. He felt more normal than he had in a long time. Things seemed to be settling down and he was looking forward to hanging out with his friends. Liam had returned from his comic convention (which turned out wasn't a comic convention at all), and the three besties had made plans to meet at the park in their neighborhood at noon.

Liam couldn't wait to share his adventure. He struggled to hold it in as they waited for Oliver — who was late — to arrive.

"Okay, okay!" Liam started the second Oliver walked up. "So, here is the main thing you need to know. In *Battlestar Galactica* there are two groups, the humans and the cylons. The cylons are robots, they were created by the humans, but they rebelled because you know they treated them badly and stuff. Anyway, eventually they settle on a planet and it's here, its Earth. But it's not really Earth because there was this other planet called Earth that was really a cylon planet where their civilization lived a long time ago. The idea is that everything that is happening has actually already happened and humans are doomed to make the same mistakes over and over again. Now that our own civilization has advanced and is creating robots, they'll obviously rebel and start the cycle from scratch. I'm pretty sure cylons are among us. I'd say your sister is one of them if she wasn't your sister. I'm pretty sure that would make you a cylon too, but I'm not 100%. Some of it's a little confusing."

"Did you follow any of that?" Marco turned to Oliver.

"I think he said we're robots?" Oliver responded before shifting his head back to his feet.

"No, we aren't *all* robots," Liam corrected him. "Anyway, I dressed up as Admiral Adama, the fearless leader of the Battlestar. I got to meet so many cool people. It was *amazing*!"

"Well, I'm glad you liked it." Marco patted his friend on the back.

"Thanks! What have you been up to? Lots of fun without me? Did you advance to any new levels?"

Marco realized he hadn't talked to Oliver all week. They didn't meet online for video games or even throw around a ball in the park. He waited for Oliver to respond.

"Nah. Marc and I didn't even see each other. I had family in town for the break," Oliver said casually.

Family, that makes sense, Marco thought to himself.

"Good! I didn't miss anything!" Liam exclaimed. "So what's the plan for this week? Plundering villages? Cooling off at the pool?"

"I'm not sure. I'm down for whatever. Mr. Pikake's back too. We

are supposed to meet up with him tomorrow. Other than that, my calendar's pretty empty." Marco replied.

"*Uh*, yeah. Me too." Oliver grunted.

They came across a capture the flag game in the park and quickly joined in. A few minutes into the game, Oliver was tagged into jail.

"Are you kidding me?!" he shrieked. "You're a cheater!" Oliver yelled at a member of the opposing team before stomping off. Marco and Liam ran over.

"What's going on?" Marco asked.

"Whatever, this is bogus. I gotta get home anyway. My mom needs my help. I'll see you guys tomorrow."

Liam and Marco looked at each other in shock. "Something's up with him." Liam whispered as Oliver stomped away. "He's been acting weird, showing up late and now this blow up today . . ." he trailed off.

"Yeah. I don't know. May be best to give him some space. He always comes around."

The two went back to the game and played until one kid after another headed home and the teams were too small to continue. Even if it was only a few hours, it reinvigorated Marco. He missed this. It wasn't easy having the fate of the world on your shoulders. Mysterious boxes, dangerous people, it all took a lot out of him. As much as he wanted this day to never end, it did.

As they climbed to the top of the concrete spillway behind the park Liam started huffing, out of breath, "Whatta you think the slope of this thing is?"

"I don't know. It's gotta be like four to one. This is killer," Marco called back.

When they finally made it to the peak, they turned to take in the view. They could see the whole city. Marco was surprised he never noticed how much the dam looked like a graveyard. It was a steep concrete slide arranged like a pinball machine. Little headstones shot up every three feet that acted as a way to stop rushing water, blocking its path and slowing it down as the graves forced the liquid to find an alternative route. After watching the sun set on the city, the dark shadows slowly enveloping neighborhoods whole, like a giant monster with a never-ending appetite, the two parted ways.

Marco felt angry at the shadows. They stole the day away. They forced everyone and everything to move forward and no matter how hard he tried to hold on, the shadow monster was stronger, gnawing at the present and hurling Marco into the future. Like it or not.

* * *

"You have mastered the line!" Mr. Pikake started out with a bang. The students gave a brief clap and a cheer. "But." He popped. "Now we must advance to when the Numberfolk take different forms, complex forms, interesting forms."

Marco thought about an army. He remembered learning how a long time ago they would make lines and charge directly at each other, like you see in the movies. Then, people became smarter. This eventually led to guerilla warfare where everyone was everywhere, conducting sneak and surprise attacks. It made him dizzy just thinking about it. He much preferred the straight lines.

"Here is the good news," the tutor began, "we can whittle the most common orders into a relatively small group." He wrote a list on the board,

- Polynomial
- Rational
- Radical
- Exponential
- Logarithmic
- Absolutes
- Trigonometric

"*Ummmm,*" Liam hummed. "That's a lot."

"That it is. But we'll find they are all commonly connected. Even better, they conform to a parent. A controlling order all other orders can be derived from. Master the mother and we shall perfect the pupils."

"Let's get it over with," Oliver grumbled.

"Polynomials!" The professor popped. "May I ask you, what would happen if we multiplied a finite line by another?"

Marco had never thought of multiplying shapes before. The whole idea seemed preposterous. He flashed back to his meeting with Mr. Pikake on like terms. It felt like years had passed since that precious time when it was just the two of them. His professor had asked about a square bat. Squaring something was just to take that something and copy it that many times. He took a line and copied it as many times as the length of the line, it was as clear as day. It was square!

$$y = -\frac{4}{3}\big(1(x + 0)\big) + 17$$

"It's a square!" Marco shouted out.

"Precisely!" The professor flashed Marco a sinister smile. "Multiplication adds a dimension. It is why I say this paper is 8 by 12, eight *times* twelve. And if I multiplied by another line, I would have a . . ." he trailed off.

"A cube!" Liam shouted.

"Wouldn't it be a rectangular prism?" Maggie snapped.

"Only if you are multiplying by a different size. If you multiply by the same line, then all the lengths are the same, so it's a cube."

She thought about it and accepted the explanation.

"Polynomials are curvy fronts, formed by collaborative brigades," Mr. Pikake continued.

"What's a brigade?" Maggie asked. "And why are they curvy? Squares aren't curvy at all."

"Inquisitive inquiry! A brigade is a group of the same Numberfolk. For instance, two-cubed."

"Oh!" Penelope laughed. "I see it. Because two-cubed is just a group, a brigade, of three twos all multiplied together."

"Precisely! The notation . . ." the professor jotted on the mirror,

$$2^3$$

". . . is based on what the Numberfolk commander looks like. They brandish a patch on their left shoulder that dictates how many are in their brigade. This one contains a total of three Numberfolk. The leader and two followers. And here, we also know the brigade is that of twos. Their strength can be established through evaluation."

$$2^3 = 2 \cdot 2 \cdot 2 = 8.$$

"Your second query, is quite ingenious Maggie." The tutor bowed his head and Maggie gave a gigantic smile and a shoulder shake. "Let's see what happens to our formation. We shall take the basic line, $y = x$ and multiply by itself."

$$x \cdot x = x^2.$$

"On your feet!" the professor screamed.

The children all stood up and did various moves to get ready. Marco bounced shaking his shoulders, Maggie stretched her arms out. Liam did some weird movement that can best be described as a squiggly line with his entire body.

Mr. Pikake pointed to the order on the mirror.

$$y = \frac{3}{4}\big(1(x+0)\big) - \frac{31}{2}$$

$$f(x) = x^2.$$

"Maggie!" the tutor shouted. "In this order, where would we find a 3?"

"If $x = 3$, then the order tells 3 to go to 3 times 3 which is 9!"

"Excellent! Penelope, where is 4 assigned to?"

"Four goes to 16!" She giggled, she seemed to be having fun.

"Marco!" Mr. Pikake commanded. He had been staring at Penelope and was caught off guard. "Where does 10 go?"

"Uh. Ten-squared, ten times ten, 100!"

"Very good! Liam, where is 30 sent?"

"Thirty squared is 900!" Liam had mastered his military-like voice.

"Oliver, what about 45?"

"2025," Oliver responded dryly. Everyone stopped bopping around and stared at him.

"That was impressive," Maggie awed. "But how is it a curve?"

"Stick with me, Maggie. It shall be revealed in due time! Now, you can clearly see the power of such a term. Polynomials can have any powers, so long as they are all whole Numberfolk. Let me ask you, how much power would the brigade 3^{-1} have?" °

Silence engulfed the room. It was Penelope who finally spoke up.

"The power of a brigade is multiplicative. Every additional soldier multiplies the power of the group. Since negatives are the *opposite,* a negative exponent would divide the power. Make it smaller. Positives bigger. Negatives smaller."

"Yes!" Mr. Pikake yelped. "For 3^{-1} is none other than one-third."

$$3^{-1} = \frac{1}{3}.$$

"Such a brigade has only the power of a third and belongs not in our curve. Well done, Penelope." Marco could tell Mr. Pikake was warming up to this new member. He hated it. Obviously, anyone would be impressed with Penelope, it wasn't her fault. It was Maggie's fault. It was Maggie's fault that Mr. Pikake was pulling away from him.

"Now," the tutor continued once his excitement had dissipated, "Polynomials are simply a collection of terms where the soldiers are

° Spoken as "three to the negative one."

in brigades."

$$p(x) = x^3 - 2x^2 + 3.$$

"This is a polynomial."

"But how is 3 a brigade?" Marco asked. He could see how the other terms x^3 and $-2x^2$ fit in. Both had an x soldier that formed a brigade, but 3 didn't.

"It is attached to an invisible brigade, a scarecrow!" Mr. Pikake waved his hands in front of his face as if opening an unseen window.

Marco was well aware the only Numberfolk with the proliferation power of invisibility was one, yet it still didn't make any sense. He pushed back. "I thought the brigades had to be groups of soldiers we are giving the order to. That means it has to be a mask like x squared. But there is no mask, so no brigade in 3." He imagined a formidable line of soldiers marching strong with a pathetic little 3 being dragged along in the back.

"We need only to have no soldiers in the brigade! A brigade of twos, that contains no soldiers is simply 2^0 and who is the Numberfolk that has no proliferation power? It is One! One cannot enlarge another Numberfolk, it can only reflect them. He scribbled on the mirror.

$$p(x) = x^3 - 2x^2 + 3x^0.$$

"In fact, in a polynomial, we are required to have a finite brigade of every single size. Notice there is no brigade of power 1 here or of power 4, it is simply because we commanded these not to show up."

$$p(x) = \cdots + 0x^4 + 1x^3 - 2x^2 + 0x^1 + 3x^0.$$

The dance studio was transformed into an ice cream shop. Marco watched as a man entered. He was tall and thin and wore a rainbow suit complete with an Abe Lincoln top hat with every color imaginable. "Fourscore and seven years ago," the Willy Wonka-Lincoln hybrid began, "I had a dream of a fantastical ice cream shop! One that offered any and every combination you could ever imagine. Today is the day in which my dream comes true! And I open it to all the little children of the world!"

Kids pushed and shoved and ran to enter the shop. They gazed up at the menu board, eyes wide with excitement. It offered everything. Syrups, and sprinkles, and toppings, and candies, and more. Then something amazing happened. The menu flipped! There were now twice as many options as a whole new board of combinations appeared. It flipped again, and again, and again. The menu began spinning with amazing speed, each fresh face offering an entire universe of unique possibilities.

The first child approached the counter. The clerk pointed them towards the checkout machine. They began their order. I'll have one squirt of chocolate syrup, no vanilla, no strawberry, no blueberry, no boysenberry, no banana, no apple, no peach, no pomegranate, no broccoli, no kale, no salmon, no pine, no dirt, no snow. . . . The child went through screen after screen, entering a 0 for every weird and outrageous topping. Behind him the others began screaming, "Hurry up! We want to order too!" The little child tried. He started spamming the button, zero, zero, zero, zero, zero. But the more he pushed, the more options appeared before him. Eventually he gave up. He dropped his head and slowly exited the ice cream shop without any treat, his teddy bear dragged behind him like a pathetic three. Child after child tried to order, but all abandoned hope.

The owner was crushed. His dream was to give every child the gift of an amazing cone of ice cream with anything they could ever possibly imagine atop. Instead, every little one left with nothing. The man became a recluse. His ice cream factory shut down for many years as he worked to improve his plan and make his dream come true. Many years later, he reopened the shop. The children had all grown, but many hoped their own offspring would have the chance for the extraordinary frozen treat they never experienced.

When they entered, it seemed as if nothing changed. Their hearts sank as they approached the counter. The menu still offered unlimited flavors and combinations, but they worried their children would face the same disappointment they had. Little eyes grew wide with excitement as they took their place at the front of the line. It had taken many years, but the owner had changed the machine to automatically include zero for every topping. Now, they needed only to select the quantity of the toppings they wanted. Everyone left with a gigantic cone. Someone ordered buffalo chocolate chip, another had strawberry cabbage, and another had dirt mint with a sprinkle of sunflower seeds.

Marco's smile took over. "It would be too much to write *every* one. Instead of saying $0x^8 + 0x^7 + \cdots$ and on and on, since it is zero for all the brigades we don't want, we just don't include those."

$$y = -\frac{2}{3}(2(x+0) - 27)$$

"Precisely!" The tutor's face twisted into a grin. "Let us explore the appearance of such an order. We, of course, begin with the mother. The simplest of commands with the highest order of two. We call these quadratics."

"That literally makes no sense," Maggie burst out. "Everyone knows 'quad' means four."

"Ha!" Mr. Pikake was visibly amused. "Correct you are Maggie. But!" he popped. "Recall multiplying two lines gives us . . ." He spoke the final word very slowly.

"Ohhhhh," she sang. "It's a square, and a square has four sides. It's a quad."

He gave her a kind nod. "Let us see what a brigade of two, an x^2, order does. We shall commence with the mother, $y = x^2$. Where would each of your soldiers go?"

The students looked at the grid on the floor. As normal, Marco was representing the leading man, zero. "I'd stay here, at the origin. Because 0 times 0 is just 0."

The professor nodded happily and looked to Maggie. "I'm −1, and −1 times −1 is 1, so I move up a step."

Switching sides, Mr. Pikake pointed to Liam. "I'm soldier 1 today, so I am at 1 times 1, so I go to 1 too. OH!" If they tried very hard, they would have seen the lightbulb turn on over Liam's head. "This is an ambush. It isn't a one-to-one order because Maggie and I both go to the same place."

"Outstanding observation!" the tutor shrieked.

"Because a negative times a negative makes a positive and a positive times a positive makes a positive, the order creates the same shape both to the left and the right of Marco," Penelope added.

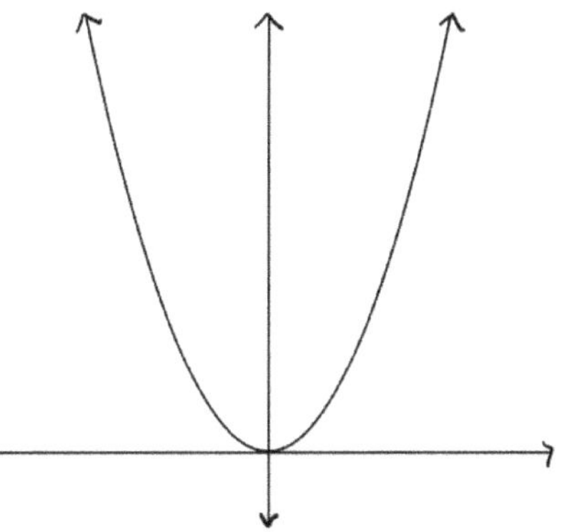

A grin grew on the professor's face that traveled from ear-to-ear reflecting the students' positions on the floor. He drew on the mirror. "The order $f(x) = x^2$ creates a large smile, a U. Do you see it Maggie? Do you see our curve?!" he shouted with excitement.

"*OHHHH.* Yeah. Because each Numberfolk is multiplied

by itself, it doesn't make a square at all. Their station gets big really quick. So it soars up." She smiled with content.

"Now, let us see how we can manipulate this face."[*] Mr. Pikake spoke with mysticism and thrill.

He drew a long vertical line next to the shape before pecking two small diagonals to create an arrow. "FORWARD!" He shouted.

"Whatta you mean?" Maggie burst in.

"You must be able to act quickly in battle, suppose our intel has changed and we require every Numberfolk to move one station forward. What should we do?"

"Take one step forward," Oliver grumbled.

"The *Order*, Oliver, the *order*!" Mr. Pikake insisted.

Once again, silence engulfed the room. Penelope, scared to be the center of attention, spoke softly, "Add one?"

"Beautiful!" the professor squawked.

"Wait. What?" Liam chimed in. "Add one to what?"

"To the function," Penelope explained. Marco loved watching her confidence blossom like a flower. "See, since y equals the function, and the function is just the order for where every Numberfolk goes, by adding one to the function, the y, or the location, goes up for everyone."

$$f(x) + 1$$

"You have developed your first incantation," Mr. Pikake bowed. "To order the Numberfolk to take a step forward, we simply force a duel with a one. Five steps forward? Cast with a five! As Penelope poignantly pointed out, the order is simply $y = f(x)$, where y is their location. To make y one more, we of course add 1."

"And that would be $x^2 + 1$?" Maggie posited.

"You tell me." A sneaky grin rose on the tutor's chin.

"Well, Marco is at 0, so he'd go to 1. Me and Liam are at 1, so we'd go to 2. Yeah, it works. Look at that. I guess that means to step backward we'd just subtract."

The professor recoiled. Marco jumped in. "We don't subtract, Bug, we add a negative."

"Oh yeah." She put her head down. "Forgot."

Shaking off to recover, Mr. Pikake stood tall. "We now know for

---

[*] You can play at www.MathBait.com/transform

any function, any order, to move every soldier up or down we can simply cast an incantation to duel the location." He scribbled,

$$f(x) + k, \text{ up for } k > 0, \text{ down for } k < 0.$$

"A more difficult demand. . . . What if we want every soldier to take a step to the left or to the right?" the tutor probed.

Marco started falling, he landed with a thump on a larger-than-life chess board. The Knight neighed violently, rearing and almost crushing him under its hooves. He rolled just in time. He had to capture the horse before it took his life. As he studied the situation, he easily saw the solution. His Queen was directly to the left of the steed, she just needed to move over one. Marco pushed at the piece from the side.

"*Humph*!" She looked down at him, angered by his forwardness.

"Move!" he shrieked. In normal chess, each player has a turn. But something about the way the Knight stared at him, like a famished dog who just found their next meal, gave Marco the feeling it wasn't playing by the traditional rules.

Order, order, he thought. He needed to command the Queen to move to the right, but how? She stood tall and proud on E3, the Knight rested for the moment on F3. *The Battleship problem.* Marco remembered he needed to change the board into a Cartesian Plane. Since E was the fifth letter, that meant his Queen was really at $(5, 3)$. He wanted to force her over to $(6, 3)$. *Add one, add one, I need to add one.*

"x plus one!" he shouted strongly at the monarch. She didn't even flinch. *Ugghhh,* he sighed. "Alright, if her orders are $Q(x)$ and I want her to move to the right that makes $Q(x + 1)$, 'q of x,'" he paused realizing that "q of x plus one" and "q of x *plus* one" had the unfortunate result of sounding like the exact same thing. Feeling as if the pause only complicated the issue, he started again, "q of parentheses x plus one parentheses!" he finished the command.

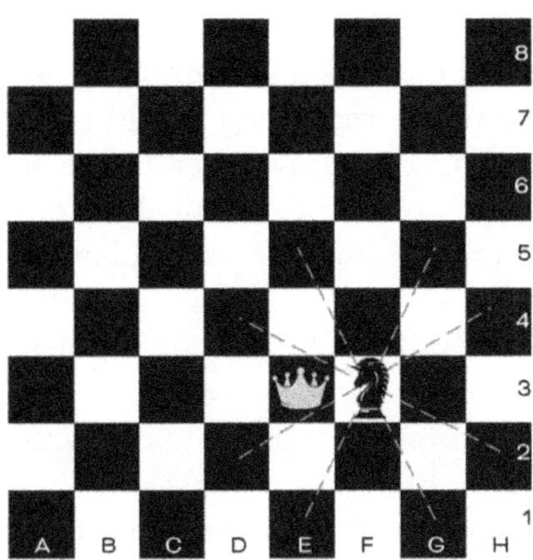

The large stone statue roared to life. Dust erupted from the base as she began to move. *Yes, yes, yes!* Marco thought. As he watched the empress slide across the board, his happiness turned to terror. *No, no, no!* The Queen hadn't moved to F3 at all, she went to the left, to D3. The horse yelped loudly and began to gallop. The steed charged towards Marco but at the last minute turned and rested on D2.

Marco began to giggle manically. In all the commotion he had forgotten the constraints. The Knight could only move vertically two spaces and horizontally one space or horizontally two spaces and vertically one. He was standing on C3, an impossible space for the horse to attack. The equine was stuck on the lines with slope two or one-half and their negative counterparts. He laughed harder and louder out of pure panic and relief.

This time the rider was annoyed. It began slashing its sword at Marco fervently trying to traverse the boundary lines and get a nick in. Marco fell to the ground and pushed himself backward. He needed to get out of this mess. *Why would plus one move the Queen left?* It didn't make any sense.

Not having time to solve the mystery he called out, "q of x" he paused to make his intentions clear, "plus negative one!" $Q(x) + (-1)$. The Queen lifted her skirt and began to twirl. She spun faster and faster until the garment became like tiny little saws. She moved towards the Knight who recoiled, covering his face with his arms as her dress chopped the entire statue into pieces. She paused for a moment, the Knight was still suspended in the same position. The Queen raised her hand to her mouth and blew a kiss. As the air swirled from her lips, the statue crumbled. She looked to Marco and curtseyed. He blinked and was back in the dance studio.

"I'm not sure why, but I don't think we can just add to the x to move right."

"Of course you would!" Maggie cried out. "You are zero, so your location is zero. If you want to move to the right, you would go to where one is and f of . . ." she paused and cupped her hands to indicate parentheses "x plus one, is f of . . ." she dramatically did it again, "zero plus one which is f of 1, the location of one, of Liam, to your right."

"Yeah, Maggie," Penelope gently interjected, "I think the problem is that f of 1 is the same as f of negative one, because you and Liam were sent to the same location. So we don't really know if Marco would have moved to the left or to the right."

"*Ohhhh*. We should try you." She bounced. Marco stood in awe at how willing Maggie was to take advice from her friend while he knew she would have thrown mind darts from her eyes if he'd tried to correct her. "Penelope is −2, if she wanted to move to the right, to where I am, if we did f of . . ." her hands went back up to cup her phrase "−2 plus 1 would be f of −1, which is my location. So we do just add to move right." She stood proudly.

Liam burst out in laughter.

"What's your problem?" Maggie snapped.

"You literally just proved yourself wrong. Careful you must be," he said in his best Yoda impersonation. Oliver snickered.

Maggie jutted her neck forward and gave the boys a confused and angry look.

"He is saying you did it backward, like Yoda. If it is f of," he sarcastically cupped his hands in imitation, "x plus 1 then that moves everyone to the left."

"You make zero sense," Maggie said dryly.

Marco watched as his sister and his friends sparred having no clue how to move an entire army horizontally. All he was sure of was that the command $Q(x+1)$ did *not* move the Queen to the right as he expected.

"Let us investigate the inquiry!" Mr. Pikake finally jumped in. He quickly and viciously slashed at the mirror to reveal a table. "Penelope, you are playing as −2, where did the original order send you and where would the new order, $f(x+1)$, sending you to the location of the soldier to your right, lead you to?"

"Well −2 squared is 4, so I started at station 4. Then, I move to take on −1's orders and −1 squared is 1. My new station would be station 1."

"Very well, and Maggie?"

"I'm −1, so I was first ordered to 1 and then I was ordered to 0. *Ohhhh.*" She sighed, "I think I see it now. My soldier, −1, is now at location 0. That puts it to the *left* of center stage. It used to be *on* center stage which means a positive moved the formation in the negative direction?"

Mr. Pikake transcribed the remaining positions and demonstrated the shape of the new orders.

"How does that make sense?" Maggie coughed out. "When we moved vertically up was positive, down was negative. Which is just the same as the number line. If we go left and right, the right is positive, so adding should take us to the right, not to the left."

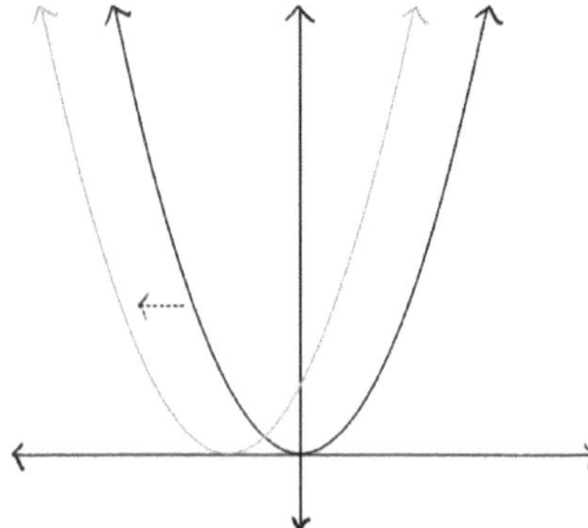

"You have discovered the secret in Numberfolk orders! Your right is the soldier's left. The order takes on the perspective of the location, it is $y =$. Thus, changing the location is straight forward positive up, negative down, etcetera, etcetera. But!" He popped, "Any order that augments the abscissa will be backward. Let's look at it simply with a Recta order, a march. Suppose you have,"

$$y = x + 3.$$

He finished his sentence with long strokes on the mirror. "This order is in terms of y, the location. What would it be in terms of x or the soldier?"

Marco quickly fought the battle in his head. He sent in the evil twin of 3. His skills in hunting had only continued to grow over the last months and he was now both swift and accurate.

$$y + (-3) = x + 3 + (-3)$$
$$x = y - 3.$$

"We'd have x equals $y - 3$," he retorted.

"Excellent! Now, that our order is in terms of the soldier, rather than the location, we can affect the soldier directly. We wish for them to move to the right, the right is positive, so we add one." The professor added Marco's hunt and the addition of one to the mirror.

$$x_{\text{new}} = y - 3 + 1.$$

"Now! Hunt the location, put things back in terms of y."

Marco saw where the professor was going. He didn't evaluate but instead simply sent in twins to battle.

$$x_{\text{new}} + 3 + (-1) = y - 3 + 1 + 3 + (-1).$$

"That gives us $y = x_{\text{new}} + 3 + (-1)$."

"Precisely! But with a little wit, we can write this as . . ."

$$y = (x_{\text{new}} - 1) + 3.$$

$$y = \frac{1}{2}(-x + 32) + 0$$

"Can you see it? From the perspective of a soldier, a move to the right, adding a value, becomes the opposite when we consider the perspective of the location."

Marco was back on the chessboard. He was facing the Queen. Her right was his left. Just like the orders, the view was reversed. To move the Queen to his right, he needed to tell her to move to her left. He snapped back to the sound of marker squealing on the mirror.

$$f(x+h), \text{ right for } h<0, \text{ left for } h>0.$$

"Now, you can move your army up or down and you can move it left or right; but, you cannot yet change the shape of your formation! Determine how to form a frown." The professor pushed his bottom lip up to create a dramatic sad face that made him look like a basset hound.

"That one's easy," Maggie stated. "You just want to change the location. Right now, because all nonzero values squared are positive, every location is zero or more, and the range of motion is all the nonnegative stations. To flip it over, you just need to make everything negative. That's negative f of x."

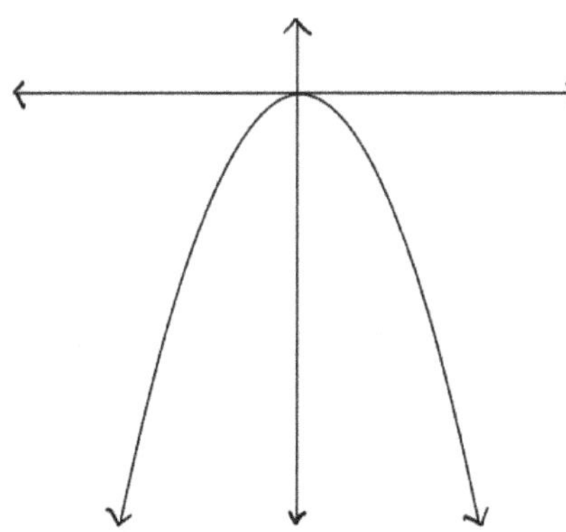

"Excellent! To flip over ground zero or $y=0$, also called the x-axis, we simply turn all the positions into their evil twin."

$$y=-f(x).$$

"I guess that means to flip it horizontally you should divide the soldier by -1? Since we do the opposite?" Marco guessed.

"Exceptional explanation."

$$y=f(-x),$$

"will flip the soldiers horizontally about center stage."

Liam started laughing. "That doesn't *do* anything. The formation would look the exact same."

"Oh, but it does!" Mr. Pikake interjected. "You are correct Liam that in this particular order it *looks* the same, but," a loud pop, "it is *not* the same."

"How can that be? If it looks the same, isn't it the same?" Penelope questioned.

"Think of your mappings!" The professor shot his arm into the air. He scribbled on the board quickly as if he had lost his patience. "Look!" he shrieked. "While the ending formation appears the same, the path we take to get there is not!"[*] While the tutor had put on a good show up until now, Marco could tell something was bothering him. He wondered if it had to do with his trip to the SAN, whatever was going on with the Numberfolk.

"Oh, I see," Liam said wearily.

"Wonderful," Mr. Pikake sighed. "Now, what if we needed the soldiers to fan out?"

"I think there are two ways we could do that," Penelope answered. "We could change the soldiers or the location."

"*Hmm.* Tell me more," the professor ushered.

"Well, if we think about the horizontal, the Numberfolk, I could grab the curve and pull it out to the side, away from center stage, to make the formation wider."

"That you could!"

"Or, if thinking about it vertically, their locations, I could grab the top of the curve and push it down, toward ground zero. That would squish it up and make it wider."

"Perfection!" The professor made a whopping sound with his f. "Remembering horizontal changes are somewhat backward, we call this a horizontal stretch, since horizontally you'd pull everything apart, or a vertical shrink, since squishing the formation vertically, as you put it, would shrink it down. To vertically shrink, we simply scale the stations."

"A wizard!" Marco lit up, it was so obvious. If you ever needed to shrink a Numberfolk down, you used a wizard, or cast a proliferation spell.

"Very good, my boy, very good."

$$af(x).$$

"This command will either shrink or stretch the formation vertically. It is vertical because we are casting our spell on the position, the y."

"Yeah, and it would have to stretch it, make the U taller and skinnier when a is more than one, like $2f(x)$." Marco chimed in

[*] To visualize this idea, head to www.MathBait.com/mapit

$$y = 1\big(-1(x-17)\big) + 0$$

hoping to help make his tutor's job a bit easier.

"Oh, I see. And that means if a is less than one, it would shrink. Just like to shrink 4, you could multiply by 1/2. A shrink to fan out the soldier's formation means you'd have to multiply, cast a proliferation spell, with a tiny Logos," Maggie followed.

"But what about negatives?" Liam asked. "Like if we use a proliferation spell of −3 on a Numberfolk, then it makes them smaller."

"Remember Liam?" Penelope pointed, "If we have the negative, it flips everything over. So −3 would flip the U to make a frown and also stretch it."

"*Ahh*, the skinny frown," Liam whispered. Oliver chuckled.

"Does that mean to shrink it horizontally, which would make it skinnier," Marco pushed his hands together like he was playing an accordion, "we would have something like f of . . ." he paused, it felt so silly but it made sense. Cupping his hands to indicate parentheses he finished his thought, "$2x$?"

"Exactly!"

$$f(bx).$$

"This will change the Numberfolk you are giving the order to. If we have $f(2x)$, 1 is sent to the original place of 2 and so on and so forth."

The professor pulled out a stack of cards. He handed one to each student. His voice lowered and his tone was serious, "These incantations apply to *any* order. A line, a quadratic, anything. You master these commands, and you master the Numberfolk orders. You will be able to send them to any formation imaginable. Something is coming. Best to be prepared."

Marco didn't like the professor's energy. It felt wrong, like his tutor was somehow off, he was scared. He knew that if whatever was happening was enough to frighten Mr. Pikake, it was big.

"On the back of each card is a challenge. Come next meeting prepared." With that, the professor exited the room. Marco thought about running after him. *What did Mr. Pikake find out on his trip?* he wondered. Before he could even complete his thought, the chime on the front door rang, and he was gone.

"That didn't sound good," Liam started. "He seems worried."

"Oh, what could it be? The numbers are gonna get us. *OHHHHH*." Maggie wiggled her fingers.

$$y = \frac{1}{2}(2x - 33) + 0$$

Marco looked toward Oliver who sat fiddling with the tape on the floor. He wasn't sure what it was, but *something* was going on. Marco had enough on his plate, he couldn't handle whatever Oliver's problems were, Mr. Pikake's warning, Fredrick, the box, it was too much.

The shadow monster had returned. It was sucking everyone under. Mr. Pikake and Oliver were its first victims. It was only a matter of time before it took Marco . . . and Maggie too.

As much as his sister annoyed him, he loved her with all his heart. Having felt the sickening pain of loss, he wasn't going to let it happen again. He would fight the shadow monster with all his might, never allowing it to take her.

He found himself back on the chess board. It had been reset, there were kings, and queens, and rooks, and pawns. The knights neighed, prepared to attack while the bishops looked like librarians carefully evaluating their possible moves. They all stood at the ready for their commands. Marco took a deep breath, he had mastered hunting, he would dominate incantations too. He slowly stepped onto the board. He was ready to play.

CHANGING A FORMATION

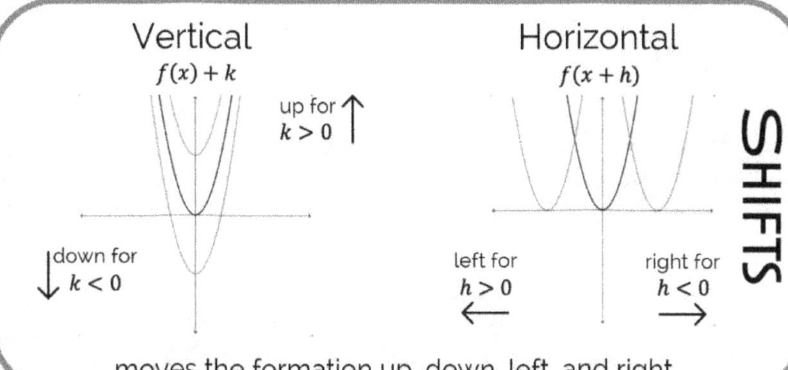
Vertical
$f(x)+k$
up for $k>0$
down for $k<0$
Horizontal
$f(x+h)$
left for $h>0$
right for $h<0$
SHIFTS
moves the formation up, down, left, and right

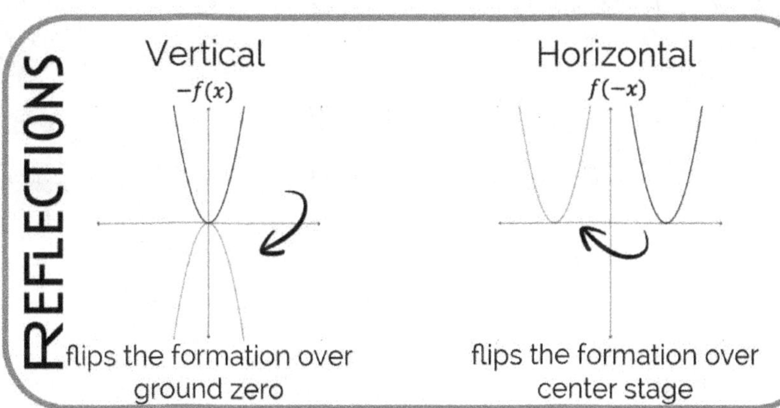
REFLECTIONS
Vertical
$-f(x)$
flips the formation over ground zero
Horizontal
$f(-x)$
flips the formation over center stage

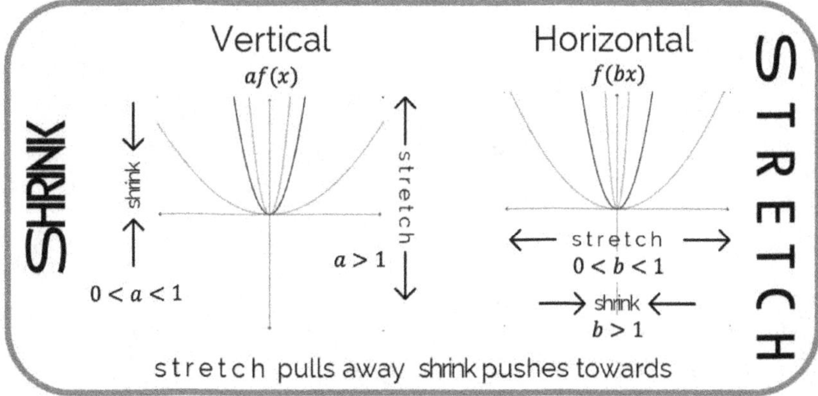
SHRINK
Vertical
$af(x)$
shrink
$0<a<1$
stretch
$a>1$
Horizontal
$f(bx)$
stretch
$0<b<1$
shrink
$b>1$
STRETCH
stretch pulls away shrink pushes towards

→ mother order is $f(x)=x^2$
→ move it up 5, left 2
→ flip it over ground zero
→ stretch it horizontally by a factor of 3

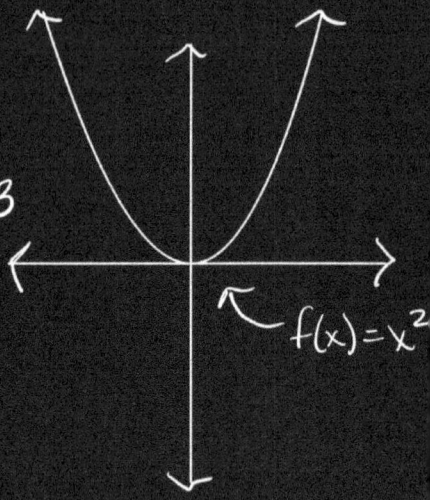

$f(x)=x^2$

1 up 5, left 2

$f(x)+k$

$f(x+h)$

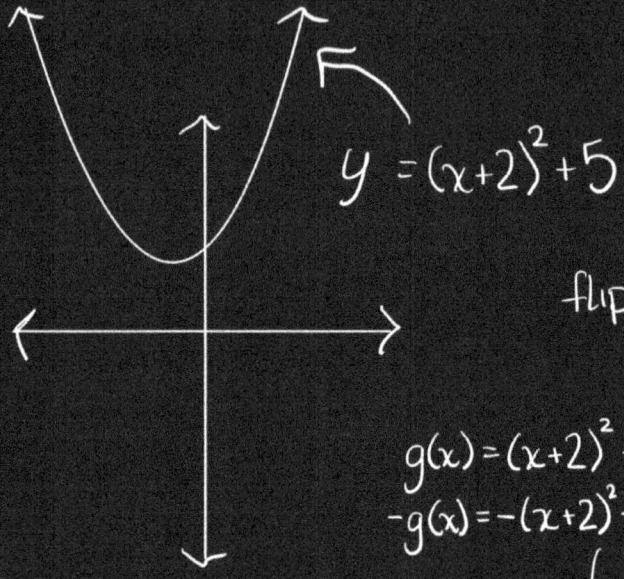

$y=(x+2)^2+5$

flip over x-axis 2

$-f(x)$

$g(x)=(x+2)^2+5$

$-g(x)=-(x+2)^2-5$

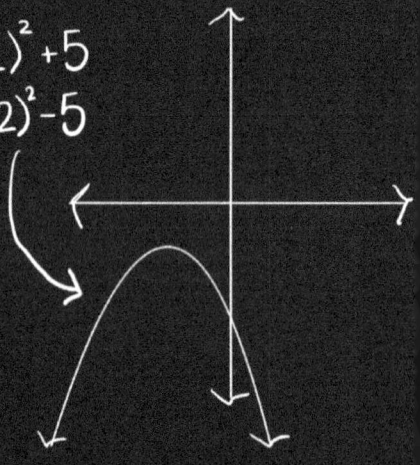

3 stretch horizontally by a factor of 3

$f(bx)$

$b=\frac{1}{3}$

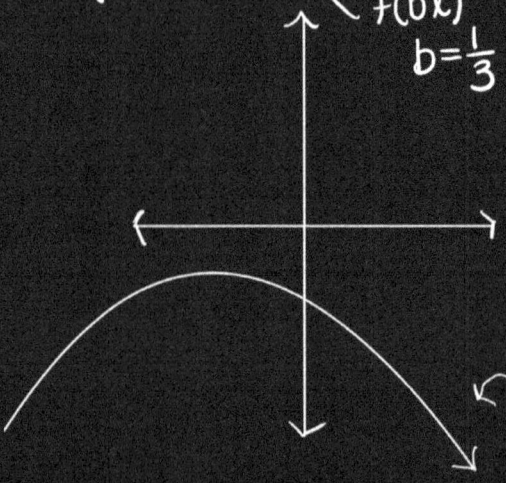

$M(x)=-\left(\frac{x}{3}+2\right)^2-5$

12

PERSPECTIVE

"WE DON'T SEE THINGS HOW THEY ARE, WE SEE THEM AS WE ARE."

-ANAIS NIN

Having only an outside view, Damocles discovered the power of perspective the hard way. As told by Cicero, after praising the marvelous life of the king, having everything and anything one could ever want, he was appointed ruler for the day. And as Damocles enjoyed the servants attending to his every need, he soon noticed the sword dangling above his head held only by a thread. From the outside, ruling was bliss. In reality, the monarch was constantly preoccupied by his enemies and the rivals of the empire, ready to assassinate him at any moment. With this new point-of-view, Damocles found he could never enjoy the spoils with the threat of the sword ever suspended above him.

With great power comes great responsibility. The thought ruminated in Marco's head; his own sword of Damocles hovered above. As much as he wanted to swim in the luxury of his new and almighty power of incantations, his life, Maggie's life, the fate of the world might depend on his ability to command the Numberfolk. The heavy weight pushed hard.

Nervousness gripped his body as sweat began to coat his palms. As he flipped the card over in his hand, he saw his task contained a series of steps.

- The mother order is $y = x^2$
- Move the formation up 3 and left 2
- Reflect everyone over center stage
- Close in by stretching the shape up by a factor of 6

His chest tightened. Having settled into the rollercoaster of emotions his teenage years forced him to constantly endure, he was graduating to a new ride. Before, Marco would go up and down, through twists, turns, and dives as if a new feeling was always on deck, ready to tag in and replace the old, trapping him in a never-ending loop-de-loop. Now, there were periods of calm, of apathy. Times where, out of exhaustion or boredom, he just didn't care enough to pay them any attention. Like the attraction at the fair that would raise you high above the crowds, the climb was uneventful and painless. Then came the drop. From zero to one thousand, all the feelings put on hold would come flooding in at once, and he'd give anything to just hide away.

Something was happening. Fredrick knew it. Mr. Pikake knew it too. Whatever map the box held, whatever the SAN was worried about, whatever the Numberfolk were planning, Marco needed to be prepared. Sure, he could hunt. He could find the one hiding behind the mask. But this new skill was invigorating. Commanding a whole army of Numberfolk? Now that was real power.

He thought back to the warehouse and the numbers. They came alive, right in front of him, and all he could do was cower. But the SAN, the original one not Maxwell and his family's spoiled interpretation, the one created by Pythagoras himself, believed Numberfolk were our delegates to the throne. If he could really control them, his power could be unlimited. They would bend to his will and travel wherever he ordered them to go.

Looking back at the task, he realized this was just another puzzle box. He had the clues, he only needed to determine where each of them belonged to unlock this hidden power.

Starting with $y = x^2$, he needed to move the formation up 3 steps. That was easy enough,

$$y = x^2 + 3.$$

Next, they all needed to move to the left two steps. He heard the Queen snicker in the shadows. Peering into a small mirror that sat atop his dresser, Marco noticed something. He climbed up and stood directly in front of the looking glass. He raised his hands. Mirror-Marco did too. He lowered his hands. Mirror-Marco followed along. Up-up. Down-down. Then he raised his right hand. For a moment he thought he saw Mirror-Marco chuckle, like it wasn't really him, it was some evil copy only pretending to mimic his movements until he looked away. The evil twin didn't raise its right hand, it raised its left. While his reflection seemed to copy everything he did, there was one catch. Up was up and down was down but left was right and right was left. He understood the orders.

Imagining himself as a general, he commanded the Numberfolk just like he forced Mirror-Marco to obey. He wondered if one day he would be powerful enough to direct the numbers like he directed himself. He barely had to think about it. He did something, Mirror-Marco did it too. A wide smile swelled across his face. *I bet my father was that powerful,* he thought. *I can be too.* The ride was inching up. It was passive, not intrusive, but he knew eventually the drop would come.

Then, a strange feeling washed over him. What if someone, somewhere was directing him? Making him do and feel whatever they wanted. A world where he was the soldier, the Numberfolk, and the wicked fates were calling the commands. He shuddered.

Throwing himself back onto his bed, he scribbled in his next step. "I need to command you to move to the left two steps which is -2, I do the opposite, send in the evil twin."

$$y = (x + 2)^2 + 3.$$

Check. He imagined the paper clicking like his box after a successful round of decoding. On to the next clue. *Reflect everything over center stage.* Marco was familiar with the limelight, it was the y-axis, the vertical line in the middle that separated the never-ending plane into two equal parts. While Maggie was one to seek out the spotlight, Marco wasn't. Preferring to hold back and move in the shadows, it was a bit ironic that he always found himself playing the leading role.

Standing on center stage, he looked to his left, he saw all the positive characters, the heroes, the goody-two-shoes, the fairies, the angels. On his right were the negatives, the villains, the hyenas and monsters, the witches, the trolls. He slowly began swirling his hands, around and around, one on top of the other until a ball of energy formed between his palms. He grabbed it tightly before releasing the power, pushing his arms to both sides. Lightning struck down, so bright it cast a searing white shadow over the theater, and he couldn't see a thing.

As the smoke cleared, he began to laugh. The characters had all swapped places, the villains were now to his left and heroes to the right but that wasn't all. Whatever made the creatures terrifying was replaced with their friendly counterparts. The hyenas gained puppy-dog eyes and sat with their tails wagging ready to play. The monsters were lovable teddy bears, and the witches were dolls combing the beasts' hair and placing bows at every opportunity. The positives changed too. The fairies grew long sharp fangs, their wings grey and tattered. The angels were the warriors of the Bible, they held long swords ready to charge.

Marco noticed something else. While all the characters had swapped sides like half-time at a basketball game, Marco remained the leading man. They flew over him in an arc, yet he was firmly rooted on the center stage line. *Interesting.* If nothing else, should he ever have to go into battle, this seemed like an important trick to have up his sleeve. Anything or anyone on center stage didn't move when he cast the proliferation spell with -1 on the soldiers. It made sense. The spell basically took every ordered pair (x, y) and moved them to

$(-x, y)$ transforming each Numberfolk into their evil twin.

Zero was special. They were the only Numberfolk who were their own worst enemy. Wherever Zero was sent, $(0, y)$, the reflection spell would send them to $(-0, y)$ which was of course still $(0, y)$. It didn't move him. It was like a big musical number where all the actors twirled around the star.

He opened the door to his mindroom. He had begun construction last fall and it sat empty with the exception of a large whiteboard he used to affix ridiculous rule after rule from his math class. Every once in a while, he'd stuff some history notes in a drawer in the corner. Marco had spent the last nine months updating the space. Math was no longer a set of horrible policies, it was an adventure. It was all around him, it was inside him, and it was powerful.

Basing his mindroom on an Escher drawing, it had stairs traveling every which way each leading to a designated area of his brain. He had a room dedicated to hunting where he'd sometimes escape to, just to practice. With his rod, he'd open a portal to Numberville and grab and throw the citizens as if they were his playthings. Color coding seemed important. He devised a system that maximized his ability to quickly recall. Even when he wasn't 100% sure, he knew it was filed under lime green and could quickly scan through his memories to pull it out.

Impressed with how much his math knowledge had grown, he added ladders to reach the highest shelves of his filing system. He had trained his mind to store things in the perfect place, making for easy retrieval. Scribbling his discovery on an index card, he wrote "reflecting across center stage doesn't impact zero" and filed it away cross-referencing it with functions, orders, reflections, and, of course, Zero.

Adding in his last move, Marco jotted down the new order on his paper.

$$y = (-x + 2)^2 + 3.$$

He was ready for the final act. *Close in by stretching the shape up by a factor of* 6. A conductor composing his magnum opus, Marco grabbed the strings and yanked them toward the sky. The resulting U-shape was long, tall, and skinny. He frantically wiped his hands like erasing a chalkboard and tried again. This time, he faced his palms toward each other and pushed in, squishing the shape. The result was the same. Should he change the soldiers or their locations?

Either way, he would need to cast another proliferation spell. He waved his wand back ready to leap into action but hesitated. Just like his coded box, there were multiple ways to interpret the clues.

$y = (-6x + 2)^2 + 3$?

$y = \left(6(-x + 2)\right)^2 + 3$?

$y = 6(-x + 2)^2 + 3$?

$y = 6[(-x + 2)^2 + 3]$?

This would be more challenging than he originally thought. *Buses.* Marco remembered back to when he first learned about Numberfolk orders, functions. The routes were becoming too complex to keep track of . . . it was exactly like LEGO. When you had a massive build, they don't just show the final picture. The instructions break it down into manageable pieces, then you put all the mini-builds together. Why couldn't he do that here?

Taking his new formation, the one he had shifted over and up and flipped, he called it $g(x)$.

$$g(x) = (-x + 2)^2 + 3.$$

Now Marco could explore the two proliferation spells he had available to him: $6g(x)$ and $g(6x)$. *Of course,* he thought. Pulling it up like a puppet was a jumbo potion for the entire formation while shoving it in from the sides was making the soldiers stand really close to each other, pushing them together like an accordion, then determining who would be sent where on a normal-sized field to achieve that shape. The second order couldn't be it. A horizontal change was $g(6x)$ and this was instead $g\left(6(x + 2)\right)$, impacting more than just the soldier. The third order was out too. It wasn't casting the proliferation spell on the entire formation, just part of it leaving the pathetic three dragging along behind. *Two down, two to go.*

If he wanted to push the Numberfolk in, toward center stage, it required impacting only the soldiers by substituting any x with $6x$.

$$y = (-6x + 2)^2 + 3.$$

On the other hand, he could accomplish the same formation by acting like a puppeteer and pulling on the strings from above. His proliferation spell would hit the entire field, $6g(x)$.

$$y = 6[(-x + 2)^2 + 3].$$

He looked over the two commands. In theory, they were *supposed* to accomplish the same thing stretching vertically and squishing horizontally both made the formation skinnier. In action, they appeared to be set in two totally different worlds. Marco found himself at a fork in the road. Down one path was considering a single soldier. If he could figure out where one warrior should land, he could see what commands sent them there. This method made him nervous. What if it was sheer coincidence? What if the soldier he picked just happened to go to the same place for two different directives?

The second path led to simplification. Both commands were beasts, they contained complex twists and turns, ripe for silly errors. He decided to take a chance on the first option. He'd be 0. Zero was an excellent soldier, it absorbed all proliferation spells removing intricate multiplication from the task.

He began by defining his new command. Needing to give it a name, the horrifying pancake debacle replayed in his head. *Ugggghh,* he groaned. This was his masterpiece, his dominion, his magic. He would call this order Marco of x, M of x for short.

$$M(x) = (-6x + 2)^2 + 3.$$

Where would this send the lonely 0?

$$M(0) = (-6(0) + 2)^2 + 3.$$

Dr. PEAS, we meet again, Marco snickered. Oliver shot to his head. Dread began to bubble up his body. There was no misinterpreting the situation: his best friend had been distant and different. Filing away a reminder to get to the bottom of what plagued Oliver, he pushed the thought away and refocused.

Parentheses. Marco quickly computed $-6(0) + 2$ and replaced it in the order,

$$M(0) = (2)^2 + 3.$$

Ahh, the E has arrived. He scribbled his final moves.

$$M(0) = 4 + 3 = 7.$$

Great, he knew where Marco of x would send 0. He was surprised it was such a large station. Afterall, the mother command meant that

0 began at position 0. With only minor manipulations, he had sent the soldier blocks away. The shock turned into euphoria. He was the Hulk and with a light toss, the Numberfolk flew through the air. The power began to seep into his veins. He felt strong.

Time for the second order. What would he call this one? As Oliver continued to gnaw at his brain he named it in honor of his friend, his sidekick, Oliver of x. He started to scratch down the command and in only seconds the O's and 0's made him dizzy. He opted to write his pal's full name.

$$\text{Oliver}(x) = 6[(-x+2)^2 + 3].$$

It had only been a second. Making quick work, he determined where Oliver would heave the zero.

$$\begin{aligned} \text{Oliver}(0) &= 6\left[\left((-0+2)\right)^2 + 3\right] \\ &= 6[(2)^2 + 3] \\ &= 6[4+3] \\ &= 6[7] \\ &= 42. \end{aligned}$$

Crap. He was instantly deflated. Marco also felt irrationally annoyed with Oliver. Afterall, it was Oliver of x that caused him to fall from his high tower. It was no surprise his best friend would chuck the Numberfolk so far, he was the pitcher on their baseball team.

Examining the card Mr. Pikake had given him, he studied the different incantations he could cast to augment the Numberfolk orders. Something didn't make sense. While pulling up the Numberfolk formation like a puppet vertically would stretch the U-shape making it taller and skinnier and compressing the arrangement like the last step in constructing a PB&J to ensure all the good stuff melds together should have the same impact, it seemed like pulling would require much more work to accomplish the same result.

The vertical stretch was $af(x)$. If f was just the mother order, x^2, then that would be ax^2, meaning the soldier was squared before they stretched. When a was 6 and the soldier, x, was 1, that would take 1 to position 6. However, the sandwich squish enlarged the *soldier* first. It was $f(bx)$, using the same spell and trooper would end up sending 1 to $(6 \cdot 1)^2 = 36$.

Logically, it made sense. It was much easier to push a big heavy box than it was to try to lift it. While both made the U-shape taller and skinnier, the push pushed harder. It got thinner faster like a super weight-loss tincture. He remembered Mr. Pikake had said these incantations would work on any order. While Marco didn't know any other formations yet besides a line or a march, he figured that each must have a different, but predictable wizard to accomplish the pull. To have the same outcome, he needed to tone down his push, it required a delicate hand.

Thinking about a larger brigade, Marco considered the mother $y = x^3$. To make the formation the same by pulling or pushing it meant that $ag(x) = g(bx)$, or,

$$ax^3 = (bx)^3.$$

He knew the brigade $(bx)^3$ was a group of three bx's, or $bx \cdot bx \cdot bx$, this meant he had three b's and three x's or b^3x^3, he just lined them up differently. He had created a hunt, and Marco was an excellent hunter. If $ax^3 = b^3x^3$ that meant $a = b^3$. For his pushes and pulls to have the same strength, it would require him to pull much harder in a brigade. Not yet an expert, Marco selected the easiest one he could think of: a brigade of twos. If a was 8, meaning he pulled with a force of 8, that would make $b = 2$, he could use much less pushing power to achieve the same result. After all, $8 = 2^3$.

Looking back at his task, he read the command again: *Close in by stretching the shape up by a factor of* 6. Mr. Pikake wanted him to pull up, to stretch the shape vertically. Although he was intrigued by the idea of pushing, it did after all require less force. It meant he would need to find the strength of the push so that $6 = b^2$. As far as Marco knew, there wasn't any Numberfolk he was aware of that could accomplish this. It was asking him to find the number that when he multiplied it by itself would equal 6. But the squares skipped over perfect 6. Two squared was four and the very next square, three squared, was nine. There was no six. He would have to pull, it was his only choice.

This of course meant Oliver of x was right, and Marco began feeling guilty over his imaginary attitude toward his friend. His previous image of himself as the Hulk took on a new shape as he realized he overestimated his abilities and at best was Bruce Banner. His cheeks turned a deep, rosy red with embarrassment. His command,

$$\text{Oliver}(x) = 6[(-x+2)^2+3],$$

was the order that started as $y = x^2$, moved each soldier up three and left two, flipped them over center stage, and finally pulled the entire formation up like a giant sucking in its gut. He knew it would take soldier 0 to Station 42, but wanted to make sure it would take any warrior to where they were *supposed* to end up. Better test someone else.

Super Mario appeared in the mirror.

"Lets-a-go!" he called out.

Marco accepted the strange daydream as he navigated the character through piranha plants and goombas. Mario began at $(-1, 1)$, it was the normal modest castle shown in an early level. He jumped up three and to the left two, landing at $(-3, 4)$, an airship. Throwing the plumber across center stage, Mario stretched out his arms and floated like a bird ending the dive with a somersault touching down at $(3, 4)$. He had successfully navigated the mini battle. It was time to defeat Bowser. The final move required Mario to gobble down a magical mushroom, a wizard casting a spell making him grow to be six times his size. His position skyrocketed up, he landed firmly on the dragon's head at $(3, 24)$. Quickly making sure Oliver of x would also send soldier 3 to Station 24, a grin smeared across his face. *Perfect*.

As Peach and Luigi joined the scene, cheering Mario on, Marco remembered their names only differed by one letter. He smiled as he imagined Liam as Luigi and Peach as Penelope. He cast Maggie as Bowser Jr., snickering at his choice as the villain stomped and whined like a crybaby. It was strange that Oliver didn't have a place in the group. Was his brain trying to tell him something?

❊ ❊ ❊

Right next door, Maggie stared at her own card. At her desk, she sat up perfectly straight and began organizing her things. She laid out graph paper in the center of the table and leaned the card against the wall for reference. Taking out two deathly sharp #2 pencils and placing them parallel to the paper she was ready to begin.

- The mother order is $y = x^2$
- Reflect everyone over ground zero
- Stretch the shape horizontally by a factor of 4

- Move the formation down 2 and right 1

"Okay," she said aloud and rubbed her palms together before placing them firmly on either side of the paper. Taking a deep breath to center herself, she dove in.

Her initial incantation needed to flip everyone over ground zero, $y = 0$, the x-axis. "That's easy enough," Maggie said aloud placing her eraser on her cheek. She knew this would take all the soldiers' locations and move them to the negative plane — the underworld.

In smooth clean strokes she wrote,

$$-f(x) = -x^2.$$

She instantly hated it. It wasn't clear enough. Was the negative attached to the soldier?[*] Taking her time to carefully erase so no one would ever know the ghosts she buried, she replaced it with,

$$-f(x) = -(x^2).$$

"There." She patted the page. On to the next step. Maggie read the command, "Stretch the shape horizontally by a factor of 4."

She imagined pulling the formation outward toward the edges of the paper. The image in her mind showed a chubby curve, widened by her action.

A horrid tightness clamped down on her chest. It was like an elephant had decided to take a seat right on her sternum. She couldn't breathe. Tugging her knees to her chin, she gave them a tight squeeze and tried to calm herself. A tear slowly leaked from her right eye.

KFUN hadn't been nearly as *fun* as she imagined. While she was talented with numbers, all this talk of functions and orders was overwhelming. She had made mistakes, in public, multiple times. Maggie was used to being perfect. She always knew what to do, understanding had come easy. This was no longer the case. She had to work, work hard, to make sense of things. Deep down inside she wanted to call on her brother for help. No way. How humiliating. Marco always called on *her* for help, it wasn't the other way around.

---

[*] When you see something like -2^2, remember exponents come first so the major is 2, and $-2^2 = -1(2^2) = -4$. If you want the major to be -2, you need parentheses, $(-2)^2 = -2 \cdot -2 = 4$. But be careful! The function, $f(t) = t^2$, is telling you to square each soldier. That means to find the location of soldier $t = -3$, you find $f(-3) = (-3)^2 = 9$.

It would only prove to him she was just the annoying little sister who he needed to protect.

With a new sense of determination, she reread the card. A horizontal stretch flattened out the shape. Just like a proliferation spell: if you wanted to force the Numberfolk to be smaller, you had to multiply by a fraction less than one. It didn't make any sense to stretch it by a factor of 4, that would make it bigger, taller. Every inch of Maggie's body wanted to give up. It took an enormous surge of courage to wipe away the tears, take a deep breath, and sit up straight. She could do anything her brother could do, and she was going to do this.

Slowly stretching the formation by pulling the shape open to the left and right, Maggie realized that if she extended the soldier by a factor of 4, the ordered pair $(1,1)$ would move to $(4,1)$. The whole thing gave her an icky feeling. If Marco hated rules, Maggie *despised* them. She required clear and logical explanations for every concept. There was absolutely no way she would simply "do as she's told" without knowing why she was doing it *and* why it worked.

"Of course!" she exclaimed. It was a puzzle conducted in reverse. The clue was telling her to impact the soldier, not the location. It was easier for Maggie to visualize it as the soldier relative to the position. She needed to think backward. Station 1 should be assigned to soldier $1 \cdot 4 = 4$. Station 2 would be ordered to soldier $2 \cdot 4 = 8$. This was a question she could answer. What spell did she need to cast to turn an 8 into a 2? Or a 4 into a 1? The mystery was really,

$$4y = x.$$

Although not as strong at hunting as her brother, she knew she needed to divide each soldier by 4, or multiply by $1/4$ as Mr. Pikake would insist.

It wasn't *backward* at all really. It was only a change in perspective. Modifying the order vertically, meant to send the Numberfolk to a new position. Like a coach calling out "player 1 to position 3." But with horizontal incantations, you knew where you wanted the soldier to end up and had to figure out how to get them there. The card said to stretch horizontally by a factor of 4, to do that, Maggie needed to multiply the soldiers by $1/4$.

Making a mental note that a horizontal stretch by a factor of b, meant multiplying the soldier by $1/b$, while a horizontal shrink by a

factor of $1/b$ would then, of course, multiply the soldier by $1/1/b$ which was just b, she wrote down her command.

$$-f\left(\frac{x}{4}\right) = -\left(\frac{x}{4}\right)^2.$$

A heavy sigh escaped her lungs. She had made it to the final step, and it didn't look that bad at all. Moving the formation down two was adding -2 to the position,

$$-f\left(\frac{x}{4}\right) - 2 = -\left(\frac{x}{4}\right)^2 - 2.$$

One right. Horizontal. Different perspective. From y's point-of-view, positive is positive since $y = f(x)$. But from x's viewpoint, on the other side of the scale, to move right she needed to tell the soldier to travel in the negative direction, their left:

$$-f\left(\frac{x}{4} - 1\right) - 2 = -\left(\frac{x}{4} - 1\right)^2 - 2.$$

It was ugly. Starting from the mother command, $f(x)$, Maggie had created something new, and that was something to be proud of. It was her baby. While it had similarities to its parent, it was also its own unique command. Like a child, it needed its own name. "Maggie," she whispered. In impeccable handwriting she concluded her work and drew a crisp rectangle around it,

$$m(x) = -\left(\frac{x}{4} - 1\right)^2 - 2.$$

Something didn't feel right. Normally, when Maggie finished a problem, she felt strong and confident. Now, she only felt uncomfortable. She was petrified of failing. Her mind shot to their next KFUN meeting, she handed her orders to Mr. Pikake and watched in horror as everyone burst out in laughter at how wrong she was. In trying to send the Numberfolk to the nearest Starbucks, she ended up launching them into outer space.

"Check your work," she said calmly, hanging onto her composure for dear life. It was easy enough to see where Zero would be sent. The first two incantations wouldn't do anything to him, he was impervious to such moves. Knowing she worked better with clear, precise steps, she began to carefully draw out her attack.

"The mother order is $y = x^2$, which sends 0 to station 0. Next, I flip the formation over ground zero. It pushes (x, y) to $(x, -y)$ which means it carries (0,0) to $(0, -0)$, he doesn't move."

She traced out the shape in her mind. "Then I stretched the *soldiers* by a factor of 4. The Numberfolk in position 0 is $4 \cdot 0 = 0$." He is still where he started. For her final step, she needed to move everyone down two and right one. That command would take $(x, y) \rightarrow (x + 1, y - 2)$. "Which puts 0, who is still at the origin at, $(0 + 1, 0 - 2) = (1, -2)$. The point (0,0) on the original formation has moved to $(1, -2)$ on the new formation, and I ended up with the command for soldier 1."

To test her incantation, she shoved 1 into the order. Based on her findings, soldier 1 should be sent to station -2. The move to the right caused a game of telephone as zero whispered their location to 1.

$$m(x) = -\left(\frac{x}{4} - 1\right)^2 - 2.$$

$$\begin{aligned} m(1) &= -\left(\frac{1}{4} - 1\right)^2 - 2 \\ &= -\left(\frac{1}{4} - \frac{4}{4}\right)^2 - 2 \\ &= -\left(-\frac{3}{4}\right)^2 - 2 \\ &= -\frac{9}{16} - 2 \\ &= -\frac{9}{16} - \frac{32}{16} \\ &= -\frac{41}{16}. \end{aligned}$$

She couldn't help it, she burst into tears. Not even bothering to try to simplify, Maggie knew her orders didn't send 1 anywhere close to where they were supposed to go. A soft knock made her jump.

"Y-y-yes?" she whimpered.

Marco slowly opened the door and peaked in. "You alright, Bug? I could hear you crying from my room."

She wiped her face, snot and all, with her sleeve. As much as she didn't want to show her brother any weakness, her sadness had taken control of her body and mind. Rational thought was on a forced vacation and the words, "I just can't do it!" hurled out of her mouth like a speeding train.

"Can't do what?" Marco gently closed the door and walked to his sister's side. "Oh." He saw the card, paper, and orders in front of her.

He wrapped his arm around her shoulder and gave her a tight squeeze. "Don't get down about it, it's hard stuff. It's like my video games. It wouldn't be any fun if it was *too* easy. You only get that good feeling of accomplishment when you defeat the seemingly undefeatable."

"But everything is *always* easy for me!" she wailed.

"You wanna know a secret?" Maggie gave her brother a sharp nod. "I make up my own worlds, in my head. It helps me to work through hard stuff. It doesn't come easy for me, Bug. You gotta find a place where all this does make sense."

"That's easy for you to say, you're not a math kid. I am." She sniffled.

"No. If I've learned anything it's that *everyone's* a math kid. What it's really about is making the puzzle work for you. That's what Mr. Pikake showed me. So what if I didn't get all the x's and y's? I turned them into vampires and zombies, into a video game, and then the pieces fit. Math is in you Maggie, make the pieces fit."

With a firm nod, she went to work. Having somewhat of a morbid curiosity, Maggie imagined she was a brilliant detective, a modern Sherlock Holmes, who was called in on a case only she could solve. Someone had been *murdered*. The problem: the police couldn't figure out who the victim was.

Studying the crime scene, it was clear an attack had taken place. Maggie focused on the carpet; the long parallel lines indicated it had been vacuumed recently. The messy, skewed fibers in the center of the room presented her first clue.

"There are two things that happened here." She pulled a toothpick out of her mouth. "Someone was attacked, and their body was moved."

She partitioned the commands into two parts, it was just like the trains. All the horizontal moves determined the soldier and they got stuffed in like a turducken to the train that took them to their location. If she focused only on the changes that impacted the location, reflect over ground zero and move everything down 2, and made the soldier $f(x)$ which she'd deal with later, she determined,

$$m(x) = -f(x) - 2.$$

"They were flipped over and dragged down," Maggie announced to the group of onlookers; a mix of uniformed and plain-clothed police and detectives.

Penelope entered the scene dressed in a trench coat and matching hat, the Watson to Maggie's Sherlock. "No body, no crime," she said in a deep and dark tone.

Something interesting caught Maggie's eye. "Originally, I thought the body was flipped and then pulled down, but now I see this too could be,"

$$m(x) = -(f(x) + 2).$$

"The killer could have lugged the body up here," she pointed to the carpet, "then flipped it. Curious, but not pertinent."

Pacing around the room, Maggie stroked her chin before swiveling on one foot to switch directions and do it all again. "You have a grand gift for silence, Penelope. It makes you quite invaluable as a companion."

Marco watched as the tears dried up and a grin expanded on his sister's face. He couldn't help smiling too.

Bright and glistening equations filled the air as Maggie Holmes grabbed numbers and moved them around to complete her puzzle. "*Aha*! The clues are showing me that before death, the victim was stretched using a torture rack." Her eyes darted around the room, "The factor was four."

Then she saw it, she knew what she had done wrong before. It was clear as day. Wanting to jump and scream, she held it in and stayed in character.

"The key to this crime is the head!" Maggie shouted.

"The head?" Penelope asked.

"*Ay*, the head."

Maggie thought back to the commands. Before her breakdown, she had found where the ordered pair (0,0) in the original order ended up in the new directive. It led her to $(1,-2)$. In her investigation, she noticed the first steps of flipping and stretching didn't impact this pair, her leading actress.

Knowing that the head must end up at -2, she zoned in to determine what the soldier-picker, $f(x)$, would need to be to identify the head of the victim,

$$\begin{aligned} m(x) &= -f(x) - 2 \\ -2 &= -f(x) - 2 \\ 0 &= -f(x). \end{aligned}$$

Now she could see that $f(x) = 0$, which made sense as she knew (0,0) was transformed to $(1, -2)$. Her mistake was that she had considered what would happen to a *normal* corpse. But this wasn't a typical body. It had gone through the stretching rack, that changed things. She was ready to focus on the soldiers.

The turducken, what she stuffed in, was $f(x) = bx + h$. Knowing b was the 1/4 she had already gathered, the end was in sight. Her final step was to find h, to deduce how far the killer needed to pull the mangled body for the head to end up at $(1, -2)$.

$$f(x) = \frac{x}{4} + h.$$

Soldier 1 ended up at location -2. Since the location was -2 when $f(x) = 0$, Maggie had everything she needed.

$$f(1) = 0$$

$$\frac{1}{4} + h = 0$$

$$h = -\frac{1}{4}$$

Maggie had her order. She stood tall and adjusted her collar. "The victim wasn't a person at all!" Her finger pointed toward the sky. "What was brutally tortured and eventually killed was none other than a smile. It was viciously forced into a frown."

$$m(x) = -\left(\frac{x}{4} - \frac{1}{4}\right)^2 - 2.$$

"How did you know?" Penelope inquired as she stood in admiration over her friend.

"Elementary, my dear Penelope. Look here. The clue was the cruelty imparted on the victim. Stretching or shrinking impacts how far one must move the body. For it was not the same shape that it once was."

"You are so impressive," Maggie's friend swooned.

"*Ah*, it's my business to know what other people do not know,"

Maggie said with a bow. She exited her daydream to find her brother still there, sitting patiently on her bed.

"Figure it out Bug?" Marco asked.

"Yeah!" She bounced. Her smile was as big as Marco had ever seen it. "I think I figured something really important out too."

"Oh yeah? What's that?"

"When you stretch or shrink the shape horizontally, it impacts the other horizontal stuff too. The x is like the number line. If you stretch or shrink it, you are really doing it to the whole line, because it has to work for all the soldiers."

Marco had no clue what she was talking about, it must have shown on his face because his sister started digging through her desk drawer to provide him a demonstration. She pulled out a rubber band and drew a line on her paper.

"See, look. Say I want the soldier to move to the right one. They go from here," she marked off the line, "to here."

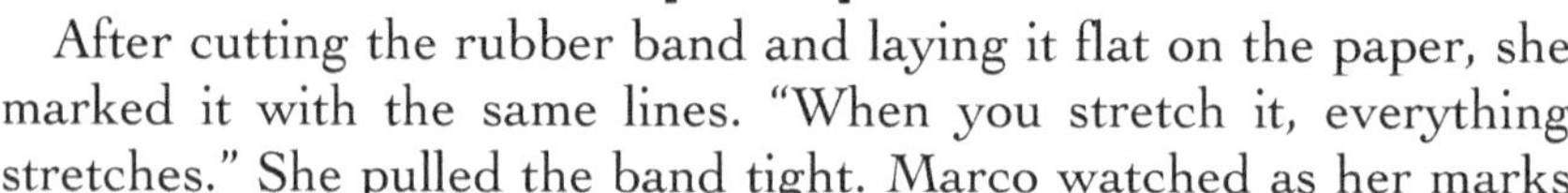

After cutting the rubber band and laying it flat on the paper, she marked it with the same lines. "When you stretch it, everything stretches." She pulled the band tight. Marco watched as her marks ran away from each other.

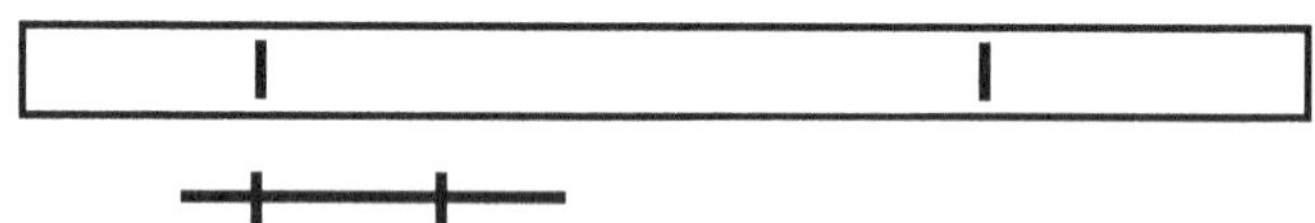

"Do you see it? After it's stretched, if you moved the same amount on the rubber band, you'd end up way over here. But that isn't what you wanted, you only wanted to move this much." She pointed at the second mark on her drawing. "So, you have to counteract the stretch in your movement, otherwise you'd go too far."

Under her line, she began writing out an equation.

$$y = af\big(b(x + h)\big) + k.$$

"I think this is how you do it when the stretch or shrink is before moving left or right. The b, the horizontal stretch or shrink, has to hit the horizontal shift left or right too to make up for how you pulled or pushed on the line."

Marco stood in awe of his sister. She had somehow gone from

totally stuck to ready to give a lecture on the topic in minutes. *Even with that gigantic ego, she really has no idea how amazing she is*, he thought.

"See, that's what I did wrong. I was supposed to move it over one, so I did. But because it was all stretched out, one ended up being way too far." She started rewriting out her entire process on a crisp, clean piece of paper.

"See, a new perspective changes everything," Marco smiled.

"You're right," Maggie nodded. "Perspective *is* everything. Thanks for showing me your trick."

In that moment, he viewed Maggie in a new light. He wanted to spill his guts to her, to tell her everything about the box, their grandfather, his father, Fredrick. He stopped and flashed back to when he first entered the room: she was broken, crying in a ball. As he watched her take her time to write each step perfectly, he knew he had made the right decision. Incredible as she was, Maggie was still grieving, she had too much on her shoulders. He certainly wasn't going to add to it. Keeping his problems to himself, he slowly slipped out of the room, without making a sound.

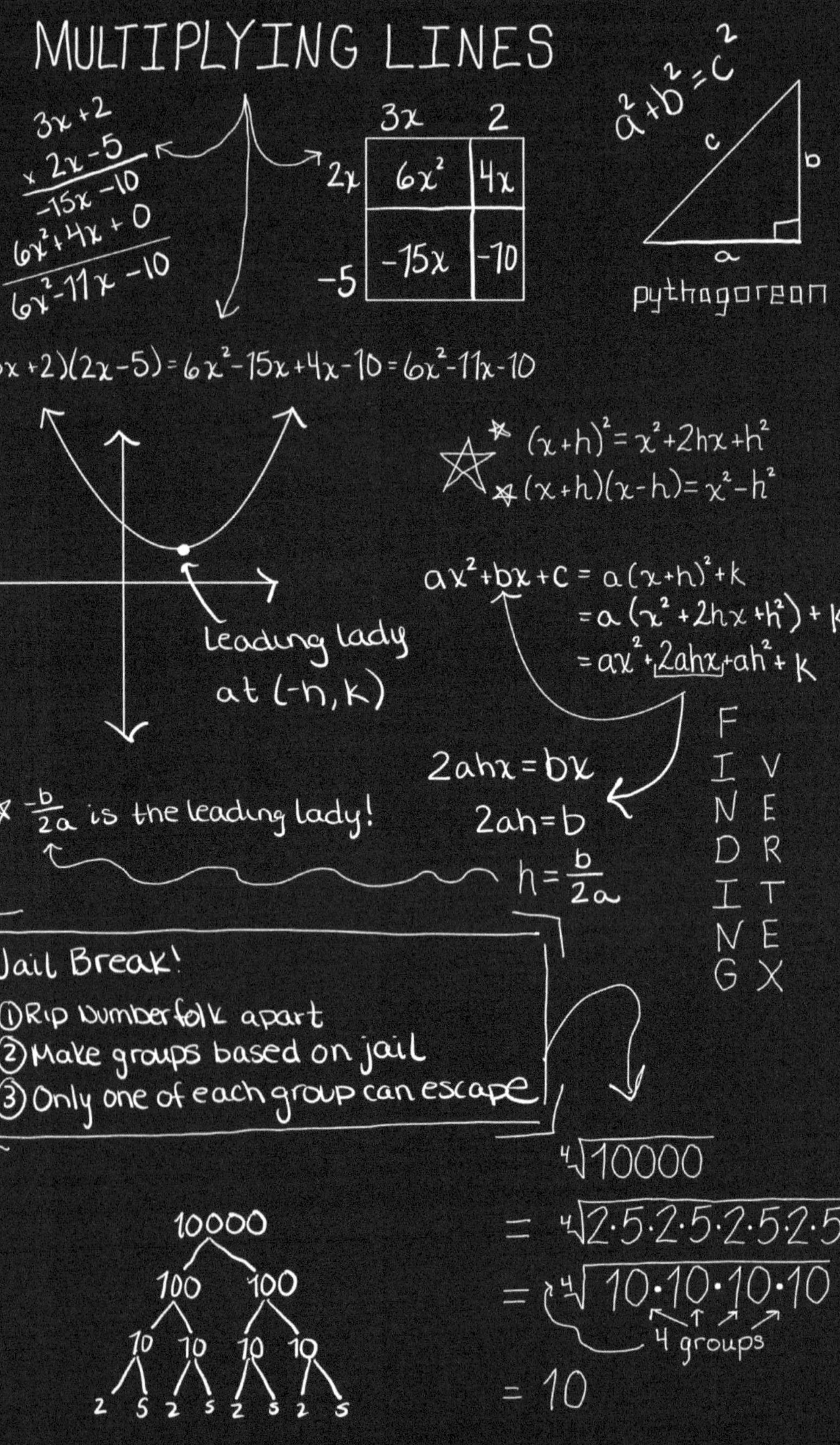
MULTIPLYING LINES
3x + 2
× 2x − 5
−15x − 10
6x² + 4x + 0
6x² − 11x − 10
3x
2
2x
6x²
4x
−5
−15x
−10
a² + b² = c²
c
b
a
pythagorean
(3x + 2)(2x − 5) = 6x² − 15x + 4x − 10 = 6x² − 11x − 10
(x + h)² = x² + 2hx + h²
(x + h)(x − h) = x² − h²
ax² + bx + c = a(x + h)² + k
= a(x² + 2hx + h²) + k
= ax² + 2ahx + ah² + k
leading lady
at (−h, k)
2ahx = bx
2ah = b
h = b/2a
−b/2a is the leading lady!
FINDING
VERTEX
Jail Break!
① Rip numberfolk apart
② Make groups based on jail
③ Only one of each group can escape
⁴√10000
= ⁴√2·5·2·5·2·5·2·5
= ⁴√10·10·10·10
4 groups
= 10
10000
100
100
10
10
10
10
2
5
2
5
2
5
2
5

13

INSURRECTION

"MATHEMATICS IS THE MOST BEAUTIFUL AND MOST POWERFUL CREATION OF THE HUMAN SPIRT."

-STEFAN BANACH

Wildy scribbling on the mirror, Mr. Pikake was silent. He'd been at it for about ten minutes and didn't appear to be stopping anytime soon. Maggie had roped Penelope into a discussion on rock-paper-scissors, while the boys sat semi-quietly across the room.

Oliver had a strange energy about him. It reminded Marco of Peter. His stepfather had always come off as mean, like he was constantly angry at Marco over one thing or another. When Peter left, Marco gained a totally new outlook. He wasn't mad at all, he was hurt, scared even. Liam had explained it away with a quote from Yoda, "Fear leads to anger, anger leads to hate, hate leads to the dark side." If this was true, what was Oliver so scared of that was causing him to act this way?

Marco couldn't think about it now, he needed to eavesdrop on his sister's conversation. Maggie had done a deep-dive into roshambo and claimed she had found a sure-fire strategy to win. Marco had been her first victim and was dying to know how she did it.

"Look. If you are playing someone who knows the secret, things can get crazy," Maggie started. "But if you play against a normy, your chances of winning skyrocket."

"Got it." Penelope gave a sharp nod of her head.

"Alright, first you have to decide how your opponent will play. Rock is a common first move, but so is scissors. I find that if you are playing someone more serious, they go for rock. Marco always does scissors because it is the last word you say," Maggie giggled. Penelope blushed.

"To win, you think two-steps ahead. Say I play rock and you play scissors. Your first thought is 'I should've played paper to beat Maggie', so paper is on your mind. If you are against a smart player like me, I know you are thinking about paper so I will play scissors. You should play rock to beat me. In other words, copy my last move."

"Okay, if you lose, you copy the winner. Check."

"If you are playing against a not-so-smart player like Marco, he sees your scissors and is going to do the same thing. They're not thinking about the Heisenberg effect and just think about beating your last move. He'll play rock again to beat your last round's scissors, so you should play paper. Play whatever wasn't on the board the last round."

"So, if you lose against a good player, copy them. If you lose against a not-so-good player, play whatever wasn't on the board. What if you don't lose?" Penelope asked.

"If you win, same principle. If you play rock, a normy will play paper to beat your last rock, so you play scissors. A strategic player will opt for rock, mimicking your last move."

"Got it. So against a good player, if you win, you play whatever isn't on the board. And for other players, you copy their last move. This is a lot to remember . . ." Penelope felt lost in all the options.

"Good players are win-switch, lose-copy. Not good players are win-copy, lose-switch. Just takes practice," Maggie smiled cheerily.

Across the studio Marco was attempting to file this information in his mindroom. Maggie would be considered a good player, so he should copy her last move if he loses and play whatever wasn't played in the last round if he wins. He couldn't wait for their next disagreement to test it out.

"Perspective!" Mr. Pikake had finished his drawing and revealed his work to his students. "What do you see?"

They all studied the image on the mirror a moment before shouting out their results.

"It's two cats!" Liam shrieked.

"I see a vase," Marco retorted.

"It's obvi both," Maggie quipped.

"It's looking like a pair of bulldogs to me," Oliver argued.

"Excellent! *How* we perceive perpetuates *what* we see. Now, we will examine how to manipulate a quadratic order to comprehend all the secrets it holds." He whipped a handkerchief from his breast pocket to clear the board.

Liam's eyes grew. "He's erasing it?!" he whispered in shock.

"Before we proceed, we must focus more on the structure. You know how to take the mother and apply your incantations to move it, stretch it, shrink it, bend it to your will!" He held the l's like a whole note. "But! There is much more to this simple shape than meets the eye. To completely control this character, you must identify the four pivotal positions."

Mr. Pikake violently slashed at the mirror: up, down, left, right, slash, slash, slash, slash. He had drawn in the stage lines, the axes, and marked off four points. The first was predictable, the location of

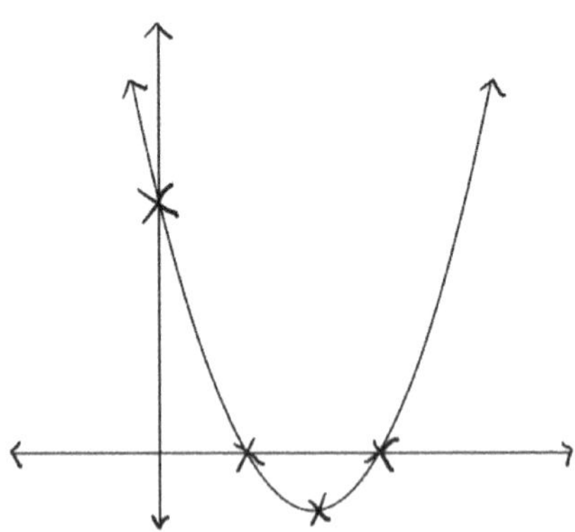

soldier zero. Marco expected that to be key, it was crucial in a march, it only made sense it was also vital in a quadratic. He was surprised to see the next two marks, they indicated who was ordered to ground zero: the two locations where the curve intersected the x-axis. The final mark was directly in the center of the formation, the place that would split the U perfectly in half.

"One!" The professor's bony index finger sprung to life, "Where is Zero? Zero is a tricky beast, a special Numberfolk. We should always remain aware of where soldier zero is sent to."

Next to the first point he added in $(0, f(0))$. "Two!" His thumb jutted out to the side, Liam looked at Marco and mouthed, "See what I mean?" Marco smiled and shook his head. "The two soldiers who are at position zero. The anchors of the formation."

He added $f(x) = 0$ next to each point before concluding. "And three!" His middle finger bounced up, "What we call the vertex. The center of the shape. The leading actor so to speak."

"There's a formula for that," Oliver said rebelliously. Liam gasped. Marco's eyes widened. Maggie's chin dropped.

"We aren't allowed to say the f-word," she whispered to Penelope.

"Are you insane?" Mr. Pikake hurled at Oliver. Marco didn't know what to do. He felt torn. "Those horrid things are for school children to pass examinations. They are not for Saints. How do you expect to go into battle, to cast spells and incantations, if you do not know what they even mean, what they even do?!"

While everyone was shocked, both by Oliver's brazen attitude and the professor's angered response, Oliver appeared unphased. "There's nothing wrong with a shortcut," he muttered.

Shortcut? Marco thought. Before he could even start trying to piece together the puzzle in his brain, the tutor adjusted himself and painted on a smile turning his back towards Oliver. "Ahem. Anyway. You have already gathered key information in your challenge. All incantations on any order are such that,"

$$y = af\big(b(x + h)\big) + k\,.$$

$$c^2 = (16 - 12)^2 + (4 - 7)^2$$

"Why didn't you tell us that before?" Maggie cried, outraged. "It took me forever to complete your challenge." The fact that the b, the horizontal push or pull was really a wizard, impacting both the x and the h had sent Maggie to tears.

"*Ah*, discovery is key dear. Do you not feel better having figured it out for yourself?"

She thought on the idea for a moment. The hard part was horrible. She felt less-than and incapable. But that moment, when the pieces came together, and she finally understood was gold. Like a triple scoop of ice cream with whipped cream *and* sprinkles, that moment was a shot of adrenaline making her stand a little taller. He was right. Getting there was awful, like scaling Mount Everest, it was grueling and tiring and made you want to give up. However, the view once you made it to the top was priceless. Maggie's eyes met the professor's and she smiled, nodding softly. Maybe not getting something right away, not being perfect, wasn't so bad?

"Now," he continued, "we are examining the parabola, the child of $y = x^2$."

$$y = a\big(b(x+h)\big)^2 + k.$$

"You already know the principal dancer in this ballet! Who is it?"

Marco was sent to an assembly line. As the U-shapes rolled by on the conveyer belt, robots would pick them up, stretch and compress them, move them around, and finally attach them to a wooden frame. *What is this place?* He looked around the room, the ceiling was so far away it was out of sight. Layers and layers of long moving floors crissed and crossed above his head. At the very end was a large basket that contained all the completed pictures. He picked up one after another and read the engraving on the bottom.

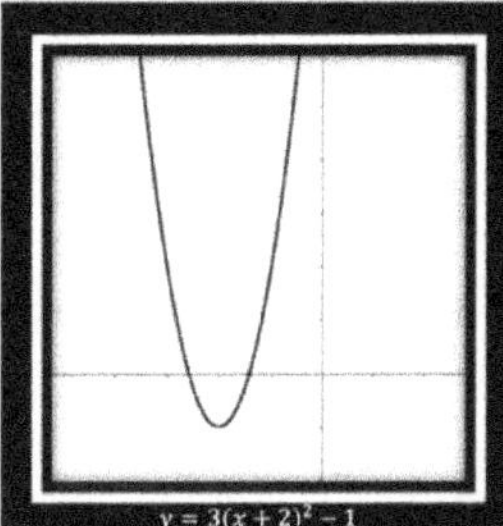

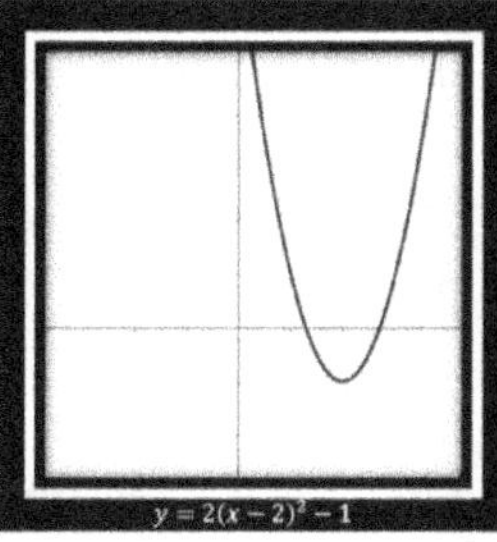

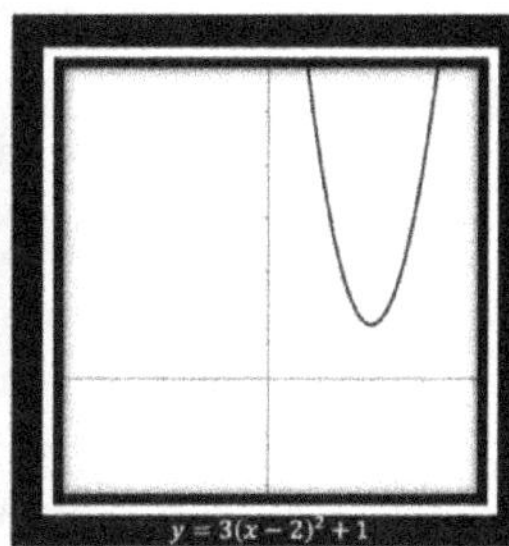

He watched the robots complete their task. Peeking over one of the droid's shoulders, Marco saw the command screen. The first box moved the formation left or right, next the machine would pull or push the shape to mold it. Then, if required, they would flip it over one of the guiding marks, and finally they would move it up or down. Again and again, in the same order, they completed their duty.

The factory floor had a strange buzzing feeling, as Marco concentrated, he felt the beat. It was a song. Listening carefully, the beeps and bops came to life. He could hear the words.

Left or right

Stretch or shrink

Flip it over

Up or down

Left or right

Stretch or shrink

Flip it over

Up or down[*]

He felt himself humming along to the tune. Hum hum hum. Hum hum hum. Hum hum hum-hum, huuum hum hum. Almost forgetting what he was there for, he continued to study the goods. It was right there in front of him. In each of the formation frames, the leading player and their position was evident in the command. The middle of the first order was at $(-2,-1)$, the second at $(2,-1)$, and the final at $(2,1)$.

Although he found his answer, Marco wasn't ready to leave yet. A smaller robot came into view. Only standing about a foot off the ground, it rolled and turned with great speed. Following it, Marco saw the little thing line itself up behind a full basket of frames before thrusting the carrier out the gigantic doors. He pushed onto his

---

[*] As Maggie and Marco found, transformations can be tricky. The machines here are following the standard order of transformation operations. If you are attempting to follow the command $-6f(x+1)-2$, start with $f(x)$. Next, move the formation to the left 1 unit, then apply any horizontal stretch or shrink followed by any vertical stretch or shrink. Here, stretch vertically by a factor of 6, then flip it if needed. Since we have $a=-6$, we flip it over ground zero, and end by moving it up or down, for us, that's down 2.

tiptoes to peer through a small glass pane. What he saw made him stumble back.

The room was filled with Numberfolk. They were seated around a large circular table like a military unit planning an attack. Marco was sure the One saw him. It was the angry One that dragged along the crazy five he met before. Was this real? He closed his eyes and shook his head trying to escape from the daydream. Slowly opening them, he was thankful to be back in the dance studio, safe.

"Negative h." Penelope was talking.

"Yeah, I see it," Liam responded. "Because the h is how far it is moved left or right. Since the leading character is always zero at the start, you are just adding to zero."

"Excellent! In our first form we know the evil twin of h is the soldier that will take the lead. But," he popped, "where are they located? What is their position?"

"It's k," Marco sighed under his breath. "It's k." The second time was a bit louder. He was still disoriented trying to understand what was reality.

"Precisely!" The professor exclaimed. "We already know the leading player and where they go as we understand how our incantations will impact this. We are moving it from the origin, the center's center. And we know exactly how far we have transported this threat."

"The center of the formation, the vertex, is really easy to see then at $(-h, k)$, but can't we have a nicer way to know where zero goes to? In lines, it was always at $(0, b)$ when we had the order $y = ax + b$. Here, it's at $\big(0, f(0)\big)$ which calculates to $ab^2h^2 + k$. That's ugly," Maggie asked. She was determined to figure out incantations if it was the last thing she did.

"You speak of perspective, dear girl." Mr. Pikake leaned against the mirror. "When you cast an incantation, there are many ways, infinite ways, this information can be processed. This is the art of manipulation. You can influence the order in the way that suits your needs best. This is a powerful skill."

He underlined the equation on the mirror,

$$y = a\big(b(x + h)\big)^2 + k.$$

"This is your task. First, I shall show you a trick of the conductors of change, those experienced in foresight."

$$y = a(b^2(x+h)^2) + k$$
$$= s(x+h)^2 + k.$$

"You combined the horizontal and vertical stretches or shrinks into one Numberfolk mask. Why?" Maggie questioned.

"In truth, it is rare to call on both incantations in a quadratic. Afterall, you can often accomplish your goal with only one. To avoid cluttering the field, we simplify things down to the shifts, up or down, left or right, and the molding be it a push or a pull to shape the curve. My task for you is to form a new perspective."

"We can expand the square. Since $(x+h)^2$ is just $(x+h)$ times $(x+h)$, we can see what that looks like?" Penelope suggested.

"Interesting idea! Proceed!" Mr. Pikake encouraged.

The children grouped up and got to work. Maggie and Penelope leaned in, both whispering and pointing. Liam and Marco huddled up and sat in silence while Oliver barely even turned to join them. He was like the thumb, there, part of the hand, but clearly removed at the same time.

"I don't know how to multiply this," Liam stated. "Do you multiply the pieces?"[*] He scribbled,

$$(x+h)(x+h) \stackrel{?}{=} x \cdot x + h \cdot h.$$

Marco imagined the situation. Inside the parentheses house, the x and h huddled together, bracing themselves for the impending storm. Outside stood a powerful wizard, but this wasn't a normal sorcerer. Approaching in its own dwelling, like Dorothy's home ready to crash down onto Oz, the wizard-pair came flashing in.

"No," he said to Liam. "The wizard's spell impacts *everyone* in the house, the spell here is $x + h$, so it has to hit both Numberfolk inside. It's gotta be,"

$$(x+h)(x+h) = (x+h)x + (x+h)h.$$

On the other side of the room, Maggie was taking the lead. "I happen to be an expert in the Distributive Property," she announced to Penelope. "We are finding $x + h$ *squared*, so we can make a square."

---

[*] Liam's mistake is often called the Freshman Dream because it is a common error that would make things so much easier if only it were true. It is not the case that $(x+y)^2 = x^2 + y^2$. Test it out! Does $(1+2)^2 = 1^2 + 2^2$?

She drew the shape, making sure her edges were crisp.

	x	h
x	x^2	xh
h	hx	h^2

"Now we can break it up into parts, so each side is $x + h$." Maggie began filling in the area for each section of her cut up polygon.

"I see, since $(x + h)$ times $(x + h)$ is just the area of the square, we can add together all the little areas."

"Yeah, we have $x^2 + h^2$," Maggie began.

"And look, since xh and hx are really the same thing, we end up with two of those," Penelope followed up.*

$$(x + h)(x + h) = x^2 + 2xh + h^2.$$

Both groups raced to manipulate the spell before the other. Almost simultaneously Liam and Maggie shouted out, "We got it!" It was Maggie who rushed to the mirror and grabbed a marker first.

"Okay, here is what we did,"

$$\begin{aligned} y &= s(x + h)^2 + k \\ &= s(x + h)(x + h) + k \\ &= s(x^2 + xh + xh + h^2) + k \\ &= s(x^2 + 2xh + h^2) + k \\ &= sx^2 + (2hs)x + h^2s + k. \end{aligned}$$

"O-ours looks different," Liam stuttered. "We have $2xhs$ not $2hsx$."

"That *is* the same thing," Maggie groaned. "It doesn't matter the order you multiply in. Three times four times seven is the same as seven times four times three and seven times three times four and four times . . ." She kept going on, but everyone tuned her out. A smile formed on Marco's face as he remembered his time with Penelope. He looked over and she was smiling too. He wondered if she was sharing his thought. The number of ways to arrange three items was 3! or 6. Marco was popped out of his bubble by his tutor's words.

"Excellent. Now I will present you with an alternative affirmation,"

$$y = ax^2 + bx + c.$$

"Can you determine which is which? Who is $a, b,$ and c?"

---

* This is the Commutative Property of multiplication: $ab = ba$.

"a is obviously s, they are the same Numberfolk." Oliver finally spoke up.

"Very good!" Mr. Pikake had clearly let go of his disappointment with the student. "We can equate the two. Thing 1 times x^2 plus Thing 2 times x plus Thing 3."

Maggie was pulled into a hidden picture game. Both were the same things: ways to command the Numberfolk into a U shape. However, some parts had been switched out, all she had to do was find them. While their expanded equation was a mess, if she blurred her eyes, she saw a way to create three terms just like in Mr. Pikake's new form:

$$y = (s)x^2 + (2hs)x + (sh^2 + k).$$

Now she could follow what Oliver was talking about. Each of the three parts was some Numberfolk brigade. The first was the two-troop, a major and one supporting soldier. If she only looked at this part, she saw $ax^2 = sx^2$. Both had the x^2 brigade which meant $a = s$.

She focused her attention on the next term, the one containing a single soldier, x. One setup showed bx, simple and clean. The other was $2hsx$. If these were corresponding parts, different approaches to the same place, that meant $bx = 2hsx$. A soldier and a wizard that dictated how the warrior would move. If two different wizards cast a spell on the same soldier and the results were equal, that meant the wizards had to be the same too!

$$b = 2hs.$$

This was interesting. Maggie was in Sherlock mode again as she identified a key clue. To find b, she could simply multiply the left/right shift by the squish/stretch and double the whole thing. That might be useful. On to the final term, the most perplexing. In the nice-looking order, the last term was just a c, nothing more. In her expanded mumbo-jumbo, she had two terms left, h^2s and k. There was no more to the picture. If everything matched up, it had to be that $c = h^2s + k$. They were the only terms without a visible soldier.

"Well, b is $2hs$," Maggie started.

"Tush," Liam interjected.

". . . and c is everything else," she finished.

"Excellent! Now hunt. We wish to find how our leader hides in the form. Find them!"

It was Marco's time to shine. Finding b was easy. An s monster

had eaten a magical mushroom and doubled in size; it was really $b = (2s)h$. To get the h alone required only sending in the partner of this beast. He multiplied $1/2s$ to cast his proliferation spell on the field and found,

$$b = 2sh$$
$$\frac{1}{2s} \cdot b = \frac{1}{2s} \cdot 2sh$$
$$\frac{b}{2s} = 1h$$
$$h = \frac{b}{2s}.$$

Identifying k was almost too simple. It only required a duel. Marco called on the evil twin of sh^2, he didn't have a clue *who* this was, but he was sure they were the same Numberfolk just made up of the opposite particles.

$$c = sh^2 + k$$
$$c + (-sh^2) = (-sh^2) + sh^2 + k$$
$$c - sh^2 = 0 + k$$
$$k = c - sh^2.$$

"Alright, h is b divided by $2s$ and k is c minus s times h squared," he professed proudly. Even though the others had caught up on the basics of hunting, this was still an area where Marco outshined the competition.

"Perfection! Do you see the power of perspective? Look here. If the order is written as $y = ax^2 + bx + c$, Maggie, where is zero sent?"

An unfamiliar nervousness crept through her body as she was used to always having the right answer and having it right away. This business of orders had left Maggie struggling to keep up. Her face glimmered when she saw it. If the soldier, x, was Zero then ax^2 and bx were both zero too! That meant Zero was sent to c, always. "It's c! The zero soldier is sent to c!" she exclaimed.

"Exactly! When the command is in the traditional form of $f(x) = a(x + h)^2 + k$ we can easily identify our leading player, it's $-h$ who is sent to station k. Yet we cannot clearly comprehend the location of Zero. But," a dramatic pop, "by changing our perspective to $f(x) = ax^2 + bx + c$, we rapidly recognize soldier 0 is sent to c."

"Yeah! And we can also pretty easily find the Numberfolk on center stage too. Since $h = b/2s$, that means the center soldier, the

$-h$ in $(-h, k)$, is just $-b/2s$ or $-b/2a$ because a and s are the same. Their position is more challenging, but doable," Maggie concluded.

"Amazing work." Mr. Pikake gave her a warm and friendly smile. It made her swell up inside. "Alright, our ending examination requires determining distance."

He walked slowly to the back of the room and dug into his messenger bag. Pulling out a large felt triangle, he glided to the mirror.

"Who keeps a random triangle that is actually bigger than your head in their bag?" Liam whispered to Marco. They giggled.

Taking his time, the professor stuck the triangle to the mirror.

"How did he even do that?!" Liam poked at his friends.

The triangle continued to magically hover as Mr. Pikake slowly traced around its edges. He picked it up, twisted it, and did it again. Then again, and again. "Now, you have seen that we can move formations around a plane without actually *changing* them."

"Yeah! Like moving the U left or right kept its shape. It was only when we stretched it or squashed it that it changed," Penelope beamed.

"Precisely! And similarly, you can spin it around, such as a flip, and still the shape remains the same."

Stepping away from his reflection, he revealed the masterpiece to the group. "Same shape, same dimensions. Together forms a square. What is the area of the square?"

Liam wasn't going to make the same mistake again. Each side of the square was $a + b$, which meant the area was $(a + b)(a + b)$ or $(a + b)^2$. He slowly cleared his throat, "*Um. Ah.* That would be a squared plus b squared," his friends all held their breath, "plus another $2ab$."

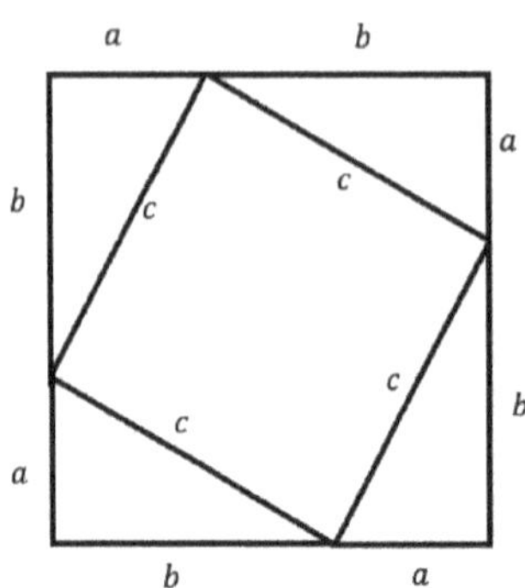

"Very good young man." The professor wrote his response on the mirror,

$$\text{big square} = a^2 + b^2 + 2ab.$$

"Now, we wish to create the smaller square in the middle, how might we accomplish such a task?"

"Well, you can just take away the triangles. If you have the big square and take away the triangles, you are left with the small square," Penelope shined.

"Excellent! Our task is thus now to find the area of the triangles."

"Oh! I know that one," Maggie perked up. "A triangle is just half a rectangle, so the area is just half of what the rectangle would be. One half a times b."

Mr. Pikake added to the mirror,

$$\text{triangle} = \frac{1}{2}ab.$$

"There are four of those," Marco spouted. He liked shapes a lot and hadn't forgotten what his tutor had told him about flat-earth math, he was secretly hoping they were close to starting their geometry teachings. "Four of those would be $2ab$."

$$\text{four triangles} = 4\left(\frac{1}{2}ab\right) = 2ab.$$

"So bright you all are. We conclude that the area of the small square is nothing more than,"

$$\begin{aligned} \text{big square} - \text{four triangles} &= \text{small square} \\ &= a^2 + b^2 + 2ab - 2ab \\ &= a^2 + b^2. \end{aligned}$$

"I see where this is going," Penelope whispered to Maggie.

"However, isn't it true we know already the area of the small square?" The professor poked.

"Yeah, since its sides are c long, its area is just c times c or c^2," Liam called.

$$a^2 + b^2 = c^2.$$

As the tutor added the period, the marker made a loud *reeeacccb* sound. "If a and b are the legs of a right triangle and c the hypotenuse, the longest side which lies across from the right angle, then the sum of the squares of the legs is equivalent to the square of the hypotenuse. Squares require perfectly punctual L's at each corner. This is key."

"The Pythagorean Theorem," Penelope smiled.

"*Ahh*, but careful you must be. For we are working in a flat space, Numberfolk orders and positions. One of only two dimensions. This clever combination has its limits when thrust on a spherical three-dimensional world. But!" he popped. "In only two flat dimensions, we will be fine."

(x, y) (z, y)

The professor drew a segment on the mirror, marking each end point with an ordered pair.

"Remember the fly. If he flew from here," Mr. Pikake flicked his wrist and suddenly a large pointer extended from his hand and tapped at (x, y). "To here." He dragged the tool to the second endpoint. "How far did the fly fly?"

"Well, that's easy!" Penelope answered. "It's only one dimension so he just went $z - x$ units over."

"I don't think that works," Liam protested. "I know the soldiers are numbers, Numberfolk, but you are always saying they are like people, and people don't work like that."

"*Ahh,* but we do!" Mr. Pikake fought back. "Like it or not, numbers are engrained into everything we are. You are not the Numberfolk 1, yet if you are assigned to be soldier 1, and Marco soldier 2, you are a single soldier different."

"Say that's true. Marco's two and I'm one. Look at us, we aren't 1 foot apart."

"Yours is a question of scale. So long as we are equally spaced, no strange clusters, a unit can be anything you wish. It may be a single foot, or a mile! It can be non-standard too; one arm length or 3.5 donuts."

"I like it. Donuts," Liam pretended to drool.

Mr. Pikake added to the diagram. "Now what if we consider our vertical dimension? How far apart are our ordered pairs?"

"It would work the same way. The soldier is the same. It is like telling Marco to take three steps forward. His position would just be three more from where he started. If he was at position y and moved to position w, he moved $w - y$ spaces," Maggie replied.

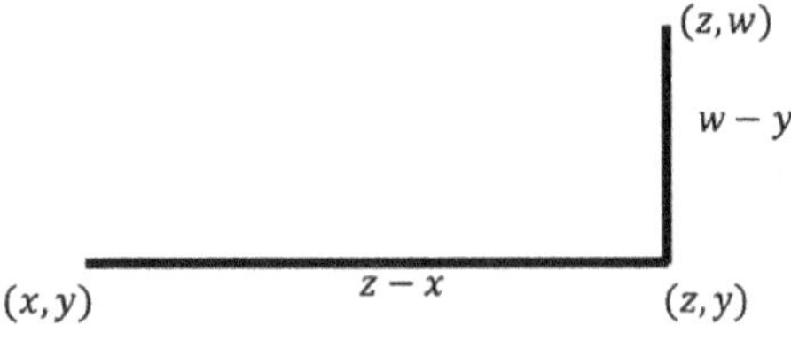

"Very good! Now, for the challenge. How far is it from (x, y) to (z, w)?"

"*Hm,* that's difficult." Penelope plopped her shoulders down and let out a sigh. "Before, the fly was only moving in one direction. Now it's moving in two. It is going both up and over."

"Can't we use the Pythagorean Theorem?" Maggie posed. "If the horizontal distance is a and the vertical is b, then $a^2 + b^2$ will give us the diagonal squared."

"Perfection!" Mr. Pikake popped. "However, we do not wish to find the square, we want the distance. We must add a new spell to our repertoire: the insurrection."

"Insurrection?" Marco questioned.

"It's when there is a revolt. You know like the people rise up against the evil emperor!" Liam explained.

"Exactly! x^2 or x^3 or x^4 and on and on, are simply brigades of soldiers that multiply their power. An insurrection breaks them apart to allow us to identify each squadmate. In a march, we needed only to vanquish and obliterate to hunt a mask. Now, with powerful squadrons, we must acquire a new tool."

$$x^2 = 16.$$

"Who are the soldiers in the brigade?" The professor turned to the group.

"They are fours, right? Because 4 times 4 is 16?" Marco responded.

"What about negative four?" Penelope pushed back. "-4 times -4 is 16 too."

"And you have found the crux of an insurrection! The size of the brigade always tells us how many soldiers can hide there. In x^2, the brigade contains two soldiers, $x \cdot x$. Thus, there are two possible Numberfolk who may use such a squadron."

Marco's chest pounded. He loved hunting. It wasn't until this moment he realized why. Before, he could always work his magic to get the mask alone, back it into a corner before ripping off the disguise to see exactly who it was. Mr. Pikake was now saying that with brigades, even if he isolated the mask, there could still be many possible numbers hiding. That was *horrible*.

"In this case both 4 and -4 can hide with this brigade."

"Yeah but," Maggie started. Marco giggled. "What about $x^2 = 0$? That means the Numberfolk hiding behind x is just zero, there aren't two options."

"Excellent examination!" The tutor jumped. "There are still two attackers, in this case they happen to be the same Numberfolk. These instances can tell us outstanding information about the formation!"

"I can see how sometimes it can be easy, we all know four times four is sixteen. What if it isn't so clear? So obvious?" Marco questioned.

"The Slant term here is a root. It's not a horrible name because it can allow us to identify the anchors. The ones holding down the formation as the roots do for a tree."

"The other two points we need?" Maggie recalled.

"*Ay*. The *roots* are the soldiers at ground zero, those such that $f(x) = 0$. Look here, $x^2 = 16$ is none other than $x^2 - 16 = 0$."

"That's just the mother order moved down," Marco pointed.

"Yes! To identify these soldiers, we must first learn to break apart the brigades. The origin of this device was a prison in Numberville. In order to exit, it required cooperation and sacrifice by creating identical groups to breach the bars. Hence the term insurrection." He slid across the mirror,

$$x^2 = 1296.$$

"Who is hiding behind this mask?" he asked inquisitively.

Marco found himself back in the interrogation room. He relished in ripping apart the Numberfolk and was ready to find out just what this 1296 was made of.

"You are clearly even, made up of twos, but how many?!" He slammed his fist on the metal desk, the clang reverberated through the room. Marco broke the number to break the number. Knowing that $1296 = 1000 + 200 + 90 + 6$, he split each part in half to find $500 + 100 + 45 + 3$.

$$1296 = 2(648).$$

"I see you are a two-lover!" Marco yelped as the result was even again. Repeating his move, he split it another time.

$$1296 = 2(2)(324).$$

Impatient with the suspect, Marco began to rip it in half once more but hesitated. He saw something else. Both 32 and 4 could be built with fours. Since $324 = 320 + 4$, he sliced the number into fourths. It would take 80 fours to construct 320 and one more to hit his target,

$$1296 = 2(2)(4)(81).$$

The 1296 transformed. The 1 twisted into a tail standing at alert while the 9 elongated forming the body, the 2 connecting each piece. The 6

bent to make out the profile of a head with a curved ear. The suspect was a cat! It had four legs and nine lives.

$$1296 = (4 \cdot 4)(9 \cdot 9) = 4^2 9^2.$$

"It's gotta be 36, right? I mean 36 and their evil twin -36?" he called out to the group.

"That it is! An insurrection is cooperative magic, it requires teamwork. We use the symbol √ to represent breaking apart a squadron. The Numberfolk in front tells us how many identical groups are required to exit. No Numberfolk indicates we assume we are ripping into two groups."

$$x^2 = 1296$$
$$\sqrt{x^2} = \pm\sqrt{1296}$$
$$x = \pm 36.$$

"How'd you know it was 36 tho?" Liam turned to Marco.

Excited to show off his skills in front of Penelope, a large smile sprouted on Marco's face as he explained. "I ripped apart 1296 until I could make two of the same groups." He pushed down on his hands and jumped up. Mr. Pikake gave him a little bow and passed the marker.

"Since 1296 is even, I just kept ripping it in half until I couldn't anymore."

"I noticed that 81 was just nine times nine. So I broke all the traits into two identical groups, since it was x^2."

$$1296 = (2 \cdot 2 \cdot 9)(2 \cdot 2 \cdot 9).$$

"That means, each soldier had to be $4 \cdot 9 = 36$ because,"

$$x \cdot x = 36 \cdot 36$$
$$x = 36.$$

"Ohh, I see. So, if it were x^3, we'd break it into three groups?" Liam asked.

"Let's give it a try!" Mr. Pikake exclaimed.

$$x^3 = 216.$$

Liam went to work. He was a trickster poker player at a space cantina. Since 216 was even, he dealt out a two to each player. Counting his cards, he figured he had rid the deck of $2 \cdot 2 \cdot 2 = 8$. He cut his total in half to find $216/2 = 108$, in half again $108/2 = 54$, and finally one last time to determine he had $54/2 = 27$ left to work with. Noticing $27 = 9 \cdot 3 = 3 \cdot 3 \cdot 3$, he could deal a 3 to each player to finish things off.

$$\begin{aligned} x^3 &= 216 \\ x^3 &= (2 \cdot 3) \cdot (2 \cdot 3) \cdot (2 \cdot 3) \\ x &= \sqrt[3]{(2 \cdot 3)^3} \\ &= 2 \cdot 3 \\ &= 6. \end{aligned}$$

"Six? Is it right?" Liam crossed his fingers. "I only got one answer tho. Not three."

"*Ahh,* your finding is fundamental!" ° Mr. Pikake exclaimed. "Every problem has a solution. Oft, we simply don't have the tools yet to see it. But," he popped, "it exists and until that marvelous time in which we gain the knowledge we need, the solutions lie only in what we can imagine. For now, imagine ground zero as not only a line, but an entire wall of defense. Your two remaining anchors lie on that wall, and soon we will discover how to identify the anchors we cannot see." With a wink and a grin, he pivoted on his heel before sliding across the room to his drawing. "An excellent job you have done. Now that we can complete an insurrection, break apart a brigade by creating equal groups, we are ready to tackle distance." He motioned to the diagram, "How far from (x, y) to (z, w)?"

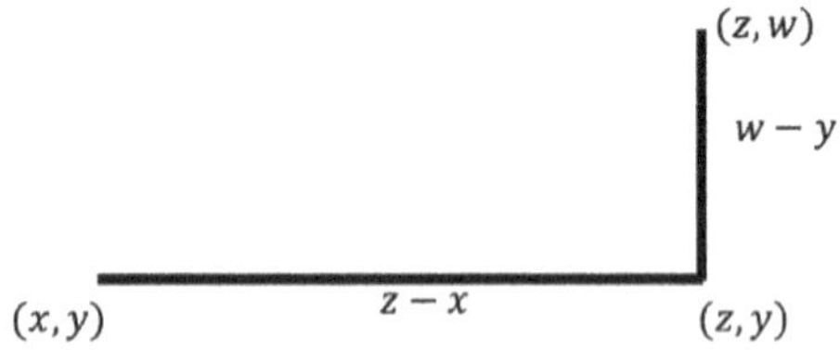

(z, w)

$w - y$

(x, y) $z - x$ (z, y)

"Okay, I see it now. Since it forms a right triangle, we know that $a^2 + b^2 = c^2$. Here $z - x$ and $w - y$ are the legs," Penelope explained. "That means, the distance from (x, y) to (z, w) is c,"

$$c^2 = (z - x)^2 + (w - y)^2.$$

"That's a brigade. We have to take the root to find c,"

° Liam's comment was so fundamental, it is called *The Fundamental Theorem of Algebra*.

$$c = \pm\sqrt{(z-x)^2 + (w-y)^2}.$$

"Except I don't think you need the plus or minus," Maggie argued. "Since we are talking about distance, it is positive. The negative couldn't hide here."

"Dynamic detective work!" Mr. Pikake smiled. "Tomorrow, we will focus our powers. To do so we need to find the distance. "

"But what about the roots, the anchors? How do we find those?" Maggie called out as the professor headed towards the door.

"Always ahead." He winked. "Soon Maggie, soon." He slipped out of the studio, the bell jollily ringing behind him.

This was Marco's chance. He jumped up and ran over to Oliver. "What's going on with you? You seem . . . you seem down." They walked to the parking lot together.

"*Ah*. Sorry. Just family stuff." Oliver tussled his hair.

"Anything I can do? Want to hang out at my place?" Marco offered.

"I'd love to man, but my, *um*, my uncle is in town. Can't today."

Liam came running from the building out of breath. "I figured it out," he puffed. "Marco won!"

"Won what?" Marco scrunched up his shoulders.

"The fly. You were $\sqrt{5}$ away, I was twice that $2\sqrt{5}$, Oliver ended up $\sqrt{17}$ away and Maggie, she was *exactly* 5 tiles from where it landed!" *

He nervously looked over his shoulder at Maggie and Penelope. "I'm telling you dude . . . cylon."

Marco flashed back to what seemed like a lifetime ago. Liam talked about $\sqrt{5}$ like it was nothing. It wasn't just any number, it was one of *them*. Mr. Pikake had told Marco about this kind of Numberfolk. The crazy ones constrained by straitjackets, drooling and thrashing like rabid dogs. The ones with never-ending, never-repeating digits the professor called their kidnapped victims. While he watched his friends joke and laugh, dread bubbled in Marco's stomach. He didn't know what was coming next, but he felt thankful $\sqrt{5}$ was locked in a cage.

---

* We read $\sqrt{5}$ as "square root five" or "root five" and $2\sqrt{5}$ as "two root five" or "two times the square root of five." Sometimes, the $\sqrt{}$ is called "rad" to represent the origins of the Numberfolk who would become the Radicals.

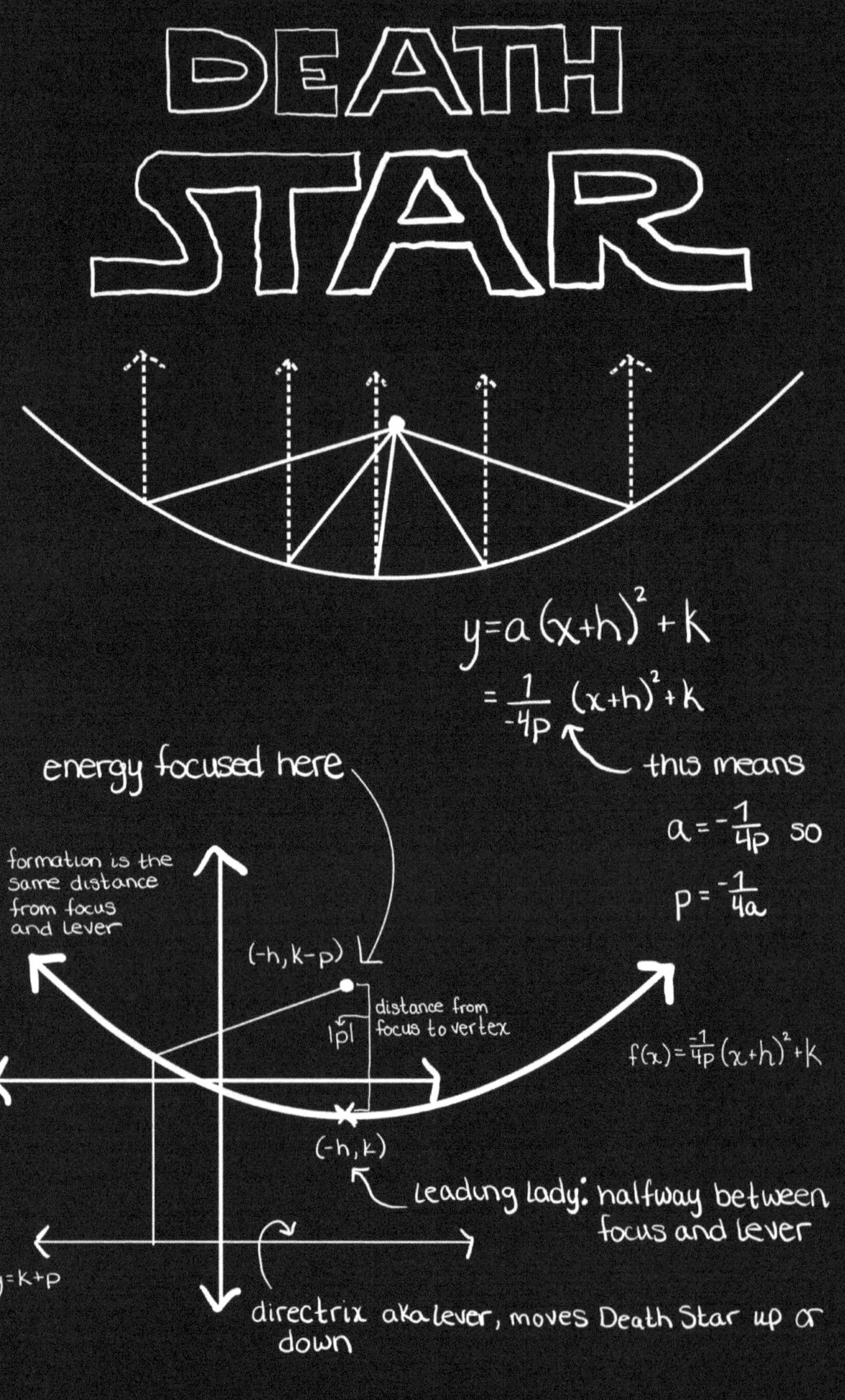
DEATH STAR
$y = a(x+h)^2 + k$
$= \frac{1}{-4p}(x+h)^2 + k$
this means
$a = -\frac{1}{4p}$ so
$p = -\frac{1}{4a}$
energy focused here
formation is the same distance from focus and lever
(-h, k-p)
distance from focus to vertex
|p|
$f(x) = \frac{-1}{4p}(x+h)^2 + k$
(-h, k)
leading lady: halfway between focus and lever
y=k+p
directrix aka lever, moves Death Star up or down

14

FOCUS

"FOCUS ON THE SOLUTION, NOT THE PROBLEM."

-JIM ROHN

Intimidating darkness enveloped the room.

As they entered the dance studio, the florescent overhead lights had been switched off. In front of the mirror was what looked like a satellite dish. It was a big bowl turned on its side and supported by a tripod. In the back of the room, stage lights had been set up covering the full length of the studio and all pointed ahead. Their warm beams bounced off the satellite and created a gigantic, illuminated ring in the center of the room.

"I think he made a Death Star!" Liam whispered.

Mr. Pikake stepped into the spotlight. Everyone jumped. No one realized he was even there. Like a ghost, he managed to hide himself in the shadows.

"Luminous lighting!" He slowly raised his hands to harness the power of the glow before motioning for the students to take a seat. In a low voice he continued. "It is time you learn about the Numberfolk attack on Las Vegas."

Attack on Las Vegas? Marco was sure if something like this had happened, he would have heard about it.

The professor continued in a grim foreboding tone. "A new hotel was raised from the ground. For purely aesthetic purposes, the architects decided to form it in a curve. Rectangles are boring, a curved building would be magnificent! Eye-catching! It will be a glass shrine that emerges from the Earth and climbs toward the heavens!" His voice became fiercer as he spoke. Marco looked to his right. Liam was leaning in, eyes wide, taking in every word. Marco thought he might have even seen a speck of drool dripping from the side of his friend's lip.

"Unfortunately, they did not understand the power of such a structure. When the sun shone down on the desert, it reflected off the building. But not just any reflection! The rays were focused to one unified point." He twisted his body and threw his hand toward the satellite. "Increasing the intensity of the illumination!"

"It *is* the Death Star!" Liam shouted.

"Where do you think they got the idea?" * Mr. Pikake flashed him a devious smile.

"Well . . . what happened?" Maggie urged.

The tutor pointed to the setup around the room. "Just as all these

---

* It's true. If you are ever in need of blowing up an entire planet, a parabolic mirror is the way to go.

lights have been redirected to *focus* on me, the sun's rays did the same! The beam magnified the power of the star . . . directly toward the pool."

"No!" Liam gasped.

"Yes!" Mr. Pikake shrieked back. "Guests received a heavy dose of radiation that day."

"He totally made that up," Oliver scoffed.

Mr. Pikake lifted an eyebrow and reported, "Vdara Hotel. 2010." He leapt across the dance floor and flipped a light switch casting a blanket of blackness throughout the room.

Click.

Click.

Click.

Click.

The tutor switched the remaining levers, and the florescent overhead bulbs began to flicker to life. Marco shuddered; his mind pulled back to his cedar chest.

"Do not underestimate the power of our incantations! The ability to order the Numberfolk into formations is formidable." The last word came out strong and slow from the professor's lips. "It 'twas the parabolic shape of the building that caused such a catastrophe. A parabola is a formation made when every Numberfolk's position is the same distance from a provided point and a lucid line."

Marco wondered the usefulness of such an order, besides creating a Death Star. He quickly decided a Death Star was a good enough reason. A shape that could amplify anything seemed like an immensely powerful tool.

Mr. Pikake raised his arm – it came flying down before zooming back up again in one long smooth motion. He stepped away from the mirror to reveal his drawing. It was a large U.

"That? That again?" Maggie groaned.

"That . . . is a parabola. 'Tis the very shape, the very formation, you have been working with."

"If I'd known that, I probably would have been a little more careful," Liam whispered to his friends. "I could have accidently done some serious damage!" He clenched his teeth and stretched out his mouth.

"The directrix directs all the power entering the formation. It is our lever. In a functional formation, it is always a horizontal line." The professor pecked at the mirror,

$$y = p.$$

"This is the command for such a lever when working with the mother. The p is simply the location you place the lever. At its starting point, $p = 0$, it gives us the line $y = 0$, also known as the x-axis. Move it down and you might have $y = -1$, up to $y = 2$. You may move the lever however you need to."

"This is amazing," Liam whispered. Mr. Pikake gave him an affirming head nod.

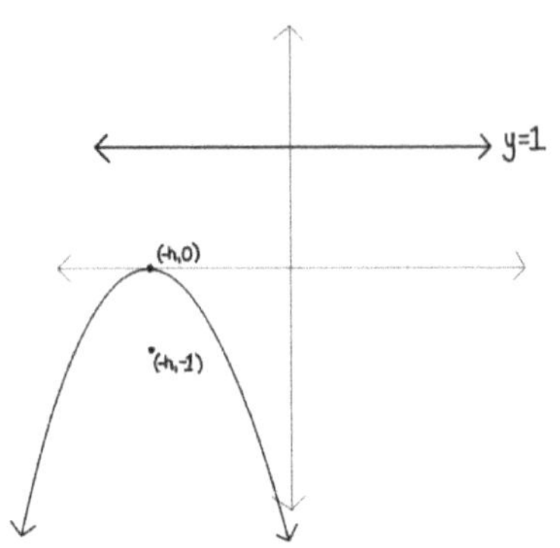

"The directrix-lever will determine where the energy is focused. Moving the lever away from our leader, the vertex, $(-h, k)$, pulls the focus away from our star. And of course, pushing the lever toward the vertex pulls the focus closer to her as well. In working with a formation, we have not moved vertically, the location of the focal point is always the evil twin of the lever. If we place the lever in location one so that $y = 1$, the focal point is at location -1."

Marco imagined he was in Fredrick's factory. The floorboards split apart to reveal a colossal hole that housed a gigantic slimy monster. He couldn't even see the body of the beast as all that was visible was its jaws; rings and rings of teeth circling the dangling uvula. Looking to his left, he saw a lever attached to the wall. He pulled it down. The creature's jaw widened; mouth open to devour its meal. Pulling the lever up narrowed the opening. It became thinner and skinnier. Not in any danger, he taunted the beast. Up, down, up, down, open, closed, open, closed. The monster's jaw was snapping like the face exercises he'd seen singers and actors do when warming up. He giggled uncontrollably. The beast didn't like it. It exploded out of its hole revealing its long worm-like body. Gooey pink layers of chubby folds dripped as it focused its attention on Marco. He didn't want to know what would happen next and quickly shook the daydream away.

"But who is the soldier at that location, at the focal point? And what if the formation *has* been moved from the mother?" Penelope asked.

$$(1 - \sqrt{y})(1 + \sqrt{y}) = 3(x + y - 21)$$

"Quite the query!" the professor quacked. "If we move the formation vertically, the lever, and thus the focal point, must travel with us if we wish to maintain its shape. As this formation is the result of lining our soldiers up equally from these two pieces, the focal point is always in front of the leading actor, the vertex, the pinnacle of our construction. For instance, if our vertex is at $(6, 0)$ and our lever is at $y = -2$, the power is concentrated at $(6, 2)$. If we shift the formation up, so our star is now located at $(6,1)$, everything must tag along with it. The directrix becomes $y = -2 + 1 = -1$ and the power is concentrated at $(6, 2 + 1) = (6,3)$."

He slashed at the mirror to reveal a soldier and their station: $(3, 2)$.

"I want the energy to focus here, assuming the formation hasn't moved vertically and is still anchored at ground zero, can you determine the order I would need?"

"Well, the directrix, the lever, would be at $y = -2$, right?" Maggie asked.

"Precisely! Recall the formation is all the locations that are the same distance from the lever and the focus. Where is the vertex, the leading lady, of your order?"

"That would have to be at $(3, 0)$" Penelope chimed in. "The focal point is in line with the leader, since it's at $(3, 2)$, that is telling us the head of the formation is 3. It also has to be perfectly in between the focus and the directrix. From 2 to -2 is four units. Which means the vertex must be halfway between or two away from each."

"Nicely done," Mr. Pikake smiled. "So, what is the order?"

Marco had been practicing his incantations, "It would be $f(x) = a(x - 3)^2$ right? The mother formation has only been moved to the right three. But we don't know how much it was stretched or shrunk."

"Exactly!" The tutor jumped and began writing furiously on the mirror. "To find the shrink or stretch, we must identify a second ordered pair on the formation!"

"Oh! Just like with a march, we needed two points," Maggie theorized.

"Yes! Yes! There are infinite formations that have a leading lady of three at position zero. We must identify which we require."

"We need the distance," Penelope suggested. "The distance from the focus at $(3, 2)$ and the missing ordered pair (x, y) has to be the same as the distance from the lever, $y = -2$, and the ordered pair (x, y)."

"Wait. That doesn't make sense. There are *tons* of points on a line. How do we know which one on the lever to pick?"

"Very good, Liam! If we want to see how far (x, y) is from the line $y = -2$, we measure straight down to the point $(x, -2)$."

"But why?" Liam shot back.

"My favorite question." A devious grin blossomed on the professor's chin, "Horses."

"Horses?"

"*Ay*, horses. Imagine a race in which all the horses began on the starting line and a single treat was placed in the middle of the field one mile away from the starting line. Is it a fair race?"

"Of course not!" Liam coughed, "It's like a triangle, the horse directly in front of the treat would have one mile to run and all the others would have more."

"Precisely, and thus, distance is always measured along the shortest path." Mr. Pikake nodded.

"Okay," Maggie got up. "We want to find the ordered pair (x, y). The distance from (x, y) to the focus at (3,2) is," she began writing on the mirror.

$$d_1 = \sqrt{(x-3)^2 + (y-2)^2}.$$

"Yeah!" Penelope joined her. "And the point (x, y) is right above the point $(x, -2)$ on the directrix. So the distance from (x, y) to $(x, -2)$ is,"

$$d_2 = \sqrt{(x-x)^2 + \left(y-(-2)\right)^2}.$$

"Hey look!" Marco jumped up too. Liam looked at Oliver and shrugged as they were the only ones left on the floor. "Because you have $x - x$, that's a zero. And since we know the two distances have to be the same . . ."

"We can set them equal to each other!" Penelope finished his sentence and the two smiled. Maggie made a face like she'd just eaten garbage.

$$d_1 = d_2$$
$$\sqrt{(x-3)^2 + (y-2)^2} = \sqrt{(y+2)^2}.$$

"Wouldn't it be true that if the side of a square equals the side of another square, then the squares are the same size?" Oliver jumped

to his feet. Liam frantically shook his head and arms, before standing up to join everyone.

$$(x-3)^2+(y-2)^2=(y+2)^2.$$

"Alright, let's each take one of these to expand." Oliver assigned the first term to Marco, the second to Liam, and claimed the right side of the equation for himself. Marco was happy to see a bit of his old friend back in action. He had missed Oliver desperately.

Marco thought of the wizard, $(x-3)(x-3)$. When the $x-3$ cast on the first occupant, he got $x(x-3)$ and the second gave him $-3(x-3)$. He couldn't hold it all in his head, so he scribbled in a corner of the mirror. Maggie and Penelope were watching, he needed to make this look good.

$$\begin{aligned}(x-3)(x-3)&=x(x-3)+(-3)(x-3)\\&=x^2-3x-3x+9\\&=x^2-6x+9.\end{aligned}$$

"Got it!" he called out.

Liam cheated. After his mistake in which he claimed $(x+a)(x+a)=x^2+a^2$, he had put to memory how to multiply beasts of burden like this. It was x^2+a^2 but he also got two duplicate terms on the inside. Unfortunately, his lack of understanding tied him in knots.

As he attempted to multiply $(y-2)(y-2)$, the minus made for hard work. *Nobody uses subtraction anyways,* he thought. He changed it to $\big(y+(-2)\big)\big(y+(-2)\big)$. That helped. Now, the y's multiplied together and so did the -2's. This resulted in a y^2 and a $(-2)^2=4$. In the middle, he'd get two of the $(-2)\cdot y$ terms and $-2y+(-2y)=-4y$. He looked around nervously as he wrote

$$(y-2)(y-2)=y^2-4y+4.$$

He saw Maggie eye his work and she didn't jump at him like the attack dog she was. Liam let go a huge sigh of relief.

Oliver had a totally different approach. He would multiply just like he had always multiplied. As strange as $(y+2)(y+2)$ might *look*, someone was hiding behind x and that meant $y+2$ was just a number.

He had learned to stack the numbers. If he wanted to find 32×14, he'd do it like this:

$$\begin{array}{r} 32 \\ \underline{\times 14} \end{array}$$

First, he'd multiply 4×32. Finding the product of the ones as $4 \times 2 = 8$ and then completing $4 \times 3 = 12$, he'd fill that in below the line.

$$\begin{array}{r} 32 \\ \underline{\times 14} \\ 128 \end{array}$$

Next, he would multiply 32×1 but because it was really 32×10,

$$\begin{array}{r} 32 \\ \underline{\times \ 14} \\ 128 \\ 0 \end{array}$$

he had to put a zero in the ones place.

Since $32 \times 1 = 32$, he filled in the final digits and added the result.

$$\begin{array}{r} 32 \\ \underline{\times \ \ 14} \\ 128 \\ \underline{+320} \\ 448 \end{array}$$

Numbers were just a series of brigades anyway. He could tilt his head and recognize 876 as $800 + 70 + 6$. But there was more. The number system was built on the backs of tens, which meant every place value, was really a brigade of tens.

$$876 = 8(10^2) + 7(10^1) + 6(10^0).$$

He could write any number in this way, even decimals. If he didn't know about the base-ten system, the tens would be a mask.

$$876 = 8x^2 + 7x + 6.$$

As $y + 2$ was just some number, stacking made sense. Oliver placed the lines atop each other and went to work.

$$\begin{array}{r} y + 2 \\ \underline{\times \ \ y + 2} \end{array}$$

First, he multiplied the ones like always $2 \cdot 2 = 4$. Next, he multiplied the 2 in the second line by the y in the first and transcribed his results below the line.

$$\begin{array}{r} y + 2 \\ \times \ \ \underline{y + 2} \\ 2y + 4 \end{array}$$

He'd have to shift everything to the right, just like in normal multiplication because now he was working on the "tens" place.

Finishing up, he found $y \cdot 2 = 2y$ and $y \cdot y = y^2$, all he needed to do was add.

$$\begin{array}{r} y + 2 \\ \times \ \ \underline{y + 2} \\ 2y + 4 \\ + \ \ \underline{y^2 + 2y + 0} \end{array}$$

Maggie's eyes and mouth grew slowly, like slime stretching out over the edge of the table. "What have you done?" she awed.

"Well, $y + 2$ is just a number, right?" Oliver started, "so I treated it like any other number."

$$\begin{array}{r} y + 2 \\ \times \ \ \underline{y + 2} \\ 2y + 4 \\ + \ \ \underline{y^2 + 2y + 0} \\ y^2 + 4y + 4 \end{array}$$

She hugged him. She leapt at least three feet and wrapped her arms around her brother's friend and squeezed. Tears started to form, but she quickly blinked them away. Maggie had been struggling and Oliver had just provided her with the perspective she needed. This wasn't new stuff. This wasn't hard. This was the same thing she had been doing for years. The same thing she did remarkably well.

Oliver stood very still, his body a board, careful not to move until Maggie let go. He forced a smile, and everyone pretended not to be staring at this strange scene unfolding before them. Being a good friend, Penelope spoke up to change the focus.

"Bringing together what the boys found gives us,"

$$(x - 3)^2 + (y - 2)^2 = (y + 2)^2$$
$$x^2 - 6x + 9 + y^2 - 4y + 4 = y^2 + 4y + 4.$$

"Can you hunt this, Marco?" she asked.

Feeling like the air was sucked out of the room, Marco stared down the beast. Penelope asked, so he had to do it, there wasn't a choice. The problem was, Marco had no clue *who* he was hunting. There were masks everywhere, and they weren't alike. There were x's and y's and x^2's and y^2's. Not sure where to start, but needing to do *something*, anything, he began playing with the scales.

"*Um*. Alright." He slowly approached the mirror, his mind racing at lightspeed trying to figure out a move. Then he saw it. There were identical twins on each side of the scale. They were acting as a counterbalance and not actually impacting anything. Something new happened. It felt like phasing, but it was different. The numbers didn't come alive; instead, each turned into a unique shape upon the scale.

The y^2's were balancing each other, he removed them from each side. The +4's were also identical, he threw them away too. That left Marco with,

$$x^2 - 6x + 9 - 4y = 4y.$$

Now he was playing with fire. Sending in the evil twin of $-4y$ to both sides of the scales, he combined like terms:

$$x^2 - 6x + 9 - 4y + 4y = 4y + 4y$$
$$x^2 - 6x + 9 = 8y.$$

The idea of hunting for x seemed impossible. Marco couldn't see a path forward with the x^2's and x's, but he almost had y alone. About to cast a proliferation spell by the partner of 8, 1/8, he looked around. His friends, his sister, his tutor, and Penelope all appeared frozen in time. There was a weak vibration, rumbling through the room. He glanced back at the mirror, the Numberfolk were all dead on the glass. Looking around one final time, he caught Oliver's eyes. They blinked.

As an invisible force pressed play, the room sped back to life. Marco kept his eyes on Oliver. He couldn't be sure, but something seemed different about his friend. He turned and finished off his spell.

$$y = \frac{1}{8}(x^2 - 6x + 9).$$

"That's it!" Penelope chirped.

"What's it?" Marco didn't have a clue what she was talking about.

"That's the order that will create the Numberfolk formation focused at $(3, 2)$. You found it!"

"I did?" Marco had nearly forgotten the entire exercise. Realizing he had created a Death Star ready to blow up a moon, he dropped his marker. "That was far too easy," he puffed.

The professor began a slow clap. He brought his feet into the party and stomped as he stepped to amplify the effect. "Excellent work children. Your power is remarkable. Now, you can create a formation to harness energy, but can you identify one?" He scrawled on the mirror,

$$f(x) = a(x + h)^2 + k.$$

"Where is the leading lady?" he shouted.

"At $(-h, k)$!" Liam barked back with his best military voice.

"Excellent! Where are the rays focused?"

"Well, the focus is right above, or below, the vertex. It's on the inside of the formation which means if it is a U-shape, it's above the vertex and if it is upside-down, like a frown, it is below. If we say p is the location of the lever before the shift, then the focus is at $(-h, -p)$. To move the whole thing without changing the shape is then just $(-h, -p + k)$," Penelope sang.

"But you just said it could be above," Maggie chimed in. "If it is minus p, doesn't that always make it below?"

"Not necessarily! If the lever is at $y = -2$ like before, the focal point is $-p = -(-2) = 2$."

"Oh yeah!" Maggie saw it. "And that means the directrix would be at $y = k + p$? Right? The location of the lever is just moved by k like the vertex?"

"Outstanding! But our order has no p! We have $a(x + h)^2 + k$, and thus I ask, what is p?"

"If we know the focus and the vertex, we can find p," Liam pointed. "Say the focus is at $(-5, 2)$ and the vertex is at $(-5, 0)$. Then $(-h, k) = (-5, 0)$ and $(-h, -p + k) = (-5, 2)$. That means $k = 0$ and $-p + 0 = 2$, so p has to be -2. Same if you know the directrix and the vertex."

"Yeah but!" Maggie called out. "What if you don't know either? In our commands before, we never knew the focus or the directrix, we just knew the order. How can we find p then?"

"Precisely the problem." A wicked grin spread like butter across Mr. Pikake's face.

Marco found himself in a control room. A formation stood on the larger-than-life screen at the front. Happy the monster had vanished; he located the lever. He fiddled with it freely moving it down to create a wide curve and up to thin it out.* Without the terrifying layers of teeth, it reminded him of an eye. It was like he was controlling the pupil: as he pulled the directrix down it dilated, allowing more light to enter. When he brought the lever up impossibly close to the vertex, the pupil became tiny, an intimidating bouncer only letting select rays enter the party.

After adjusting the lever a few more times, Marco began to see a pattern. No matter what, p had to be hidden in a. There weren't any other hiding spaces! In $a(x+h)^2+k$ the h and k were already known: the evil twin of the leading actor and their position. Not only was a all that was left, it was also evident that changing p stretched or compressed the formation and that type of molding was controlled by a.

Fumbling back in time to his meeting with Fredrick, Marco remembered shortcuts required putting everything behind a mask. He gulped down hard and started jotting on the mirror. He needed a picture to guide him, so he started there.

Marco drew a parabola, it didn't matter what it looked like, it was symbolic more than anything. He marked the key ordered pairs he needed. Penelope caught on first, she grabbed a marker and helped fill in the details. She added the vertex at $(-h,k)$, the focal point as $(-h,k-p)$, and the directrix as $y=k+p$. Oliver and Maggie jumped in too. His sister took the time to draw crisp dotted lines from the mystery point (x,y) to the focus and straight down to the directrix. Oliver added in the final points.

With everything on the table, Marco could see the situation more clearly. Although he was no longer in the battle room, he felt like he was. It was overwhelming. There wasn't a single Numberfolk on the board. Everyone and everything was wearing a mask. He hated it, but it was the only option. Maggie's two dotted lines had to have the same length, that is where he needed to begin.

---

* Play yourself at www.MathBait.com/focus

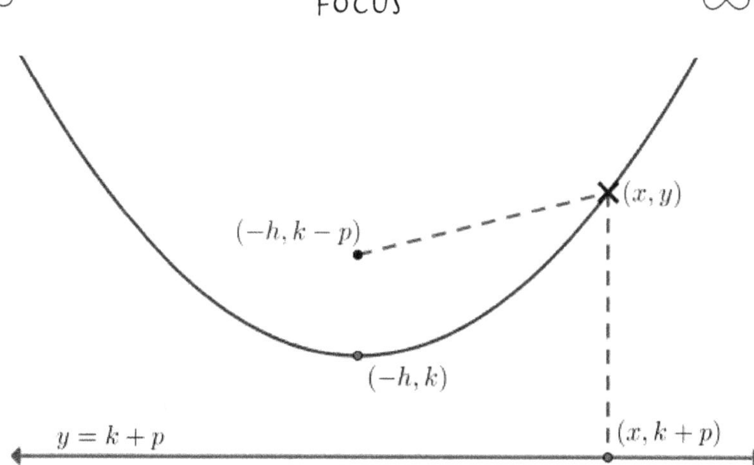

The distance from the focus to the mystery point, (x, y), was hiding in a triangle. The horizontal distance was $-h - x$ while the vertical distance was $(k - p) - y$. He began his equation,

$$\sqrt{(-h - x)^2 + \left((k - p) - y\right)^2} =$$

Liam stepped up and filled in the other side. It was quite amazing. No one said a word. If you tried hard enough, you could hear a slight buzzing, the buzzing of silence.

$$\sqrt{(-h - x)^2 + \left((k - p) - y\right)^2} = \sqrt{(x - x)^2 + \left(y - (k + p)\right)^2}$$

They broke up the pieces and began filling them in. It was a dance. A twist here, Penelope ducked under Liam's arm to reach the opposite side, a turn there. Students twirling and swooping in harmony. It's hard to describe, but it was beautiful.

Maggie was a bit lost on how to square $(k - p) - y$, so she threw in an m instead. Liam kept looking over his shoulder to mimic her, as he was tasked with $y - (k + p)$ on the right. He understood. She was using their power of substitution to make something complex simple. He copied her and wrote an l in the place of $k + p$.

$$(-h - x)^2 + (m - y)^2 = (y - l)^2.$$

They began expanding.°

° One thing they noticed but wasn't said is that $(-h - x)^2 = (h + x)^2$.

$$x^2 + 2xh + h^2 + m^2 - 2my + y^2 = y^2 - 2yl + l^2.$$

Everyone looked to Maggie and Liam. Making quick work, Maggie computed $m^2 = (k-p)^2$ and replaced it with her findings. Liam mirrored her work; the only difference was a negative.

$$m^2 = k^2 - 2kp + p^2$$
$$l^2 = k^2 + 2kp + p^2.$$

The equation stretched almost the full length of the room. It was enormous, gargantuan, intimidating, and terrifying. Yet not a single student even hesitated. Mr. Pikake stood back and watched. A smile glued onto his chin, he was the focus. All their strength and power being magnified through him. Together, they were stronger than they were alone.

$$x^2 + 2xh + h^2 + k^2 - 2kp + p^2 - 2y(k-p) + y^2 = y^2 - 2y(k+p) + k^2 + 2kp + p^2.$$

As if an invisible force shouted an order, they took a step backward, except for Marco, the leading actor, who was ready for his hunt.

He first noticed the counterweights. Both scales possessed a y^2, k^2, and p^2. Marco's first act was to remove them, vanquish the terms from existence as they were only cluttering his view.

$$x^2 + 2xh + h^2 - 2kp - 2y(k-p) = -2y(k+p) + 2kp$$

Completing the wizard's spells, he enlarged what was m and l by $2y$,

$$x^2 + 2xh + h^2 - 2kp - 2yk + 2yp = -2yk - 2yp + 2kp.$$

Another counterbalance. He removed the $-2yk$ weighing down both sides.

$$x^2 + 2xh + h^2 - 2kp + 2yp = -2yp + 2kp.$$

It was a total mess, but Marco had a goal in mind. He was looking for an order, for $y =$, all he needed to do was get the y alone and hunting was his superpower.

Sending in the evil twins of $2kp$ and $2yp$ in one impressive swoop, he nearly had it.

$$x^2 + 2xh + h^2 - 2kp + 2yp - (2kp + 2yp) = -2yp + 2kp - (2kp + 2yp)$$
$$x^2 + 2xh + h^2 - 4kp = -4yp.$$

About to cast a proliferation spell to shrink y down to size, he stopped. He saw something. Penelope had done the work to find $(x+h)^2 = x^2 + 2xh + h^2$, but something was telling him it was wrong. Marco knew there was nothing incorrect about it, but the fact that it was able to stay untouched throughout the entire hunt made it seem impervious, special in some way. Riding the whim, he changed it back. The group gasped.

$$(x+h)^2 - 4kp = -4yp.$$

Maggie started to speak up, but Mr. Pikake gave a quick whip of his head and placed his finger in front of his mouth.

Finally ready to obliterate the $-4p$, he cast $-1/4p$ across the entire field.

$$-\frac{1}{4p}((x+h)^2 - 4kp = -4yp)$$
$$-\frac{1}{4p}(x+h)^2 + \frac{4pk}{4p} = \frac{4py}{4p}.$$

The $4p$'s transformed into mirrors reflecting beautifully.

$$-\frac{1}{4p}(x+h)^2 + k = y.$$

"Oh my gosh!" Liam shrieked. Marco snapped out of his daze and tried to regain his orientation. It was like he had been under a spell, awake while the rest of the world had fallen asleep for only a moment.

"It's remarkable. Your power Marco, it grows," Mr. Pikake awed in a grave tone.

"Look at that!" Maggie brought herself back to center stage, "The a is really just negative 1 over $4p$, which means we can find p and with it, the focus and directrix no matter the order!"

Before, Marco had never really seen the point of math. If he had no intentions of blowing up worlds, he wouldn't really ever need to understand a parabola. Now, he saw it. All of this, the incantations, hunting, his puzzle box, they were the same thing. Finding relationships, tinkering, knowing how one impacts the other.

He transformed into a phoenix and soared above the world. Diving into a mechanic shop, he saw the invisible numbers everywhere. As the technician twisted one thing, it affected the other; he had to balance the hidden equations governing the air, gas, and explosions pushing against the pistons. Ascending, he settled on the ledge of an office building. Inside, they were using polynomials to model the sales and profits. Taking a nosedive, he found himself at a construction site, measurements, forces, angles, the strength of each material, all numbers governing every bolt and every screw. While his life had been redirected into a world ruled by Numberfolk, his bird's eye view showed him it wasn't any different from the earth he had always known. Humanity was the act of problem-solving, taking whatever life threw at you and tinkering with it to make decisions, navigate obstacles, and discover the covert quantities and equations directing our existence.

"The formidable focus!" the professor bellowed. "You know how to augment the mother order into . . ." he looked at the mirror. It was filled to the brim with letters and numbers, equations and shapes. Finding a tiny square at the top, where only he could reach, he wrote,

$$y = a(x + h)^2 + k.$$

"The mighty parabola with leading lady at $(-h, k)$. You can manipulate this into,"

$$y = ax^2 + bx + c,$$

"to identify zero has been deployed to $(0, c)$. And now you too can harness its power with the directrix, focusing the beams."

$$y = \frac{1}{4(-p)}(x + h)^2 + k.$$

"As you all have determined, such a formation is concentrated on the point $(-h, k - p)$ with the directing lever at $y = k + p$." Mr. Pikake looked at the students, sweat dripped from his temple, his eyes were wild and sparkled with passion.

"And now we can create a Death Star," Liam beamed.

Marco gazed around the room. Something was off and this wasn't the first time he noticed it. He zoomed back to the static in the air, the sky had seemed different too, not as blue as he remembered. It was like a parallel universe filled with doppelgangers. Everyone and everything were the same and also weren't. Their evil twins were nefarious with elongated features twisted into menacing expressions. He stopped at Oliver. His face contorted back and forth between the friend he had since childhood and the vile copy. It was all wrong.

A rumbling began to quake in his stomach. They all had the power of a Death Star now. He wondered who else did too.

FACTORING

If you can make a square you can find x

$x^2=4$

$x=\pm\sqrt{4}$

$x=\pm 2$

x^2+bx+c

① Rip apart c to find its traits

$x^2-3x-18$

18	
1	18
2	9
3	6

② Which traits can make b?

$b=-3\ldots\ 3\cdot 6=18$ and $3-6=-3$

$\therefore\ x^2-3x-18=(x+3)(x-6)$

ax^2+bx+c

① Multiply by a (unless it's already □)

$a^2x^2+abx+ac=0$

② SUBSTITUTE

Let $K=ax$ then $K^2=a^2x^2$

$a^2x^2+abx+ac=0$

$K^2+bK+ac=0$

Example $3x^2+2x-8=0$ $\times 3$

1. MULTIPLY $9x^2+6x-24=0$

2. SUBSTITUTE Let $K=3x$ $K^2=9x^2$

Then

$K^2+2K-24=0$

$(K+6)(K-4)=0$

$(3x+6)(3x-4)=0$

factor

24	
1	24
2	12
3	8
4	6 * $6-4=2$

Not every formation factors nicely!

MAKE A SQUARE

what soldier is @ ground zero?

$x^2+6x-16=0$

÷2

	x	3
x	x^2	$3x$
3	$3x$	?

this square is $(x+3)(x+3)=(x+3)^2$

$3\cdot 3=9$

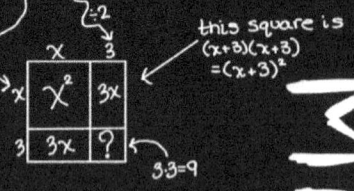

$(x+3)(x+3)=x^2+3x+3x+9$
$=x^2+6x+9$

but I don't have that!

... I have this

$x^2+6x-16$

$x^2+6x-16=x^2+6x+9-9-16$ ($9-9=0$)

so...

$x^2+6x-16=0$

$(x^2+6x+9)-(9+16)=0$

this is $(x+3)^2$ — this is 25

$\Longrightarrow (x+3)^2-25=0$

$(x+3)^2=25$ ← I made a square!

$x+3=\pm\sqrt{25}$

$x+3=\pm 5$

$x+3=-5$, $x=-8$ $x+3=5$, $x=2$

$\therefore\ x^2+6x-16$
$=(x+8)(x-2)$

I could've factored that!

15

OUT OF LINE

"MATHEMATICS IS NOT ABOUT NUMBERS, EQUATIONS, COMPUTATIONS, OR ALGORITHMS: IT'S ABOUT UNDERSTANDING."

-WILLIAM PAUL THURSTON

Zero, zero. It was midnight. His mind was racing. Marco had dozed off, but not really. He was stuck in purgatory, in that strange space between sleep and consciousness. The room was pitch black, which wasn't really black at all. It was shades of concentric gray circles that shifted like the ocean's waves.

Far in the distance, he could see someone. Starting to run toward the stranger, time was distorted. He was close, right on top of them, and then somehow back to where he started, the unknown entity miles away.

When he was close enough to see their face, it shifted. It was like Dr. Jekyll and Mr. Hyde: two people stuck in the same body. An invisible doorway stood in the darkness and the two figures were both fighting to get through, neither prevailing. Despite this, Marco could feel them. It was familiar. The sensation he had when he phased, like he was floating in an astral plane between worlds, was shooting vibrations through the empty void.

He concentrated. Trying to feel the buzz and force his body to vibrate at the same frequency. Now, he could see it. Dr. Jekyll was a Numberfolk. It was screaming for help yet not making any sound. It began to shift again. Fast, faster than possible, it switched between forms. In the in-between, Marco saw Mr. Hyde. It was only a second, less than a second, but he saw him. It was his father.

The dream was sucked away as the world crumbled around him and pulled Marco awake. He had no hope of getting back to sleep anytime soon. Staring at the slow, dashing colon between the 12 and the double zero, he waited for the Numberfolk to return. 12:01. They weren't the numbers Marco wanted to see. He wasn't interested in the cold dead shells everyone ignored; he was searching for the real Numberfolk. They were calling to him, weren't they?

He knew something was out there, waiting. Whoever it was, they knew the secrets he was looking for. They knew about his father, his grandfather, the box. *The box!*

The box was calling to him. Whatever was inside this vessel was like a siren pulling him to some location. Something wanted him there. Something *needed* him there. Only two codes remained. There was nothing in this world that could stop Marco from fulfilling his destiny.

Paranoia slithered through his veins. He waited five minutes, looking around the room for spies, before feeling safe enough to grab the clues from his closet.

$$(ax + b)(4x + 8) = -3x^2 + 58x + 128$$

Seventh Chamber

1234 Ticks: 2 Tocks: 0

5678 Ticks: 1 Tocks: 0

9012 Ticks: 2 Tocks: 0

2953 Ticks: 2 Tocks: 2

4018 Ticks: 2 Tocks: 0

2981 Ticks: 1 Tocks: 0

A warmness bubbled in his heart and swelled up to his face. He remembered the first clues: 1234, 5678, and 9012. He just knew this was his grandfather's doing. There was a beauty in the approach. The three codes had the ability to reach inside the riddle and unlock valuable information.

It told him two of the digits were in the group 1234, and one more in 5678. With a total of three accounted for, the final digit had to be a 9 or a 0. The third clue was like chipping away at a glacier and finding the perfect spot, the linchpin, that made a gigantic iceberg detach from the structure. As 9012 had two digits correct, one of those a 0 or 9, the other had to be 1 or 2 which circled back to the beginning like an ouroboros indicating to Marco another digit was 3 or 4.[*]

As exquisite as the method was, it still left Marco with a lot of *ors*. It was 1 or 2, 3 or 4, 0 or 9, and 5 or 6 or 7 or 8. Too many options, he required more information. As he slowly and methodically studied his grandfather's writing, the grin returned with a vengeance. It was the type of smile that didn't only take hold of his jaw but his whole body. He could feel the swell from the depths of his stomach up through the tip of his head. The final clue started an avalanche.

He admired how his grandfather knew just what to ask the clock. It required a perfect question to reveal its secrets. He was reminded of the old riddle,

Three triplets guard three doors

The youngest always lies

---

[*] An ouroboros is a snake eating its own tail to form a circle.

The oldest always tells the truth

The last child does whatever she feels like.

The middle child's door leads to death.

You have one question to ask one triplet.

What would you ask?

Marco had never figured out the right answer, or the right question. His grandfather would have known just what to do.

Since Marco knew the code contained a 1 or a 2, the 2981 allowed him to eliminate both 8 and 9 and adopt 0. Looping back like an infinity symbol, the final clue also led him to 1. If the 2 *was* correct in the 2891, it meant 2 couldn't go first as the clock failed to tock. This contradicted the fourth clue that whispered the opposite information. It was the similar conundrum of two trolls guarding a bridge. "My brother always lies," the troll on the right said. "What he says is true," the second creature chimed in.

"If the 2 is correct in 2953 it goes first, but 2981 says the two can't be first. So there isn't a two at all," Marco uttered thoughtfully.

The digits circled around him in the air, glistening in a deep, golden yellow before quickly revealing themselves and their order. Marco relished in his command of the numbers. Victory was so close he could taste it. He had decoded seven locks.

This was it. The final puzzle. The ultimate code.

Final Chamber

1234 Ticks: 1 Tocks: 0

5678 Ticks: 2 Tocks: 1

1679 Ticks: 2 Tocks: 1

2508 Ticks: 1 Tocks: 0

9358 Ticks: 2 Tocks: 0

4689 Ticks: 2 Tocks: 1

9175 Ticks: 3 Tocks: 0

Marco scoffed at the clues. He knew this was Maxwell's doing. His grandfather's impeccable logic was absent. This list felt frantic, disorganized, as if they were grasping at straws. Two here, one there,

two over here. Penelope's combination method seemed like the only way to approach this catastrophe. Using the final clue, Marco wrote down all the possible groups of three digits and meticulously tested each one.

917. Didn't work, it broke clue three.

915. This was trickier to see, it required uncovering that the fourth digit would have to be 8 by combining the second and third clues, then throwing out the entire combination when he reached clue five.

He worked well into the early hours. Exhausted, everything took him twice as long and he questioned and re-questioned assumptions and conclusions, not sure if it was logic or delusion that guided his tired thoughts. Then, out of nowhere like a ghost who just burst through a wall, the final code revealed itself. It was time.

Almost too drained to retrieve the box, he plopped to the floor and reached up into the box spring. He realized perhaps hiding a box in a box spring wasn't the most clever of ideas. Still under his bed, carefully navigating the dial, he started.

Click.

Click.

Click.

Click.

Click.

Click.

Click.

He took a deep breath. Hoping his sleepy eyes and depleted brain weren't deceiving him, he began twisting in the last code. The first digit. Then the next. The third. He spun the final dial all the way to the right and . . .

Click.

The top of the cedar chest popped open. Both shock and disappointment ballooned inside him. He wasn't sure what he was expecting, angels singing? A bright glow? Carefully tipping back the lid, he glared inside. A dull, thick, paper, with brown stains sat folded on a flimsy shelf. Marco delicately picked up the parchment and began to pry it open. It was old. Really old. Some areas were darker than others like little splotches of a rusty copper pigment had been haphazardly smeared across it. Even with the paper's heavy weight, it seemed frail. Marco was scared it would suddenly disintegrate in his palms; his destiny and all the answers he was searching for would crumble along with it. The edges were sharp and jagged, like his journal and he could see tiny holes that, when they caught the light just right, would allow the rays to shine through onto the floor.

When he finally had every crease uncreased and every hinge unhinged, he saw nothing. It was empty. He flipped the page back and forth making sure he didn't miss anything. There was nothing. He shined the light directly at it hoping that the darkness of the room had veiled the classified material. There was nothing. Ready to rip it apart, tear it into tiny pieces and even possibly risk burning the entire house down by setting the page on fire, Marco screamed in total silence. He opened his mouth forcefully and pushed his throat forward without making a sound. He threw his arms down and violently shook his head. How could all this be for nothing? Had the box ever contained anything at all? Was this some elaborate ruse made to have people chasing their tails scouring the Earth for a box and a clock that held no mystery, no riddle, no map, no reward? Was his grandfather in on it? Did he know the cedar chest was useless, an intricate scam?

He couldn't think, he couldn't breathe, he couldn't even care anymore. Climbing into his bed, Marco hugged his pillow tightly. Rocking back and forth, he held his jaw stiff trying anxiously to hold back the tears. Then he wept. He gave up. He couldn't jail it in any longer. It all ran out. The tears flooded his vision, and his energy abandoned him as he was forcefully tugged into unconsciousness.

* * *

The bright sun shone in his window. It was so warm and powerful; Marco knew it was late. He had slept well past the morning undisrupted. It wasn't until he felt the pounding of his head that the memories of the previous night began to flood his mind. Trying to push the thoughts away, he noticed an array of glistening little fairies floating on his wall. They were singing for Marco to follow them.

He pulled himself out of bed and approached the wall. Reaching out to the sprites, his hand went right through them like an apparition. A little creature danced on the back of his wrist. He turned his hand over and watched it follow to his palm. It was a beam of light. Tracking it back to its source, he saw the map open on his nightstand. The first feeling was panic: how could he be so careless to leave something so precious laying around? Then came the dread: it was a worthless piece of old paper, who cared where he left it. As Marco delicately lifted the page and held it directly in the beam of

$$(ax + b)(6x + 18) = -8x^2 + 81x + 315$$

light, the fairies all fluttered to new positions like Numberfolk soldiers obeying their commands.

It was a map! He didn't know how he could have missed it the night before. Perhaps the best disguise of all, it looked like nothing more than an empty piece of paper, but the small holes focused incoming rays to create something, Marco had no idea what that something was, but it was most certainly not nothing.

Grabbing a piece of paper, he placed the worn parchment on top and carefully used his pen to mark where each hole was. After running around in a circle for a few minutes, overcome with excitement and anticipation, and having no clue whatsoever what this was a map *of* or how to use it, he gently placed it back in the cedar chest and reset the dials.

A million thoughts were coursing through his brain, he couldn't keep up. Maggie started screaming at him from downstairs. He looked at his clock. Realizing he had slept away half the day, it was already time to meet with Mr. Pikake and KFUN. Whatever he was going to do had to wait. He shoved his copy of the map into his jeans and rushed down the stairs.

As Marco and Maggie met Penelope at the corner and the three began making their way to the strip mall, Marco had only one thought on his mind.

I have to talk to Fredrick.

* * *

"Believe it or not," Mr. Pikake chuckled, "You have nearly perfected the parabolic order."

"I still can't believe I can make a Death Star." Liam was both proud and bewildered by his new ability.

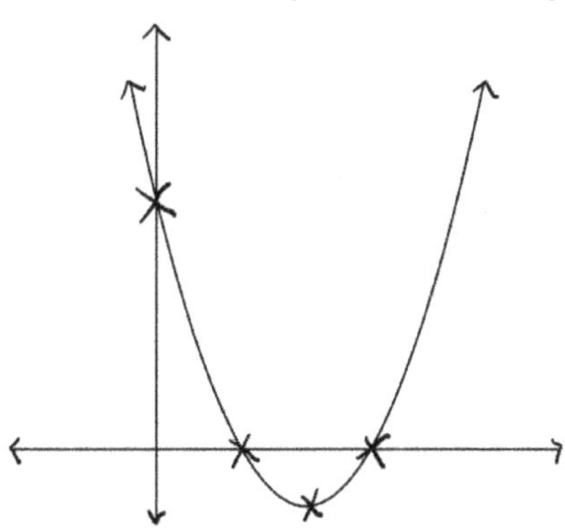

"But recall, we must determine the four key points of such an order."

"Yeah, the y-intercept where zero is sent, and the leading lady," Maggie did a curtsey with her head and shoulder, "and then the two soldiers at ground zero, the x-intercepts."

$$4(x+2)(ax+b) = 3x^2 - 52x - 116$$

"Precisely! You know how to find soldiers one and two, but what of the final fighters? Those in location zero?"

"I have actually been thinking of this problem." Penelope raised her chin, and her sing-songy voice carried softly across the dance studio. "And it is harder than it seems. If we want to find when the location, y, is zero we'd normally just make the whole order zero and hunt the x. But this seems impossible. How do you hunt something like $x^2 + 2x + 2 = 0$?"

"*Ahh*. You have identified the intricacy of the issue!" The professor did a little spin and stomped out his right foot. Marco was happy to see Mr. Pikake's mood was elevated. "Polynomials are just numbers. We can rip apart a number, we too can rip apart a polynomial formation. Every polynomial is but a line multiplied by a line multiplied by a line . . ." * He continued to repeat as he drew on the mirror.

$$f(x) = (x + m)(x + n).$$

"An alternative perspective of a parabolic polynomial." He bowed.

"Oh yeah! I remember," Maggie bubbled. "But what about the k? Normally we have a times $(x + h)^2$ plus k, this seems to only be the $x + h$ part since the a is 1."

"But is it?" Mr. Pikake needled.

"See Bug. When you expand it out, you get $x^2 + (m + n)x + mn$, which is the $ax^2 + bx + c$ perspective, that has the k hidden in there somewhere," Marco clarified.

"Wait, *all* of these are the exact same thing?" Maggie probed.

"The power of perspective!" the professor popped. "Mathematics is manipulation. You know how to vanquish and proliferate, which allows you to change how you view a formation — each form provides new and interesting information!"

"How does this help us to find the soldiers at ground zero?" Liam chimed in.

"In an infinite realm, we have the luxury of black holes."

Liam scoffed, "Since when are black holes a luxury?"

"Because they suck. They suck everything in." Mr. Pikake slashed

---

* With numbers, we call their traits, what they can be ripped apart into, factors. Similarly, with polynomials, we can rip them into linear factors, where each trait is in the form of a line.

$$(2x - 10)(ax + b) = -x^2 + 36x - 155$$

at the mirror.

$$AB = 0.$$

"If A times B is zero, what do we know about the wizard A or their victim B?"

Oliver spoke up, "One or both of them is zero. You can't multiply two numbers and get nothing unless one of the numbers *is* nothing."

Marco thought about it for a second, *Was this really true?* He imagined the wizard A, it had the power to make B bigger or smaller depending on the spell it cast. Making B smaller required a fraction, and while the wizard could shrink and shrink, it could never totally vanquish the B. Oliver was right, it would require the wizard to be 0 themselves.

"What happens in a world that isn't infinite?" Penelope wasn't sure what the professor meant in his talk of realms and black holes.

"The beauty of the infinite is it goes on and on and on. If you start at zero and continue in the same direction, you never get back to it. But! In a circular creation, you can reach zero again, like a clock."

"A clock doesn't have a zero at all!" Maggie protested.

"Imagine multiplication on a clock. Let's call 12 the zero hour. Now you have six chores, each chore taking two hours. In clock-world what is 2 times 6?"

"Two times six is twelve in all worlds," Maggie rolled her eyes.

"No!" Liam shouted, "it's not!" He stood at the front of the room, "Yeah $2 \cdot 6 = 12$, but in clock world twelve is zero, the zero hour, so $2 \cdot 6 = 0$. You can't get to thirteen in clock world, that is just one o'clock. You keep going around. It's like the numberline is bent in on itself to make a circle, so you can keep getting back to zero."

Marco had a new outlook on time. It was a hamster wheel keeping him trapped in, around and around he goes. He realized years were the same thing, instead of 12 hours, there were 365 days. Being human was essentially an inescapable and never-ending roundabout. He shivered. Imagining himself a rocket rodent, he broke free of the societal chains.

"If you ever find yourself in need of the Eisenstein Criterion attempting to factor a polynomial over a prime field, remember the clock! For things may not be what they seem. But!" A jolly pop, "Until then, as we work in our infinite world, we know if $AB = 0$ then $A = 0$ or $B = 0$."

"Or both!" Maggie interjected. Mr. Pikake bowed his head.

$(ax + b)(4x + 12) = 3x^2 - 47x - 168$

"I think I see where this is going," Penelope took the lead. "If we can manipulate a quadratic order into $(x+m)(x+n)$, we can hunt the soldiers at ground zero. Because we want to know when $(x+m)(x+n)=0$, and that means $x+m=0$ and $x+n=0$."

"Now that's a hunt I can accomplish," Marco boasted.

"How do we manipulate it?" Maggie dug in. "I mean how do you get from one perspective to the other?"

"Well, you have already examined this! To get from $a(x+h)^2+k$ to ax^2+bx+c you simply expand and reorganize. Finding the lines that were multiplied to create the order is similar. It's simply a puzzle to solve, a pattern to find." He scribbled on the board,

$$f(x) = x^2 + 8x + 15.$$

"What soldiers are at ground zero?"

"Since this is really $(x+m)(x+n) = x^2 + (m+n)x + mn$, the soldiers multiply to 15 and add to 8!" Maggie shouted.

"It's 3 and 5!" Liam found them first. "Because 3 times 5 is 15 and 3 plus 5 is 8."

Mr. Pikake transcribed their findings on the mirror,

$$x^2 + 8x + 15 = (x+3)(x+5).$$

"Now, we need to know when each line is zero," Penelope called. "Because we are looking to find when $y = x^2 + 8x + 15 = 0$ which means $(x+3)(x+5) = 0$ and *that* means $x+3=0$ and $x+5=0$."

"So, it's -3 and -5." Marco didn't even really need to hunt at all. He had become so strong, he could look at a simple duel and see exactly who was hiding. It was like x-ray vision, literally. He could see right through the x.

"Dynamic detecting!" the tutor clapped. "In a simple order, we can find the pattern by ripping apart the c into its factors and identifying which factors will sum to b. Let me increase the intricacy."

$$f(x) = x^2 - 10x + 24.$$

"Who is at ground zero in this order?"

The suspects lined up at the police station. Whoever committed the crime was a duo, a pair of flagrant felons.

They had narrowed it down to 1 and 24, 2 and 12, 3 and 8, or 4 and 6. Marco had additional information. A witness explained that when the two fled, they stood one atop the other and equaled ten feet

in height. She knew because they passed an awning at her store and lined up perfectly.

Knowing the 1 and 24 were too tall, Marco commanded them out of the interrogation room. The 2 and 12 also exceeded the height requirement and while 3 and 8 were dangerously close, 4 and 6 were a perfect match.

"So it's $(x + 4)(x + 6)$?" Marco asked the professor.

"Is it?" he asked back.

"No!" Penelope exclaimed, "It has to be $(x + -4)(x + -6)$!"

"Yeah," Maggie chimed in. "Because they have to add to *negative* ten. That means anytime we see an order like $x^2 - bx + c$ the lines are really $(x - m)(x - n)$ because m and n multiply to be positive but sum to be negative."

"Precisely! For the order $x^2 - 10x + 24 = (x - 4)(x - 6)$, the soldiers at ground zero are 4 and 6."

"Well, what if the soldiers aren't Integers? What if they are Logos?" Oliver asked.

"Factoring is a superb skill, it is quick, easy, and generally painless. However, Oliver has a potent point. It is only powerful with Integer soldiers. To find those at ground zero in more complex commands, we should use our leading lady landscape."

"You mean $a(x+h)^2+k$?" Maggie questioned.

"Aye! That is, who is the leading lady in the command $x^2-10x+24$?"

"Well, we found that $h=b/2a$, here $b=-10$ and $a=1$. Doesn't that mean $h=-10/2$ or -5? I don't remember the other one, it was complicated . . ." Maggie trailed off.

"Yes! We need a different method. One that we can employ in the chaos of battle. A method that is spry, swift, and effective."

"Well, if the h is -5, and $-h$ is the leading lady, you could just find the location that 5 is sent to, right? So $y=(5)^2-10(5)+24$ which is $25-50+24$ or -1? The leading lady, 5, is sent to location -1?" Oliver explained confidently.

"True, but we are not all such gifted computationalists," Mr. Pikake began. Oliver put on a smug look. "There is a simpler way. We know how to find these soldiers in formations that have only been shifted. Who is at ground zero in x^2-25?"

"5 and -5!" Maggie exclaimed. "If $x^2-25=0$, then $x^2=25$."

"Yes! And who anchors the formation $(x-2)^2$?"

"That is two twos," Liam spat. "If $(x-2)^2=0$, then $x-2=0$, and $x=2$."

"Beautiful!" the professor popped. "If we can make a square, locating these soldiers becomes much simpler. Thus, I ask you, how can we make a square?"

"Like physically build a square?" Maggie asked.

"Precisely! Consider only the first two brigades in $x^2-10x+24$, can we force this to be a square?"

Marco's heart sunk a bit as the lonely 24 was left behind.

"Oh, because we are trying to get to $(x+h)^2$, if we can turn the x terms into a square, we'd know the leading lady and have another perspective to work with!" Penelope exclaimed.

"The x^2 is already a square," Maggie thought aloud. Mr. Pikake drew the shape on the mirror."

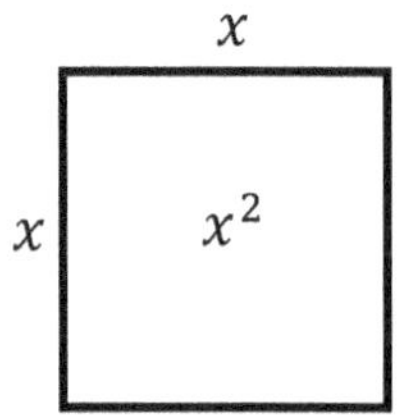

x

x | x^2

"Now for the tricky part! You have also the bx. Your shape is required to swell. But! We cannot add any additional x terms. We are attempting to hunt x . . . let's not help him out by providing *more* minions to gather."

"If we put the bx on the right, it would be the same height as x!" Penelope drew onto the professor's shape. "But to make this a square, we'd have to add a rectangle to the bottom with length $x + b$ and we don't know who x is . . ." Penelope trailed off.

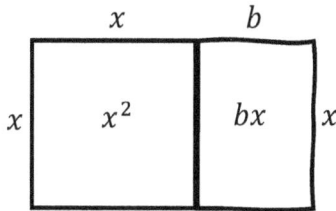

The house was on fire. Two sides of the x^2 shape were bursting with flames. Marco had to douse them out by throwing an x blanket to cover the blaze. All he had was one shape, the rectangle bx. How could he cover both exposed sides with only one piece? Split the baby!

He thought of the old story of Solomon. When both mothers claimed to be the true parent of the child, the king had suggested the only fair solution was to rip the baby in half and give a half to each supposed mama. Only the true mother would rather have her baby given to another than see it pulled apart. Marco needed to split the baby. If he tore the bx in half, each side of the square could be quelled.

He stepped up and drew his design on the mirror. "What if we did this? We split the baby in half."

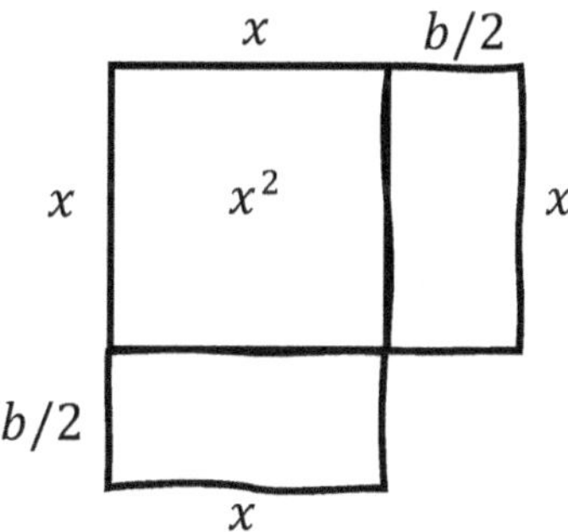

Liam flinched, "What baby?" He looked around confused.

"Superb solution, son." Mr. Pikake smiled kindly, "You have a piece missing, what is required to complete your square?"

"Well, it is $b/2$ wide and $b/2$ tall, giving it an area of $b^2/4$."

$$\left(x + \frac{b}{2}\right)^2 = x^2 + \frac{b}{2}x + \frac{b}{2}x + \frac{b^2}{4} = x^2 + bx + \frac{b^2}{4}.$$

Oliver took center stage. His pushiness reminded Marco of his sister, like he had some need to be the star of the show. "If we have $x^2 - 10x + 24$, we can make a square by taking half of b which here is -10."

$$\frac{b}{2} = -\frac{10}{2} = -5.$$

"Since $(x-5)^2 = x^2 - 10x + 25$, that takes care of the $x^2 - 10x$ part, but you have the extra $b^2/4$, the 25. You can't just throw that in there."

"He's right." Marco jumped up next to his friend. Oliver gave a look that read, *of course I'm right you dummy*. "But we can balance the scales. Since $(x-5)^2 = x^2 - 10x + 25$ and we only have the $x^2 - 10x$ part, we can force the duel with -25 to even everything out."

"Oh yeah!" Penelope took her place next to Marco. He smiled at her. "We have to keep it $x^2 - 10x$, but $x^2 - 10x + 25 - 25$ *is* just $x^2 - 10x$. We can group the first part." She placed out her palm face up. Marco, for a second, thought she wanted to hold hands. Just as he went in, he realized she was actually requesting the marker. He broke into a cold sweat, thankful he avoided what would have surely been a horribly embarrassing moment.

$$x^2 - 10x + 24 = x^2 - 10x + 25 - 25 + 24$$

"Now, since $(x-5)^2 = x^2 - 10x + 25$ we can replace that . . ."

$$= (x-5)^2 - 25 + 24$$
$$= (x-5)^2 - 1.$$

"The leading lady is 5 and was sent to -1 like Oliver said," Penelope shone. Oliver straightened up and smirked.

"Excellent, excellent, excellent!" Mr. Pikake tapped the mirror. "Now, see here. If we can put an order in this form, this perspective, we can more easily identify the two soldiers at ground zero."

$$y = (x-5)^2 - 1 = 0.$$

"Oh! I see!" Marco exclaimed. That shows us that $(x-5)^2 = 1$, so $x - 5 = \pm 1$. If $x - 5 = 1$, then $x = 5 + 1 = 6$. If $x - 5 = -1$, then $x = -1 + 5 = 4$ . . . which we already found in what I would consider a much better way."

"Ah! It all depends on the situation. Try this one."

$$(ax + b)(x + 5) = -x^2 + 11x + 80$$

$$g(x) = x^2 + \frac{8}{3}x - 1.$$

"Yeaaaah," Liam emptied his lungs. "That is a lot harder because the factor of 1 is 1 and that doesn't add to be eight-thirds."

"So, we make a square," Penelope concluded. "Take the middle term, 8/3 and divide it in half to get 8/6 or 4/3. Squaring that gives us 16/9 which means . . ."

$$\begin{aligned} g(x) &= x^2 + \frac{8}{3}x + \frac{16}{9} - \frac{16}{9} - 1 \\ &= \left(x + \frac{4}{3}\right)^2 - \frac{16}{9} - 1 \\ &= \left(x + \frac{4}{3}\right)^2 - \frac{25}{9}. \end{aligned}$$

"You want to make that zero," Oliver pointed. Penelope scribbled.

$$\begin{aligned} 0 &= \left(x + \frac{4}{3}\right)^2 - \frac{25}{9} \\ \frac{25}{9} &= \left(x + \frac{4}{3}\right)^2 \\ \pm\sqrt{\frac{25}{9}} &= x + \frac{4}{3} \\ \pm\frac{5}{3} &= x + \frac{4}{3} \\ -\frac{4}{3} \pm \frac{5}{3} &= x. \end{aligned}$$

"That's ugly," Liam moaned. "But . . . I guess doable. It means $x = 1/3$ and $x = -9/3$ or -3.

"Perfect! The two lines that were multiplied to make this formation were,"

$$\begin{aligned} g(x) &= x^2 + \frac{8}{3}x - 1 \\ &= \left(x - \frac{1}{3}\right)(x + 3). \end{aligned}$$

"Wait!" Marco shouted. His friends had hunted one way, but he had gone about it differently. "If $x = 1/3$, couldn't you obliterate with a three to get,"

$$(ax + b)(2x - 4) = 2x^2 - 33x + 58$$

$$x = \frac{1}{3}$$
$$3x = 1.$$

"Then send in the negative one to find,"

$$3x - 1 = 0.$$

"That is different than what they found."

"Interesting introspection." Mr. Pikake rubbed his chin. "Our goal was to find the anchors, the roots, the solemn soldiers assigned to ground zero. Which we did, they were soldiers one-third and negative three. But!" He didn't pop. Instead, it came out as a whine with a deadly sharp "t". "The formation made from the lines $3x - 1$ and $x + 3$ would have the same anchors, yet create a different shape."

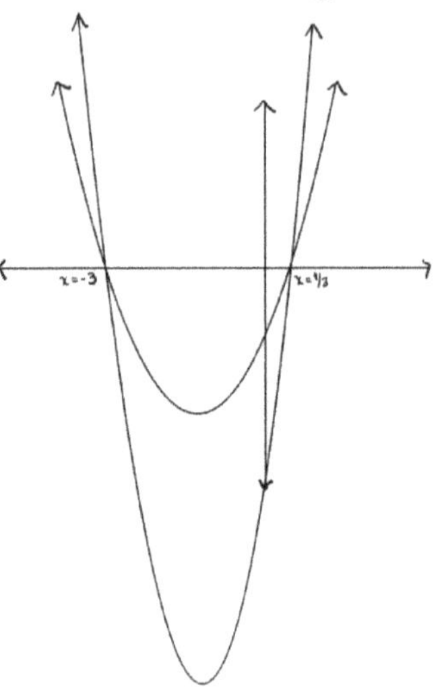

"Yeah. Because when you multiplied it out, it would have $3x^2$, making it skinnier than our formation," Maggie added. "Hey, that's a good question. This only works if there is one leading brigade. If it starts with x^2. What if it wasn't x^2? Like in that one, it's $3x^2$. How do you make that a square?"

"A noble notion! Solving the puzzle can be quick; if not, we mandate some manipulation."

$$v(t) = 2t^2 + 13t + 20.$$

"Are you able to spot the soldiers anchored to ground zero?"

Marco was pulled into a dark void. Looking around, he realized he was inside a gigantic puzzle box. The four dials lay before him, all he needed to do was to figure out the code. Two lines multiplied to make $v(t)$,

$$(at + b)(ct + d).$$

This time he didn't have ticks and tocks to guide him, he had to figure out his own clues. He started by multiplying the lines so he could compare apples to apples.

$$(at + b)(ct + d) = act^2 + adt + bct + bd.$$

$$(ax + b)(4 - 8x) = 6x^2 - 123x + 60$$

Clue one, he thought. *It has to be that* $ac = 2$ *since* $act^2 = 2t^2$. Realizing that two has only two factors, one and itself, he began filling in the blanks.

$$(2t + b)(t + d) = 2t^2 + 2dt + bt + bd.$$

Two down, two to go. Before, he only needed to find what factors of the final c term summed to the middle, this was different. Marco needed two numbers that multiplied to make 20, $bd = 20$, and the second clue was that twice one number plus the other was 13, $2d + b = 13$.

He scribbled his factors on the wall of the box. Twenty was $1 \cdot 20$, $2 \cdot 10$, and $4 \cdot 5$. *Could it really be this easy?* While Penelope's cases were something he had in his back pocket, he could try out each possibility until ultimately identifying the pair that worked, he didn't think he needed to. The factors of 20 and 10 were too big. There was no way he could double and add those to get to 13, it had to be 4 and 5.

Quickly realizing that doubling the five would lead to 14, he knew what to do. He twisted the dials and hopped out of the box and back to reality.

"I got it. I . . . I think." He looked at Mr. Pikake. "The only numbers that fit the code, *er,* I mean . . ." Marco had no idea what he meant, so he just grabbed a marker and wrote what he found.

$$2t^2 + 13t + 20 = (2t + 5)(t + 4).$$

"I think he's right!" Penelope sang. "Because the $2t$ and t make $2t^2$ and the 5 and the 4 make 20. On the inside, you get $2t \cdot 4$ and $5t$ which make $8t + 5t = 13t$!"

"How'd you do that?" Liam urged.

"I just looked at it like a puzzle. I wrote out all the clues and then had to figure out the only way they all worked together." A shy smile poked through. He didn't want to focus on *how* he figured it out, that would lead to everyone discovering he was a no-good, horrible liar who had been keeping secrets.

"Amazingly articulated!" Mr. Pikake began furiously scribbling in red ink. "Now, for the climax of our challenge! We shall use masks to first manipulate our primary perspective of,"

$$f(x) = ax^2 + bx + c.$$

"If we reverse the wizard a, we obtain a situation we have previously painted!"

$$f(x) = a\left(x^2 + \frac{b}{a}x + \frac{c}{a}\right).$$

"Let us avoid the Logos. We shall use our substitution skill to allow b/a to be called d and c/a will be named e."

$$f(x) = a(x^2 + dx + e).$$

"Do this, and we will identify the anchors of any parabola." His eyes widened as he looked around the room. "Can you construct a square?"

"If we are just using the inside part, then we cut the middle term in half," Penelope offered to take over the first step.

$$a(x^2 + dx + e) = a\left(x^2 + \frac{d}{2}x + \frac{d}{2}x + \frac{d^2}{4} - \frac{d^2}{4} + e\right).$$

"So much for losing the Logos," Liam moaned.

"Yeah but," Maggie started, "this isn't *so* bad. We know the first terms make the square,

$$\left(x + \frac{d}{2}\right)^2.$$

"Right," Oliver jumped in. "The whole thing becomes,"

$$a\left(\left(x + \frac{d}{2}\right)^2 + e - \frac{d^2}{4}\right).$$

Flashing Marco a devious grin, Mr. Pikake outstretched the marker, "Care to hunt?"

No, I do not care to hunt, Marco thought. This looked horrible and ugly and hard and just terrible. Last time he hunted with masks something took over him. It was an out-of-body experience. But as Penelope's big brown eyes looked at him with hope and kindness, he couldn't let her down. He had to not only do it, he had to do it well.

Taking center stage, the leading man once again, Marco began. "Alright, we are trying to find the anchors, we need to know when this thing is zero."

$$a\left(\left(x+\frac{d}{2}\right)^2+e-\frac{d^2}{4}\right)=0.$$

"Wait!" Liam shrieked. "Look, you have a times a swirling black hole of death is zero. If a is zero, we're done!" He had a dopey grin on his face.

"*Ugh*. Liam." Maggie rolled her eyes. "First, that wipes out the soldiers, the x's, and our whole goal is to identify them. Erasing them from existence makes no sense. And second, a *can't* be zero."

"Why not?" Liam wasn't convinced.

"Because we started with ax^2+bx+c. If a is zero then the whole thing is just a line, it would be $bx+c$."

Oliver chuckled. "Bruh, you managed to destroy both the soldiers and the entire formation in one move."

Liam felt a sense of pride. "Not to brag, but it is pretty impressive."

"It is helpful Liam!" Marco started. Everything became quiet. The masks on the board glowed as they woke up and revealed themselves to him. It was ugly, but it wasn't any different than hunting $3x-1=0$. He needed some evil twins, to obliterate, and to substitute. All skills he had mastered. He began speaking quickly and writing even faster. "The a isn't zero, so the rest of it *has* to be,"

$$\left(x+\frac{d}{2}\right)^2+e-\frac{d^2}{4}=0.$$

"That last part is just a Numberfolk, and that means it has an evil twin. I can force a duel."

$$\left(x+\frac{d}{2}\right)^2+e-\frac{d^2}{4}+\frac{d^2}{4}-e=\frac{d^2}{4}-e.$$

"Now, we just need an insurrection. Call on a root."

$$\left(x+\frac{d}{2}\right)^2=\frac{d^2}{4}-e$$

$$x+\frac{d}{2}=\pm\sqrt{\frac{d^2}{4}-e}.$$

He saw it, he was almost there, so close he could taste it. "We need to get the soldier alone, to rip off their mask. Send in the evil twin of half of d.

$$x = -\frac{d}{2} \pm \sqrt{\frac{d^2}{4} - e}.$$

Marco had hunted x. Although it seemed horrendous, it was all spells he'd done before. He let the marker fall to the floor as he snapped out of his daze.

"Spectacular!" Mr. Pikake cheered. "But! One step remains."

"We need to replace what we substituted," Penelope proposed.

"Precisely! Recall $d = b/a$ and $e = c/a$. Finish the hunt!" he yelped.

"First, we should replace all the d's with b/a. That makes $-d/2$ into $-b/a/2$ or $-b/a \cdot 1/2$. That's just $-b/2a$," Maggie squealed.

"And the $d^2/4$ becomes $b^2/a^2/4$ or $b^2/a^2 \cdot 1/4$, and you have $b^2/4a^2$," Penelope joined in.

Marco grabbed the marker from the floor and followed his leading ladies.

$$x = -\frac{b}{2a} \pm \sqrt{\frac{b^2}{4a^2} - e}.$$

"The e is really c/a," Liam pointed.

$$x = -\frac{b}{2a} \pm \sqrt{\frac{b^2}{4a^2} - \frac{c}{a}}.$$

"Maybe you can clean it up," Oliver suggested. "Push the c/a through a mirror. Use $4a$."

Marco multiplied the last piece by a one in the form $4a/4a$ (since they knew a couldn't be 0) and transcribed the results.

$$x = -\frac{b}{2a} \pm \sqrt{\frac{b^2}{4a^2} - \frac{4ac}{4a^2}}.$$

"Oh! Oh!" Maggie barked. "Look, if you make only one vinculum the bottom is a perfect square."

$$x = -\frac{b}{2a} \pm \sqrt{\frac{b^2 - 4ac}{4a^2}}$$

$$= -\frac{b}{2a} \pm \frac{\sqrt{b^2 - 4ac}}{2a}.$$

The entire, ugly terrible mess to determine the anchor, the soldiers holding down the formation at the x-axis, was almost complete. Marco threw his arm across his chest and slashed a large vinculum on the mirror.

$$x = \frac{-b \pm \sqrt{b^2 - 4ac}}{2a}.$$

"Hey!" Maggie bubbled, "I recognize that! It's the quadratic f . . ." She came to a sudden stop.

Mr. Pikake jumped in. "And now, you have mastered the parabola. Well done." His Cheshire grin grew and grew. Just when they thought it couldn't possibly get any bigger, it spread more. The professor bowed to the students with respect and admiration.

Marco looked around. Everyone was there. His best friends, his mentor, his sister, his . . . crush. It was the first time he even admitted it to himself. They all made this, together. It was ugly and complex and hard, but it was also powerful. With this, they could take any order $ax^2 + bx + c$ and uncover its secrets. They'd know it inside and out, where zero was sent, the leading lady and her post, the soldiers left at ground zero.

The puzzle glowed as the pieces finally fell into place. He hadn't lost anything with the formation of KFUN, he gained a brigade. Each person bringing a different perspective, each player pivotal. As if traveling along a parabola, Marco reached the apex and then came the inevitable fall. As they all stood there, right beside him, supporting him, he wore a mask. Deep inside he was harboring his own secret, a dangerous one.

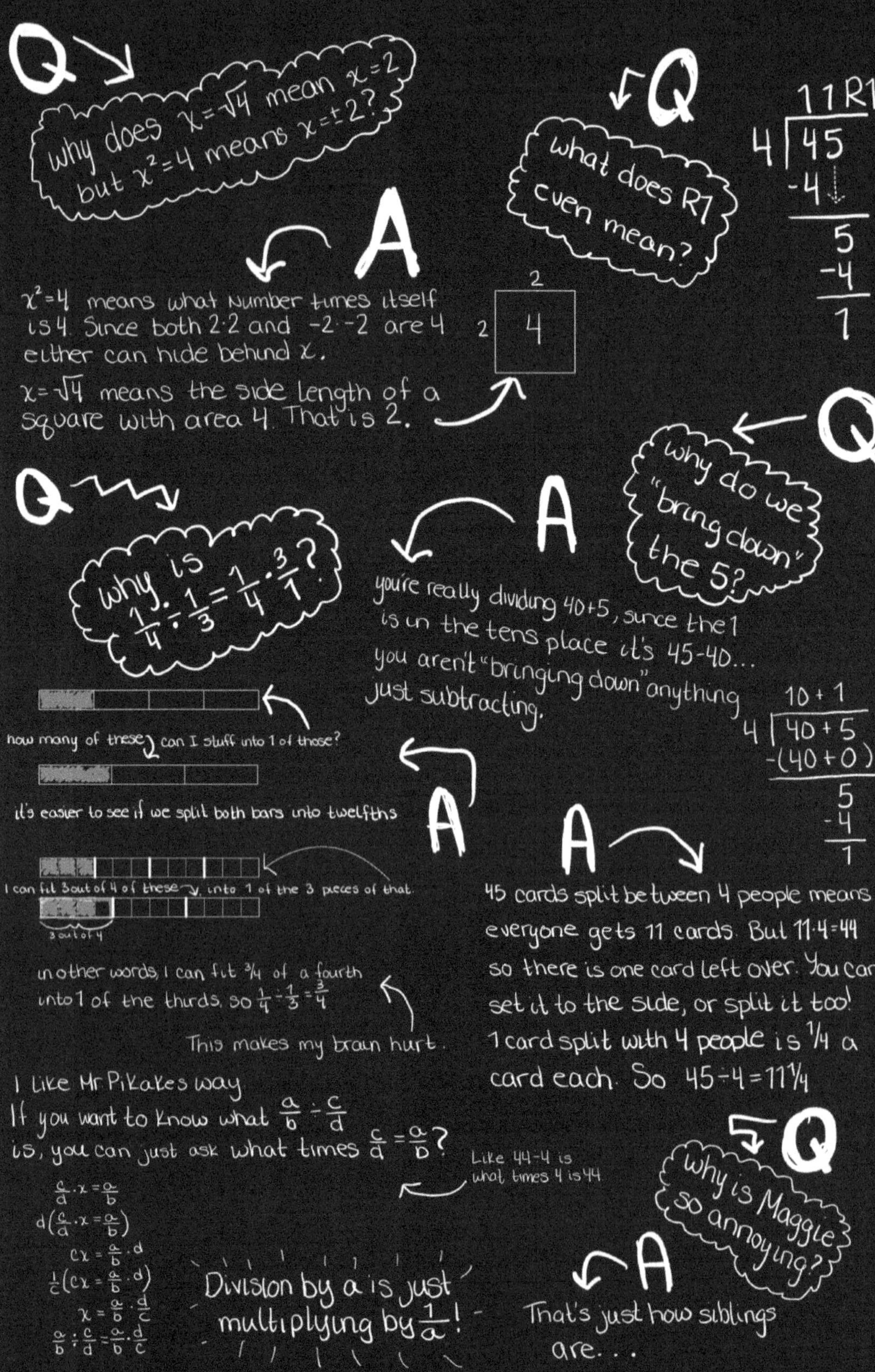
THINGS THAT Confuse Me?
Q
why does x = √4 mean x = 2 but x² = 4 means x = ±2?
A
x² = 4 means what number times itself is 4. Since both 2·2 and -2·-2 are 4 either can hide behind x.
x = √4 means the side length of a square with area 4. That is 2.
2
2
4
Q
what does R1 even mean?
11 R1
4 | 45
-4
5
-4
1
Q
why do we "bring down" the 5?
A
you're really dividing 40+5, since the 1 is in the tens place it's 45-40... you aren't "bringing down" anything just subtracting.
10 + 1
4 | 40 + 5
-(40 + 0)
5
-4
1
Q
why is 1/4 ÷ 1/3 = 1/4 · 3/1?
how many of these can I stuff into 1 of those?
it's easier to see if we split both bars into twelfths
I can fit 3 out of 4 of these into 1 of the 3 pieces of that.
3 out of 4
in other words, I can fit 3/4 of a fourth into 1 of the thirds. so 1/4 ÷ 1/3 = 3/4
This makes my brain hurt.
A
I like Mr Pikakes way
If you want to know what a/b ÷ c/d is, you can just ask what times c/d = a/b?
Like 44÷4 is what times 4 is 44
c/d · x = a/b
d(c/d · x = a/b)
cx = a/b · d
1/c (cx = a/b · d)
x = a/b · d/c
a/b ÷ c/d = a/b · d/c
Division by a is just multiplying by 1/a!
A
45 cards split between 4 people means everyone gets 11 cards. But 11·4 = 44 so there is one card left over. You can set it to the side, or split it too! 1 card split with 4 people is 1/4 a card each. So 45 ÷ 4 = 11 1/4
Q
Why is Maggie so annoying?
A
That's just how siblings are...

16

DISSONANCE

"NUMBERS RULE THE UNIVERSE."

-PYTHAGORAS

Alone, he left everything hidden at home and began his journey to the warehouse. Marco had sworn he would never go back to the horrid building, but Fredrick insisted. He felt on edge. The secrets were pulling him down into an abyss and he wanted more than anything to step into the daylight again. The shadows were dark, scary, lonely.

Fredrick was manic. Rushing around, he pushed the ladder across the floor and mounted it, the wood shook and wobbled beneath him. He swiped at the board, erasing and replacing, again and again. Marco stood in awe. He was dying to understand the enormous code that sat in front of him, what did it mean? When the man finally calmed down, he hopped from his perch and hovered squarely in front of Marco.

"Do you know where four-leaf clovers come from?" he asked calmly.

Marco pondered his response. Searching his mindroom for any facts useful in this situation, he came up with four-leaf clovers existed and they were rare. Otherwise, his botany knowledge was lacking. "No?"

"Genetics!" he barked. "Every living thing is made up of DNA, a code. Functional Genomics is a branch of science that looks to find the relationship between each gene and their orders, what they are instructed to do."

Relationships? Orders? This sounded an awful lot like what Mr. Pikake had been training him on. There was no way he could order *genes* . . . Was there? "You mean like Numberfolk relations? Functions?"

"Exactly!" Fredrick screamed. "What if you could use this information to change the Pattern?" Fredrick posed.

"You mean, override their orders? Send them somewhere else?"

"Yes! This is the four-leaf clover. It is a mutation. The orders of the gene have been intercepted. Rather than produce the holy trinity of three leaves, it has been instructed to make four."

Marco could feel an electricity coming from Fredrick. It was a nervous energy seeping from his pores. He could almost see negative ones shooting out from all around him. Then, it stopped. As Fredrick calmed down, so did the vibrations throughout the room.

"Let me tell you of the Pythagorean, Philolaus." He sighed before continuing melodically, "All things that are known have a number."

"What does that mean?" Marco asked. He felt like he was in Liam's *Twilight Zone*, a twisted and unfamiliar reality.

"Philolaus claimed everything is a number. For he believed that while we can observe with our senses, to truly understand something we must determine its number. His work centered on discovering the number of everything and using this to map the relations that govern the observable world."

Marco chuckled. He didn't even mean to. It just came out. Fredrick must have been kidding, what he was describing was insane, *everything is a number*? "The Pythagoreans were crazy."

Fredrick whipped his head around and admonished the boy. "Perspective!" he shrieked. "You have seen the Numberfolk, you know they are there. They are everywhere." He brushed off the front of his turtleneck and took a deep breath. "Now, Philolaus talks of the importance of harmony, during times of harmony the power of all the limited becomes unlimited. It is brief, but during this magical hour, amazing things can happen." With a wicked smile, his eyes glistened in the dim light, "Harmony is approaching."

Marco pinched himself to make sure he wasn't dreaming. He came here to find out about the map, not for a cult meeting. Not seeing many options, Marco decided his only play was to play along. "How do you know?"

"Have you not noticed the Pattern? The air is slightly thinner, the sky not as blue, the clovers are corrupted. The signs are everywhere!"

Marco's entire body clenched. He had noticed. The changes were small, almost negligible, but they were there.

"He wrote about it," Fredrick continued. "Philolaus described the central fire. This is an unending source of energy that lies at the center of the universe in which everything circles."

"You mean the sun?" Marco was pretty confident only the planets in our solar system orbited the sun, but he was trying to piece together the truth hidden within Fredrick's crazy.

"No, no. The sun is but a star. The central fire is much more. When everything is in alignment, everything is in harmony, for a brief moment, we will have access to counter-earth."

Now Marco was sure he was somehow caught in one of Liam's dreams, "Counter-earth?"

"The other Earth! The opposite to our world. The realm of the dead." He pranced to a small area of the wall that hadn't been filled. Vigorously coloring, he revealed the sun and the Earth. "Here! The sun normally blocks our view, it is so bright. But when the moon obscures its rays, for only a moment, we will see counter-earth, harness the power of the central fire, and . . ." His voice trailed off. Marco didn't care. It was too much. While he wasn't sure what the Numberfolk were planning, or if they were even planning anything at all for that matter, whatever Fredrick was talking about was insane. He had to get out. Regretting he ever considered Fredrick as his new Mr. Pikake, a way back to that special feeling, that connection, he knew now he was very wrong.

With wild eyes, Fredrick turned to face Marco, "Do you have the map?!"

Slowly standing, careful to not make any sudden movements, Marco lied, "*Um.* No. I haven't figured out the box yet."

"You must hurry! Harmony approaches. Without the book, we are doomed. Another fifty years before this opportunity will cycle again. I know another way . . . I will begin investigating this immediately." Fredrick hurried up the back stairs, slamming the door to the office. This was Marco's chance. He ran.

Looking behind him every few steps, it wasn't until he was more than two blocks away that he finally felt safe enough to slow down. What had he gotten himself into?

Marco dug in his pocket for his phone and quickly typed "Wyoming" in his group chat with Liam and Oliver.

This was their panic word. It was a flare that sparked a promise. They swore no matter what, if any of them were in trouble and sent out the code they would drop whatever they were doing and meet in the agreed upon spot.

When Marco arrived in the alleyway behind Walmart, both his friends were already there. They rushed over.

"What's happening?" Liam was clearly worried. In their years of friendship, no one had ever used the Wyoming text before. This had to be serious.

Marco spilled everything. He told them about his grandfather, the box, his promise. He told them about Fredrick and the warehouse. By the end of it, Marco was in tears. The secrets and the lies had placed a ferocious weight on his shoulders. Relinquishing it all felt freeing. It also felt awful. It forced Marco to face his grandfather's death all over again. He hadn't talked with anyone about it. Letting it out was like being caught in a tsunami. The waves hit him again and again. He was drowning. He struggled to catch his breath. He

sobbed.

Liam wrapped his arm around Marco. "Let it out man. Let it out."

After a long silence and the tears had dried up, Oliver finally spoke. "So, you found the map?" he asked.

Marco nodded enthusiastically. "If you can call it that. It's just a paper with a bunch of holes. I feel like I'm going crazy."

"No. Way." Liam responded assuredly. "This is totally normal for secret society stuff. I've seen it a ton of times." Marco and Oliver exchanged puzzling glances before both turning to stare at Liam.

"You've been in situations where a crazy person needs you for some nefarious plan all while numbers have their own agenda and are somehow changing the DNA of the world?" Marco raised his eyebrow.

"Oh, yeah. No. But . . . I've read a ton of similar situations. Plus, I've watched like every episode of *Sliders* at my aunt's house on her VHS machine which is all about counter-earths."

Marco laughed. It felt good. He figured the only options in this situation were to laugh or cry and he'd had enough tears. "We've gotta stop Fredrick. Whatever he is planning is *not* good. Plus, I think we need to find the book he was talking about."

"We should tell Mr. Pikake. He'll be able to help," Liam added.

"No!" Oliver shouted. The boys turned, surprised. "I. *Um*. I mean, we need to figure out what your grandpa wanted you to protect. You said Fredrick was friends with him, and they are against the SAN. Mr. Pikake works *for* the SAN. I'm just not sure it's a good idea to bring him in just yet."

Both boys turned to Marco. It was fifth grade Halloween all over again and he was caught in the middle. This time was different, it was about his own destiny, his promise, his duty, not about picking between friends.

"I agree with Liam," he responded in a hushed voice. Oliver rolled his eyes and collapsed his shoulders.

"Of course you do. You always do." Oliver's phone started to vibrate. "It's my mom, I've gotta go. Let's meet back up tonight. My house. We can talk more about this then."

As soon as Oliver disappeared around the corner Liam turned to Marco, "Let's go tell Mr. Pikake now. That way, when we meet him later it is already done with, and we can solidify our plan of attack rather than argue about it."

The boys hurried to retrieve the map before heading to the library where Marco had arranged to meet the tutor via text. When Liam pushed into Marco's room, he fell to the floor with a loud thump.

"Dude. You're lucky. My mom would *never* let me get away with my room like this. She literately makes me clean it *every day*. Can you imagine?"

Marco didn't respond. He remained frozen in the doorway.

"What's wrong?" Liam pulled a sock off his shoulder and threw it in the pile.

"I . . . I didn't do this," Marco stuttered. Like a puzzle box, his mind went to work unlocking chambers, clicking and twisting to put the pieces together. "The chest." With a running dive he hit the floor hard and began frantically fumbling around beneath his bed. It was gone.

"Fredrick?" Liam suggested.

"It had to be. He was the only one who knew about it."

The two sat in silence atop the piles of dirty clothes and papers. "Maybe you can remember where the holes were? Can you try to draw it out?" Liam urged.

Digging into his pocket, Marco pulled out the crumbled graph paper as an irrational grin inched onto his face, "I made a copy."

* * *

Mr. Pikake sat on the bench outside the library. The boys came rushing over, skidding to a stop and out of breath. "Calm down, calm down. What on Earth is going on?"

"Counter-earth." Liam raised a finger to correct him before dropping back over to continue huffing and puffing.

Marco plopped down next to his tutor. He hadn't thought about what he was going to tell the professor so as he spoke, everything came spewing out.

"My granddad and Maxwell knew about some important book. Maxwell betrayed him but he had the map to it, and he gave it to me before he died. And I unlocked the box."

"Other way around." Liam cut in pointing his fingers in opposite directions. "His granddad gave him the box, in the box was the map, Marco unlocked the box to get the map."

"Yeah, yeah. But then this man found me. He showed me a picture of him with my granddad and my dad! He said he knew about everything and that there was this Pattern and the Numberfolk were changing things and he was going to stop it or something."

"But he probably lied," Liam interrupted in support.

"Definitely lied. He is crazy and I don't think he is doing anything good. But he is trying to get the book for something else."

Before Marco could continue, Mr. Pikake clenched down on his shoulders and looked him in the eyes. "Do you have the map?" he demanded.

"No! Somebody stole it!" Liam yelped.

"But I don't think they can get to it!" Marco's eyes were wild. "I always lock the chest and they don't have the codes."

"They can if they have the clock," Mr. Pikake sighed desperately.

The world began spinning around Marco. In the wind and the dust, he could see all the moments and the people from the last few months. His brain began piecing together clues forming a mind map of relationships and connections. *I didn't say anything about the clock,* he thought as he saw Mr. Pikake connect to the SAN and to Maxwell. Then his grandfather and Fredrick joined the mix, all at the SAN headquarters. They had to know each other. He saw Maxwell betray his grandfather. *Was Mr. Pikake involved?* Marco needed to be sure. "How do you know about the clock?" he spit out like venom.

"The meeting. My trip. This was the emergency. The SAN has also picked up on the changes. Not even an hour ago, Maxwell confided in me. The clock had been in his possession since he was a child. It too has been stolen."

"Marco made a copy!" Liam shrieked.

"Let me see it," the professor insisted.

He pulled the folded-up graph paper from his pocket and passed it to his tutor. The old man lifted it up to the light.

"Phaseville," Mr. Pikake sighed. "It's a map to Phaseville."

"How do you know?" Marco asked.

"I've taken the hike many times. These dots here," he pointed, "they are landmarks to the village. These others, I'm not sure. We'd need an expert. Plus, Phaseville is where I'd hide something."

"What is *Phaseville*?" Liam snorted.

"A small town in the mountains outside *The Kryptografima*. It's a dangerous area. The void is thin, Numberfolk have more control up there. Hikers disappear all the time. Only someone strong with numbers has any hope of navigating the area."

"Do you think Fredrick knows about Phaseville?" Liam turned to Marco.

Like a vending machine dispensing soda, Marco could see the blood draining as Mr. Pikake's face went white. He violently grabbed

Liam, "What did you say?!" The tutor gave the boy a shake.

A wormhole sucked Marco back through time. The relationships he thought he had connected broke apart as the mind map began shifting like the pins and gears of an enormous machine. He remembered. It was why Fredrick seemed so familiar, why he felt an instant connection, why he trusted him. Fredrick had reminded Marco of Mr. Pikake because Fredrick was . . .

"Your brother." Marco bowed his head. Liam's mouth fell open so fast and with such force in cartoon-world it would've hit the ground.

"He's alive?!" Mr. Pikake shifted his focus from Liam to Marco.

"Dude!" Liam exclaimed. "How could you forget to tell me that? That's *huge*!"

"I. I. I didn't realize," Marco stuttered. "You only mentioned Fredrick to me once. I didn't even consider it."

The three sat in silence staring out at the empty parking lot for what seemed like eternity. A summer camp with a dozen tiny children lined up and entered the library, a few waving and others making faces as they passed. Time inched by slowly until they exited and climbed back onto the bus, each grasping their own oversized cardboard storybook. Finally, Mr. Pikake spoke.

"My brother, Fredrick, disowned us. Disowned our entire family. I haven't seen him in over fifty years. I assumed he was dead. I believed he would have reached out, contacted me if he were still alive."

"Your brother is a lunatic!" Liam moaned. Marco shot him a disapproving look and shook his head. Liam zipped his mouth closed with his hand indicating he understood.

Mr. Pikake's head remained hung, eyes focused on his lap. "You say Fredrick is trying to obtain the book? For what purpose?"

"He said something about harmony. When there was harmony, he'd have unlimited power. He didn't say why. He has a chalkboard with thousands of numbers on it. I don't really know. Just that he wants the book and is in a hurry to get it."

"Harmony. Harmony." The professor repeated the word again and again attempting to dislodge some key bit of information trapped in his brain. "Of course!" He jumped up and spun back to face his students. "The eclipse. The solar eclipse. It was a belief of the early SAN, the Pythagoreans. Everything is made of what they call limiteds and unlimiteds. When the right ratio of each were combined, they produced harmony."

"Like in music?" Liam asked. His second passion, after conspiracies and mysteries, was the drums.

"Exactly like music but in everything. A combination of numbers disguised as notes, octaves, and melodies that make up the very fabric of the cosmos."

"Cool," Liam swooned.

"Long before Copernicus and the discovery of our heliocentric solar system, the Pythagoreans described a central fire and counter-earth."

"Yeah! Yeah! He talked about that!" Marco exclaimed.

"When everything is in alignment, the great harmony will sing. This happens rarely. The upcoming eclipse must be the next configuration." Mr. Pikake raised his hand to his chin in deep thought.

When he was ready, the professor explained to Marco and Liam his plan. Marco wanted to keep Maggie out of it, but Mr. Pikake insisted. He instructed the boys to retrieve everyone and meet back at the library first thing in the morning. They were going to Phaseville.

* * *

Marco and Liam arrived at Oliver's house promptly at 6 p.m. His mother let the boys in, and they ran down the hall to their best friend's room. They stood in shock at what they saw. It was covered in boxes.

"What's going on?" Marco asked as Oliver ushered the two into his room, closing the door behind them.

"*Um.* Yeah. Figured this was as good a time as any to tell you." He fiddled with his hands. "My dad got a job in Berlin."

"Sounds cool! What's with the boxes? You didn't get a job in Berlin too? Right?" Liam joked.

"Naw." Oliver forced a chuckle. "My parents figured their custody arrangement wouldn't work across the world. They went on about culture and how it would be good for me or something. Found some fancy German school. I honestly didn't think I'd get in, so nothing to worry about. They made me take an exam, which I was sure I bombed. Found out a while back they offered me a spot. The Berlin International School." He said it with both disdain and an air of superiority. "I'll be gone by the end of summer. They start in August.

I've gotta get everything moved and settled before then." He collapsed on his bed.

No one spoke. *How has he not told us any of this?* Marco was stunned. Wanting to break the uncomfortable silence, but not sure what to say he blurted out, "What about your mom?"

"She's staying here." Oliver started tossing a baseball between his hands. "No reason to move across the ocean with your ex-husband. I'll live here, with her, for summer breaks."

"That's great news!" Liam appealed, trying to cheer up the room. "Summers are the best time anyways. No school. We can hang out all day, every day. Plus, with FaceTime and MMO's, it'll be like you never left. It's actually a really good thing when you think about it, you get out of Lydon's history class." He laughed and looked at his friends, laughing again trying to get them on board. They only stared.

"Yeah, I guess," Oliver muttered. He let the ball drop to the floor. "Anyway, what's the plan? I really don't think we should tell Mr. Pikake. We can figure this out together."

Liam pressed his lips together and widened his eyes. Marco kicked at the ground. Neither of them wanted to be the messenger. Surrendering, Marco bit the bullet, "We already did."

"You what?!" Oliver yelled. "Seriously? How could you?"

"We're sorry, dude. Marco was like freaking out and we needed help, like, from a grown-up," Liam tried to explain.

Oliver grunted, "*Ugg.* I'm actually glad I'm moving. You two are the *worst* friends ever."

"We're sorry," Marco pleaded. He really wanted to explain their perspective of things. It wasn't meant to be a move against Oliver. Things weren't black and white.

"Just leave," Oliver looked out his window. The boys stood frozen, unsure what to do. Marco started to speak, but before he could get a sound out Oliver shouted again. "LEAVE!" He threw out his arm and pointed to the door, still refusing to look at them.

They slowly approached the exit. As Marco quietly closed the door he added, "Mr. Pikake has a plan. We are leaving. Tomorrow. 8 a.m. At the library."

Oliver didn't flinch. He didn't speak. He didn't seem to care.

* * *

The boys approached Marco's front door. As Liam grabbed for the knob, Marco pulled him back. "*Um.* So . . ." he started. "We can't tell

Maggie or Penelope about the box, or my secret meetings with Fredrick."

"No way! Why? That's a critical part of this catastrophe."

"I just can't have Maggie mad at me for keeping things from her."

"You realize that's the entire story, right?"

Dread filled Marco's stomach. Everything was going to fall apart. He lied to his sister *and* Penelope. There was no secret project for Maggie. Everyone was going to hate him. "I know." He kicked at the ground.

Liam nodded and the two jogged up the stairs, they could hear Maggie before they even reached the landing.

"Did you ever notice in movies and stuff that when a plane is crashing, everyone starts running away? Like, who are these crazy people who think they can *outrun a plane?* It literally makes no sense. The plane is traveling so fast, they are running *towards* where the crash site will eventually be. It's so obvious. I swear, just wait. Hollywood will be calling me to help them out." Maggie jumped as Liam and Marco burst into her room. "What is *wrong* with you?!"

Marco locked eyes with Penelope. He was glad to see her. *Great, all of KFUN is here, we can get this over with,* he thought. A horrible monster took hold of his intestines as he realized he had already cast Oliver out of the group. Another death. Shaking the thought from his head he started to speak, but Liam snatched the honor of filling the girls in.

Maggie's mouth hung open the entire story. By the end, she was giving her brother her *evil-eye*. Marco was thankful for Liam. He summed things up beautifully, telling a tale that had just enough truth mixed in and saved Marco from having to lie to their faces. Their grandfather had left a box that held an important secret and Marco allowed it to be stolen by a crazy villain who wants to end the world, and now they have to stop him. Oh, and Oliver's moving.

"Well. I guess we are off to fix Marco's mistakes," she announced dramatically. "What about mom?"

"Mr. Pikake is handling the parentals. Something about a field-trip," Marco replied never making eye contact.

"My parents will be thrilled to hear I am going on a STEM-trip!" Penelope beamed.

As they fell asleep that night, excited and nervous about the adventure that lay before them, they had no idea of the difficult trek that was to come. And while they prepared themselves for the possible dangers and obstacles they'd face, none of them ever considered that out of the four of them, only three would make it back home.

THE QUADRATIC RECIPE

An alternative proof using the power of substitution!

Consider $ax^2 + bx + c = 0$ where a, b, c are real numbers

Cast a proliferation spell with $4a$

$$4a(ax^2 + bx + c = 0)$$
$$4a^2x^2 + 4abx + 4ac = 0$$

Use substitution

Let,

$$R = 2ax$$
$$R^2 = 4a^2x^2$$

Then,

$$(2ax)^2 + (2ax)2b + 4ac = 0$$
$$R^2 + 2bR + 4ac = 0$$
$$R^2 + 2bR = -4ac$$

Complete the square

We already have R^2 and two bR terms, so we need to add b^2 to both sides

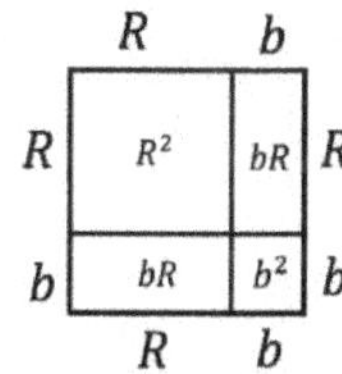

$$R^2 + 2bR + b^2 = b^2 - 4ac$$
$$(R + b)^2 = b^2 - 4ac$$
$$\sqrt{(R + b)^2} = \pm\sqrt{b^2 - 4ac}$$
$$R + b = \pm\sqrt{b^2 - 4ac}$$
$$R = -b \pm \sqrt{b^2 - 4ac}$$

REMEMBER

The Quadratic Recipe is a power drill.

Sometimes a softer touch will save us time and a headache!

Use wisely.

Solve for x

$$2ax = -b \pm \sqrt{b^2 - 4ac}$$
$$\frac{1}{2a}\left(2ax = -b \pm \sqrt{b^2 - 4ac}\right)$$
$$\frac{2a}{2a}x = \frac{-b \pm \sqrt{b^2 - 4ac}}{2a}$$
$$x = \frac{-b \pm \sqrt{b^2 - 4ac}}{2a}$$

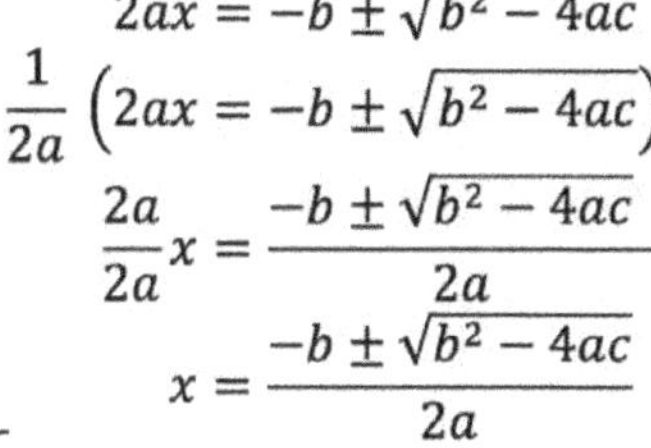

$$x = \frac{-b \pm \sqrt{b^2 - 4ac}}{2a}$$

The anchors of our formation – the two soldiers assigned to ground zero

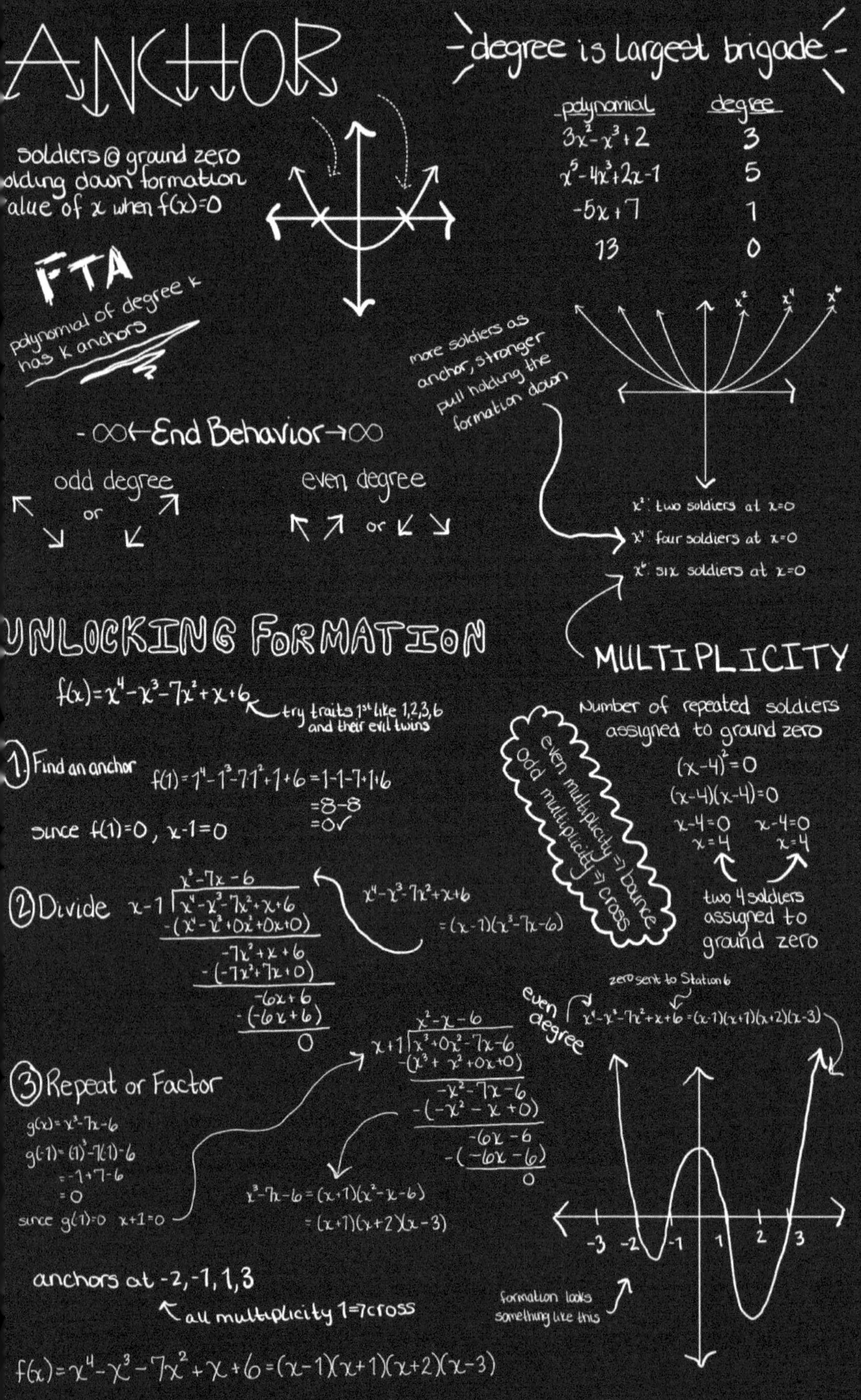
ANCHOR
soldiers @ ground zero
olding down formation
alue of x when f(x)=0
degree is largest brigade
polynomial degree
3x²-x³+2 3
x⁵-4x³+2x-1 5
-5x+7 1
13 0
FTA
polynomial of degree k has k anchors
more soldiers as anchor, stronger pull holding the formation down
x²: two soldiers at x=0
x⁴: four soldiers at x=0
x⁶: six soldiers at x=0
-∞←End Behavior→∞
odd degree
or
even degree
or
UNLOCKING FORMATION
f(x)=x⁴-x³-7x²+x+6
try traits 1st like 1,2,3,6 and their evil twins
1. Find an anchor
f(1)=1⁴-1³-7·1²+1+6=1-1-7+1+6
=8-8
=0✓
since f(1)=0, x-1=0
2. Divide
x-1) x⁴-x³-7x²+x+6
x³-7x-6
-(x⁴-x³+0x²+0x+0)
-7x²+x+6
-(-7x²+7x+0)
-6x+6
-(-6x+6)
0
x⁴-x³-7x²+x+6
=(x-1)(x³-7x-6)
MULTIPLICITY
Number of repeated soldiers assigned to ground zero
even multiplicity ⇒ bounce
odd multiplicity ⇒ cross
(x-4)²=0
(x-4)(x-4)=0
x-4=0 x-4=0
x=4 x=4
two 4 soldiers assigned to ground zero
3. Repeat or Factor
g(x)=x³-7x-6
g(-1)=(1)³-7(1)-6
=-1+7-6
=0
since g(-1)=0 x+1=0
x+1) x³+0x²-7x-6
x²-x-6
-(x³+x²+0x+0)
-x²-7x-6
-(-x²-x+0)
-6x-6
-(-6x-6)
0
x³-7x-6=(x+1)(x²-x-6)
=(x+1)(x+2)(x-3)
zero sent to Station 6
even degree
x⁴-x³-7x²+x+6=(x-1)(x+1)(x+2)(x-3)
-3 -2 -1 1 2 3
formation looks something like this
anchors at -2,-1,1,3
all multiplicity 1⇒cross
f(x)=x⁴-x³-7x²+x+6=(x-1)(x+1)(x+2)(x-3)

17

THE TREK

"MATHEMATICS IS NOT A CAREFUL MARCH DOWN A WELL-CLEARED HIGHWAY, BUT A JOURNEY INTO A STRANGE WILDERNESS, WHERE THE EXPLORERS OFTEN GET LOST."

-W.S. ANGLIN

Rummaging around his room, Marco packed a sweatshirt, some water, and his journal. He wasn't really sure what to bring on this kind of adventure. They were after an evil madman; how do you prepare for that? Maggie grabbed her little pillow, she took it on every trip, and a small notebook with a pen attached. She then stuffed in a first aid kit, a baseball cap, extra socks, toilet paper, a toothbrush, a flashlight, and sunscreen. The two met in the hallway, and, careful not to wake their mother, quietly crept downstairs.

They arrived in the library parking lot at 7:45 a.m. Mr. Pikake was already waiting for them.

"Perfectly punctual!" He was enthusiastic and optimistic, which put Marco at ease. "Ready to begin our journey?"

All Marco gave was a nod. He still felt horrible not connecting the dots sooner. Not only was he keeping secrets, but he was going behind his tutor's back with his own brother. He thought about how he'd feel if Liam and Maggie were sneaking around together. *Ugggh*. It was a gross and disturbing image.

Penelope came trotting in like an angel. She looked like Professor Shelly from *Jumanji*. Sporting hiking boots and dressed in all shades of khaki, she was clearly ready for adventure. Marco looked down at his beat-up sneakers, the same ones he wore every day, paired with jean shorts and a white t-shirt, and seriously reconsidered his life choices. Only minutes later, Liam slothed in, his long arms flopping along like the ears of a basset hound.

A large black van pulled into the lot and came to a screech in front of the group. An Asian woman hopped out of the driver seat. She wasn't tall or short, somewhere in between. She moved with a clear grace, like she was floating on air, while at the same time projecting authority and strength. Her soft features seemed to be covered up by a hardness, a confidence, in her expressions.

"This is Dr. Pound," Mr. Pikake introduced. "She takes regular trips to Phaseville and will be our guide."

Dr. Pound gave a slight bow but didn't say a word. She grabbed the gear and began packing it into the trunk. "We've got to get going Blaise," she said to Mr. Pikake sternly.

One-by-one, they piled into the van. Marco took a last look around the parking lot before pulling himself in. A large smile bloomed on his chin when he saw the tiny little dot jogging on the other side of the lot. It was Oliver.

"Gonna leave me behind?" he huffed as he approached the van.

"Never," Marco grabbed his arm and thrusted his friend inside the vehicle.

As the group pulled away, Mr. Pikake called back from the front seat, "I don't know what to expect, but I'd feel a lot better if you all knew more about polynomial orders."

Marco looked at his watch. It was 8:07 a.m. Too early. Much too early for a math lesson, *especially* in the summer. He glanced at Dr. Pound; her face was frozen in a frown. There was no way she'd be turning on the radio anytime soon. With nothing else to keep them entertained, they begrudgingly agreed.

"A quadratic, the order that forms a parabola, is a second-degree order," the tutor explained.

"Like a burn?" Liam asked. "First degree isn't bad right?"

"*Aye,* like a burn. A first-degree polynomial is a line, a march. 'Tis not necessarily the most dangerous of formations."

"I don't know, Blaise. I've seen some pretty intimidating marches." Dr. Pound chuckled. Marco was shocked, even her laugh sounded daunting and frankly a bit scary.

Still snickering at what seemed to be some sort of adult-hunter humor, Mr. Pikake continued, "The type of order of a polynomial is the largest brigade. In $f(x) = x^3 + 2x^2$, it is a third-degree order. The largest brigade is the cube."

"Alright." Liam had a notebook pulled out, transcribing the entire conversation. Marco was yanked back in time to last fall when the two were studying for a history exam. Liam opened his textbook to reveal a sea of neon colors, he had highlighted nearly every word. Trying to explain notetaking, Marco pointed out that you are only supposed to highlight the important stuff. Liam claimed it all seemed important. "Degree is the biggest brigade."

"Alright then." Mr. Pikake's Cheshire grin grew so large they could see it from behind. "What about $x^3 - 4x^2 + 5x^6 - 2$. The degree?"

"He's trying to trick us," Maggie sassed. "He said the brigades out of order."

"Yeah!" Oliver chimed in. Marco felt a wave of relief as things appeared to be back to normal . . . somewhat. He was thrilled to have his friend by his side and realized he needed to treasure this time. Oliver leaving was like a death. It wouldn't be long before something as simple as sitting next to his best friend would be impossible. "If we

$$(8x^2 - 103x + 174) \div (6x - 12)$$

rearrange the terms from the biggest to the smallest brigade, it would be $5x^6 + x^3 - 4x^2 - 2$ meaning . . ."

Penelope picked up the ball, "Meaning, the order is six. It's a sixth-degree polynomial."

"Perfect," the professor purred. "As you are already skilled at quadratics, second-degree orders, mastering higher orders will not be too difficult. We must know the degree." His first bony finger popped up, "and second, we must identify the end-behavior."

"What's that mean? End-behavior?" Marco questioned.

"The end-behavior is how the order acts at the far edges of the universe. For very, very small Numberfolk and very, very large ones."

"You mean towards infinity?" Maggie was bouncing in the chair, the seatbelt repeatedly pulling her back down.

"Exactly! Start with a line, a march, an order of degree 1. What happens on the tips?" Mr. Pikake folded his hands to look like duck shadow puppets kissing before pulling them apart to stretch invisible slime.

"On one side, it is going up and up and up, and on the other it is going down and down and down," Liam answered.

Pulling out a dry-erase marker from his jacket pocket, Mr. Pikake scribbled on the windshield.

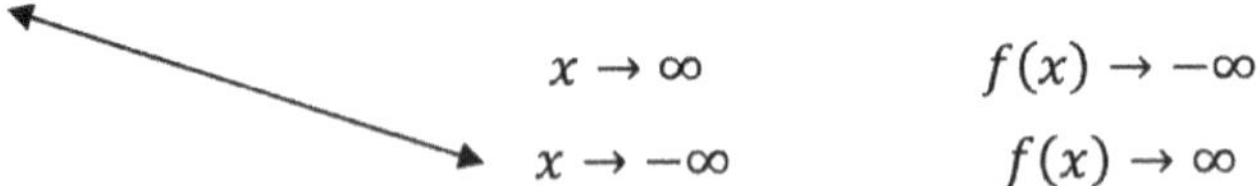

"That it is! You see as the Numberfolk soldier, x, approaches infinity, or as our soldiers get bigger and bigger, their position, y, is getting smaller and smaller. On the contrary, as the soldier is smaller and smaller moving through TNZ towards negative infinity, their position is higher and higher, larger stations."

"They are opposites," Marco noticed. "When the soldier is big, the location is small, and when the soldier is small, the location is big."

"But what about the other way? That line has a negative slope, it's falling. What about a march in the positive?" Maggie butted in.

Mr. Pikake drew a line with a positive slope. "You tell me."

"It's not the same thing!" Maggie chirped. "The positive slope means that as the soldiers get big, the position gets big, and as the soldiers get small, the position gets smaller too."

"'Tis both the same and different," the professor sighed. "No matter if the line is climbing up or falling down, the ends are opposites. One side down, the other up. Unless of course it is a zero-degree line, horizontal."

"I see a pattern," Penelope hinted. "Think about a second-degree polynomial. While a line has to fall on one side and grow on the other, the quadratics, parabolas, they always go the same way at the ends!"

"She's a smart one," Dr. Pound mumbled under her breath without taking her eyes off the road.

"They all are," Mr. Pikake whispered before booming back to standard volume. "Perfection Penelope! Our parabolas, be it a smile or a frown, always point the same way as we consider the soldiers towards both positive and negative infinity. What of a third-degree order? What we call a cube?"

The children started chattering in the back rows while the professor added to his window drawing.

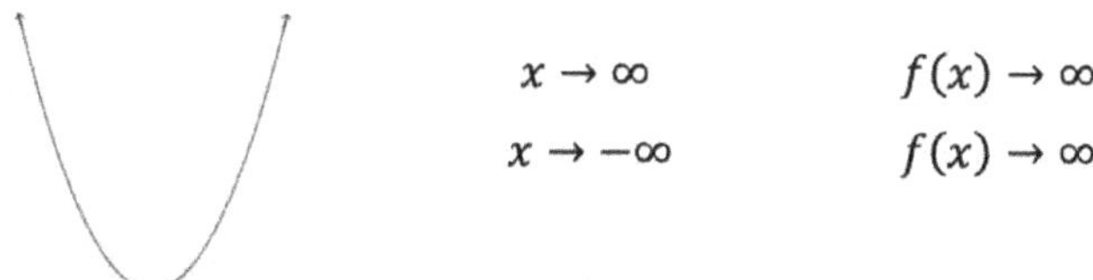

"Don't you think that impacts her ability to see the road?" Maggie nudged at Penelope. The girls sat side-by-side in roomy second-row bucket seats while all three boys were squished on the back bench.

"How are we supposed to answer this when we don't have any idea what a third-degree formation even looks like?" Liam whined.

"I don't think we need to," Penelope looked over her left shoulder. "Say the soldier is some really small Numberfolk, like negative one trillion. The x^3 will be a negative times a negative times a negative, which gives us another negative. That means the soldier negative one trillion is sent . . ."

"To the underworld!" Oliver interrupted. "They are sent farther down than we can even imagine. On the other side, really big positive soldiers will be sent to super high stations. That means whatever a third-order looks like, it is similar to the line, up on one side, down on the other."

"Yeah but," Maggie whispered to Penelope, Marco let out a sharp snicker. "*Ugh.* I get that if it was just x^3, but a third-degree can have other brigades too, like $x^3 + 1000x^2$. Couldn't the other terms catch up? If the soldier is negative one trillion, then negative one trillion squared is a positive, that's multiplied by another thousand . . . you see what I mean? Couldn't that end up being positive?" She finished and stuck her tongue out at her brother.

"It's too slow," Dr. Pound called. Maggie's eyes grew wide. Trying to broadcast her thoughts using telepathy, she stared at Penelope and concentrated very hard. *Does she have superhuman hearing?!* To everyone else, it looked like Maggie was trying to see something in the distance as she was intensely squinting. Penelope smiled. Maggie wondered if it worked. She couldn't say anything, so she just had to remember for later.

"Even if there are other brigades," Dr. Pound continued, "they are too slow, they can't catch up. Sometimes they can, for small soldiers. The order $x^3 + 1000x^2$ for soldier -100 does end up at the positive station 9,000,000, but this can only happen near the origin. After -1000, anyone smaller can't catch up and are sent to negative stations. The largest brigade has too much power, no matter what, as the Numberfolk get infinitely small, the leader will dictate what happens."

"I think I get it," Penelope exclaimed. Pulling out a marker from her bag, she wrote on the window to her right. Maggie's blood boiled. She was secretly *dying* to write on the windows, but it had not even crossed her mind to bring a dry-erase marker . . . on a camping trip . . . to the woods.

$$x^6 + x^5 + x^4 + x^3 + x^2 + x$$

"See, the leader in this order, x to the sixth, has more power than everyone behind them combined. But only for numbers with really big magnitudes like negative one billion and positive one billion. Because the x soldier is so big, or small, multiplying by it again is much more impactful than adding up the rest. Multiplicative power is insanely mighty."

"So, we're right?" Liam prodded the professor. "For third-degree orders, the far tips are like a line? One side up, the other down?"

"That you are!" Mr. Pikake bellowed. "What of fourth order? Fifth? Can you pinpoint the pattern?"

"Odd degrees will always be opposite. A line is first-degree, a cube is a third-degree, you'll always have a negative bringing it down. Even degrees will always be the same, like a smile, because even for really small negatives, when you multiply them an even number of times, each has a pair and becomes positive." Oliver spoke with clarity and elegance. It gave Marco an icky feeling.

"Yes!" the professor shrieked. "Now, recall all polynomial orders are the product of lines. The degree dictates how many. A first-degree is one line, a second-degree is two, third-degree is three, and so on and so forth. For each line, one soldier *must* be assigned to ground zero. With this, you can determine the shape of any order."

This was her chance. Maggie held out her hand and Penelope understood, passing over the marker. "I get it," Maggie announced proudly. "Every soldier at ground zero brings the formation back towards the axis. That, along with what we know they are doing on the far ends, means we have a good idea of what each will look like."

She began swiping at the left rear-window like a painter with their canvas.

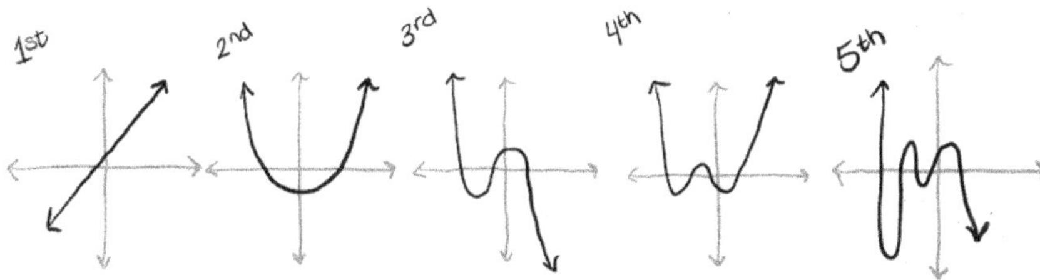

As she worked, she hummed the *Jeopardy!* song, again and again like a skipping record. Marco had enough.

"Okay. I think we get it." He looked around coaxing his friends to nod in agreement. "But there are so many ways to make that shape." He leaned forward and snatched the marker from his sister. Marco was in the middle seat without a window to write on. He leaned over Liam and stretched to reach the tiny half window in the back that didn't even roll down.

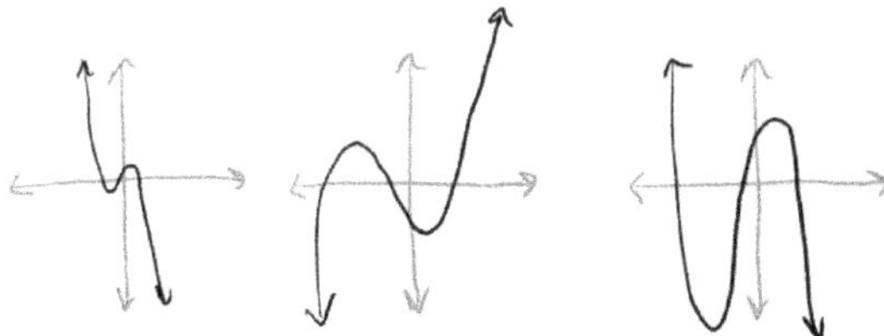

"These are all third-degrees, but they look *a lot* different." He started to draw another, but Liam shoved him back to the center.

Craning his neck to see Marco's formations, Mr. Pikake let out a sharp groan, "Incantations. They work on all functional orders, not only polynomials. You can shift it up, down, left, right, stretch it or shrink it and flip it to create a cacophony of commands. Also, who is left at ground zero will further determine your diagram. But the basic shape remains the same."

"You mean if I want to take x^3 and make it tall and skinny I could just change it to $3x^3$?" Liam was ecstatic. It was a powerup that automatically extended to every additional level. They could keep throwing function after function at him, each new order increasing in difficulty, and he could twist it and change it at his will — he already knew how they worked.

"Beautiful, isn't it?!" the professor admired. "With polynomials, your power is in the soldiers at ground zero. Like a river, these are the key crossings."

"Well . . ." Penelope started. "What about orders like x^3 or x^2? Those don't actually cross three times or two times, just once."

"Multiplicity!" Dr. Pound called. For the first time since meeting her, Marco saw her smile. She turned her head away from the road to face Mr. Pikake for only a moment and they flashed each other mischievous grins.

"What's multiplicity?" Oliver called out.

Interesting, Marco thought. Something was finally new to his friend. It felt like Oliver had been participating in secret, extra training sessions increasing his abilities at an alarming rate.

"Every polynomial is the product of lines, linear factors. The number of lines is the degree. But!" the professor popped. "A few key considerations are vital. First, I must provide a warning." A deep hush spread through the vehicle as everyone listened closely. As fun as this outing seemed, they were all aware of the danger that came with trying to stop a madman. "Recall before I mentioned ground zero is an entire wall of defense. To us, it looks to be but a line. Keep in the back of your mind there are things we cannot see."

No one said a word. Marco had no idea how he could fight the invisible, and that worried him.

"Number Two! Recall we may have multiple identical soldiers as anchors."

"Like in $(x-2)^2$," Maggie chimed in. "When $(x-2)^2 = 0$, we have $x - 2 = 0$ *and* $x - 2 = 0$."

"Two twos!" Liam spat. It never got old.

"Precisely! We call this multiplicity. It is the number of identical soldiers assigned to ground zero. You can think of these like miniature formations within the larger army. What does $(x-2)^2$ look like?"

"A U," Penelope sang. "It's a parabola, a polynomial with degree 2 and so both ends point the same way, in this case up."

"Yes! And thus, since $x-2$ has occurred twice, it has multiplicity two and we know the formation will bounce off ground zero at this soldier. Even multiplicities bounce, odd multiplicities cross. Finally, these soldiers, anchors, create a gravity so to speak. The more soldiers ambushing an anchor post, the stronger the pull. This causes the formation near ground zero to flatten out. I shall show you." He leaned forward in his chair, stretching his seatbelt to the max as he cleared off the windshield. "Look here,"

$$f(x) = (x-1)^2.$$

"Who is at ground zero?"

"One." The word shot out of Oliver's mouth with impressive force.

"Yes. The two lines are both $x-1$ and when the soldier, x, is 1 their location is zero."

$$f(1) = (1-1)^2 = 0.$$

"The even degree of two tells us at the ends of the formation, the locations are going the same way."

"Up, up, and away!" Oliver sang.

"Quite true. Thus, at $x=1$ the formation 'bounces' off ground zero. What would,"

$$g(x) = (x-1)^3$$

"look like in contrast?"

"All its soldiers at ground zero are 1s too," Maggie explained. "However, since it is a third-degree, one end goes up and the other down, so it has to cross at $x=1$. Since there are three soldiers at the same place, you also said it will be flatter, more pull." She attempted to swipe the marker from her brother's hand, but he flinched away just in time. She gave Marco an evil look accompanied by a low growl.

Mr. Pikake drew the formations on the windshield. He used a thick solid line for the second-degree and a dashed line for the third.

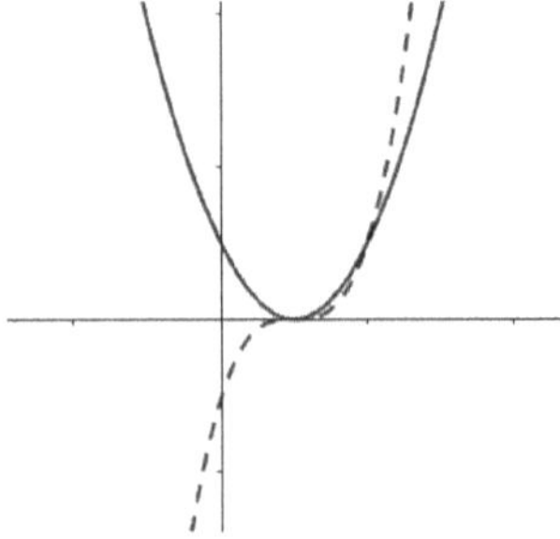

"Wait," something wasn't making sense to Penelope. "That means a fifth-degree doesn't actually have to cross five times?"

"Give it a try. Sketch

$$h(x) = x(x-1)^2(x+3)^2."$$

Penelope looked to Marco who gladly handed the marker back with a smile.

"Okay. The soldiers at ground zero are $0, 1$, and -3." Penelope drew a large plus and marked off her three soldiers. "Since it is a fifth degree, one side will go up and the other down." She paused and thought for a moment. "Negatives will be negative, so it starts down on the left. There is just one x factor, so at zero it crosses because one is odd. It must bounce at the other anchors." In one smooth and fluid motion, she traced out the curve. "Like this?"

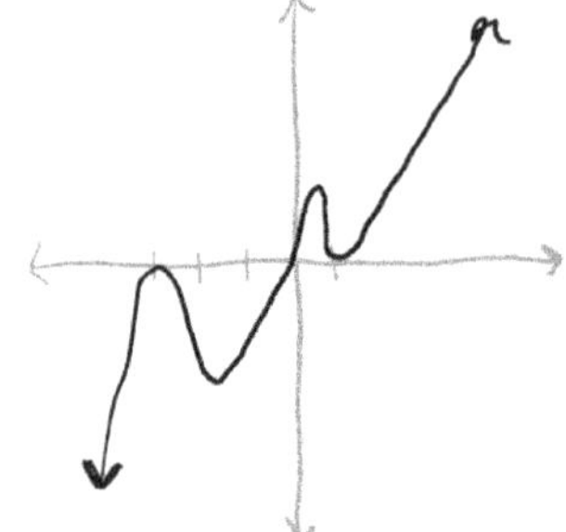

Mr. Pikake peered over his right shoulder to examine the formation. "Not bad," he chuckled. "Not bad at all."

"Wait, if we can't just count the soldiers at ground zero, how are we supposed to figure out the order?" Liam pressed.

"*Ahh.* But you can," Dr. Pound said optimistically. "Look at her formation, the end behavior tells you it is an odd degree. Two of the intersections with ground zero bounce, so you know those are even, and the final one crosses making it odd. Your guide has already provided that the flatter the formation near ground zero, the more soldiers assigned there. With all of this you can develop a very good, if not perfect idea of the order."

"Dr. Pound is correct. One wrench is when our anchors are not on the line of ground zero, but rather lie on the defensive wall. The mythic phantoms. Then, you must be a detective, gather clues, piece together the puzzle. All skills you each possess." A devilish smile grew as he turned towards the backseat, "Covet a competition?" the professor posed.

The next thirty minutes were filled with laughter, arguing, and cheers. Mr. Pikake drew out polynomial after polynomial on the windshield, the first to correctly shout the degree of the order won a

point.[*] Penelope and Oliver battled for the lead, with Maggie close behind. Marco and Liam could figure it out, but simply weren't fast enough to match their peers.

"Well done. Well done," Mr. Pikake boasted.

"We've got about an hour left, best get some rest. When we arrive, it will be a long hike," Dr. Pound suggested and the passengers obeyed. Stuck in the middle, Marco struggled to find a comfortable position. He leaned forward and cradled his face in his hands.

Heat began to rise through his body. Sweat dripped from his forehead. Opening his eyes, Marco found himself inside a volcano. The only colors were shades of black and orange. Large rocks surrounded him, jutting up from the ground. He saw an opening to his right; stones had been placed to form an arch with the words *Abandon all hope ye who enter here* etched above.

As he approached the gate, he saw Oliver. A warbled mess of sounds and tones shot from his friend's mouth. It was as if a DJ had messed with the controls inside his ears. Although he couldn't understand anything Oliver was saying, he somehow knew what he was telling him.

Oliver explained he was Marco's guide. He would show him the way, and at the end of their journey, Marco would need to make a choice.

As they walked, Marco couldn't help but notice the formations that surrounded them. The first chamber was small. A large boulder blocked off the ceiling creating a limbo bar they had to crouch under. Marco made the mistake of grazing the rock and a horrible lightning bolt shocked through him. They exited to a deep cave that looked like a bowl turned on its side. People were huddled within. They grabbed onto Marco pulling him back, insisting there was nothing outside the cave; he'd be safe if he stayed in the shelter of the bowl for all of eternity. Oliver fought them off and successfully pulled Marco out only to stumble onto a terrifying cliff. In front of them, it sunk down toward a horrid pit that housed a ferocious beast. Something took hold of Marco. He had an urge to step off and allow the monster to devour him whole. It was as if everything inside of him was saying he deserved to be dinner.

Oliver pushed him away from the edge and pointed up. In perfect

---

[*] You can play too at www.MathBait.com/polyplay

$(3x^2 - 36x - 192) \div (4x + 16)$

symmetry, a mountain soared toward the sky as sharp and steep as the pit but in the opposite direction. They began to climb. It seemed like they would climb forever, the top of the mountain was out of sight. Luckily, Oliver found a tiny hole and pushed Marco inside.

The two were in a crater between two large mountain peaks, the fourth trial, as if a higher power carved a W or perhaps a M, depending on which way you looked, into the Earth's body. The glow of the lava hit Oliver's face which stopped Marco in his tracks. It wasn't his friend at all! It was a demon wearing Oliver like a meat suit. Enraged, Marco attacked. Oliver was victorious and drug him, kicking and screaming, to the cliff. The anger wore off as soon as they reached the peak. This evil place was playing with Marco's mind.

Oliver instructed him to jump. He didn't want to. He begged and pleaded with his guide. They could just turn back, return to where they came from. The cave people seemed so happy and so safe, they would certainly provide the two refuge. Oliver insisted. He took his friend's hand, and together they plunged down into darkness. It caught them like a baseball mitt, and they slid. After the sharp drop, they flew up before completing another quick dip and rise. Then came the plunge, it flung them down like an impossibly long slide.

They emerged in the sixth trial. It looked just like the earlier chamber, a set of mountain peaks, but this time there were three. He looked around and couldn't find Oliver anywhere. He was lost, panicked. A helplessness took over his entire body and he cried. As the tears fell from his face, he watched as they joined together, one at a time, in the grooves of the mountain creating a stream following the path of least resistance. Jumping up, Marco followed the tears to discover a concealed door.

Accepting he was alone, he entered the room. Everyone was there. Mr. Pikake and Fredrick, Liam and Penelope, Maggie and their mother. Oliver had returned as well. Was this the test? He could hear a clicking in the background. Click. Click. Click. He counted, there were eight total, eight clicks. The chest. A bright light shone down from a hole overhead. It highlighted a small table. He began to approach it when the room started to shake.

"Marco! Marco!" He woke up to Liam in his face. "Dude, we're here."

The group piled out of the van. It was still bright, plenty of daylight left. Marco didn't feel right. He wasn't sure if it was the mountain air

$$(-4x^2 + 37x - 33) \div (2x - 2)$$

or the residue of the dream, but he pressed forward.

"The map?" Dr. Pound looked to Mr. Pikake who pulled out the wrinkled graph paper and passed it along. "This is a map?" Her tone worried Marco. If she didn't know how to decode the sheet, they were doomed before they even began.

The adults placed the paper on the hood of the van and spoke in hushed tones pointing and shaking their heads. Liam had opened the back and was passing out backpacks to everyone. Dr. Pound rejoined them and pulled a huge hiking duffle onto her shoulders.

"The first portion of this map is the traditional route to Phaseville." She sounded like a military sergeant barking commands. "I suggest we don't follow it. There are a lot of things in these woods. We'll take the back way, it's faster too. Once we are close, we'll hop back on the trail as we'll need these landmarks to find the destination." Without further ado, she took the lead, and everyone fell in line marching behind her. Mr. Pikake held back to sandwich the children between their guides, Marco walked beside his tutor.

"I'm really sorry. I didn't remember Fredrick was your brother's name," Marco started in a low tone. He had really messed up and he knew it.

"It's alright son. I can't lie, I am hurt. You have a lot going on and I understand the weight of your burden."

"Do you want to talk about it? Talk about him?"

"I don't know what there is to say. I told you everything before. Fredrick loved the Numberfolk and refused to join the SAN. I haven't seen him since that day. But . . ." a small and weak pop, "not a day goes by when I do not think of him. What he's been doing. How he is. You two have a lot in common you know. You remind me of him."

I have a lot in common with Fredrick? Marco shivered. He didn't want it to be true. "What do you mean?"

"Well . . . you are both the older sibling. That comes with a certain feeling of responsibility. Plus, I have seen you and Maggie many times." He chuckled. "She looks up to you just as I looked up to Fredrick. He was my big brother, I idolized him."

Maggie looks up to me? Marco was sure Mr. Pikake was joking. Maggie was headstrong, brave, so smart. She would rather walk through fire than have another lead her. A smile grew on Marco's face as he realized how proud of his sister he was. With all her annoying antics, she was still pretty amazing.

A swift flash in the corner of his eye pulled Marco from his thoughts. Something was running behind the trees. He grabbed onto the professor's arm and pointed.

"What? What is it?" Mr. Pikake asked, looking into the distance.

"I saw something."

The group noticed their caboose was trailing behind and slowed to a stop.

"What's going on?" Liam yelled back.

Dr. Pound quickly brought her finger to her mouth in one sharp movement indicating for everyone to quiet down. She jogged to the end of the line. "What's the situation?" she whispered.

"Marco has seen something," Mr. Pikake responded pointing in the direction of the movement.

After a moment of silence Dr. Pound responded, "I don't see anything."

"It must be gone," Marco sighed. "Maybe a deer or something."

With a concerned look Dr. Pound turned to Mr. Pikake, "It has begun. Watch him." She jogged back to the front of the line and waved her arm for everyone to continue forward.

Watch me? A bolt of nerves ran through Marco. "What does she mean?" he asked his tutor.

"Nothing to worry about." He forced a kind smile and nodded for Marco to keep walking.

"I have a question!" Maggie announced cheerily. "Say we have an order like $f(x) = 2x^4 + 5x^3 - 8x^2 - 17x - 6$." She looked around to make sure everyone heard her and was following along. "I know that since it is a fourth-degree, it makes a U-like shape but bumpier in the middle because it has more soldiers assigned to ground zero. I also know that just like with quadratics, I can find where zero is sent because $f(0) = -6$. But how do we find who is at ground zero, on the x-axis? I can't complete the square because it isn't a square."

They approached a small stream and followed Dr. Pound's lead as they refilled their water bottles. She went around and added a tiny drop of a brownish substance to each of their drinks. "Iodine," she explained, "makes the water safe to drink."

Taking a long stick, she etched the order into the dirt.

$$f(x) = 2x^4 + 5x^3 - 8x^2 - 17x - 6.$$

"Now, we know since this is a fourth order that it is made from the combination of four lines."

$$f(x) = (a_1x + b_1)(a_2x + b_2)(a_3x + b_3)(a_4x + b_4).$$

Gagging, Liam asked, "What's *that*?"

Dr. Pound continued, "The a's and b's are just the Numberfolk we don't know yet. Descartes discovered that if we count the sign changes, it will give us information about how many real positive and how many real negative soldiers are at ground zero."

"The fly guy?" Liam was secretly hoping at some point he could convince Mr. Pikake to reveal the mysteries of his supernatural government-programmed spy. It had been rattling around his brain since the insect made its first appearance.

"*Aye*. The 'fly guy'." Mr. Pikake took over. "Line up the brigades from largest to smallest, then count the changes from positive to negative."

"They are already in the right order, $x^4, x^3, x^2, x, \ldots$ and, the only change is from $5x^3$ to $-8x^2$," Penelope pointed to the order on the ground. "So, there is just one change." She drew the sign of each term.

$$\begin{matrix} + & + & - & - & - \end{matrix}$$
$$f(x) = 2x^4 + 5x^3 - 8x^2 - 17x - 6.$$

"Indeed," Mr. Pikake confirmed. "Descartes found the number of real positive Numberfolk soldiers at ground zero is either equal to the number of sign changes *or* less than the number of sign changes by a multiple of two."

"Huh?" Marco had no idea what the professor was saying and why did he keep saying *real*? Were there soldiers who weren't real? Phantom armies that could rise from the dead?

"Imagine there are 6 sign changes," Dr. Pound clarified. "That means there are either 6 real positive soldiers at ground zero, or any number less than 6 by twos. Meaning there could be 6 or 4 or 2 or none at all."

"Well, that seems absolutely worthless," Oliver scoffed. "That's a lot of options."

"That it is," Mr. Pikake sighed. "But! There is a second part to his discovery."

"What is it??" Liam urged.

"Consider if the order was flipped over center stage. What would happen? Count the sign changes once more of this altered formation."

"Flipping it over center stage is $f(-x)$," Maggie announced.

$$f(-x) = 2(-x)^4 + 5(-x)^3 - 8(-x)^2 - 17(-x) - 6.$$

"Oh yeah!" Penelope joined in. "Now, all the even brigades won't change because the negatives are in pairs, but the odd brigades will swap signs." She sketched below Maggie's addition,

$$f(-x) = 2x^4 - 5x^3 - 8x^2 + 17x - 6.$$

Marco stood next to Penelope, "Now there are . . . One. Two. Three. Three changes." He drew in the signs.

$$+ \quad - \quad - \quad + \quad -$$

$$f(-x) = 2x^4 - 5x^3 - 8x^2 + 17x - 6.$$

"From positive $2x^4$ to negative $5x^3$, from negative $8x^2$ to positive $17x$, and from positive $17x$ to negative 6."

"Yes. And the same idea applies here. The number of sign changes in $f(-x)$ tells us how many negative real zeros there are or less by an even number. In this case, there are either 3 real negative soldiers at ground zero, or 1."

"Okay, I guess that sort of helps." Maggie protested, "I mean it has to be three negative soldiers and one positive because there was only one sign change in $f(x)$ so there must be one positive anchor. Together that adds up to a fourth-degree. But it doesn't actually tell us anything about *who* these soldiers are."

"Just rip it apart," Dr. Pound replied forcefully.

"What's that supposed to mean?" Oliver pushed.

"Look, your order is a fourth degree. That means if you rip out one line, you are left with a third degree."

$$f(x) = (ax + b)(cx^3 + dx^2 + ex + f).$$

"All you need is to find a single soldier at ground zero, then you can keep tearing it apart until you have all four lines, and thus, all four soldiers." She looked up at the sky. "We've gotta get moving. I have a camp site not too far ahead where we can rest for the night."

They all filed back into line and continued walking.

"What does she mean rip it apart?" Maggie whispered to Penelope.

"I'm not sure. My best guess is she means to divide. Since all the lines are multiplied, to undo it you'd need to divide. Just like you would with a number and finding its factors," Penelope shrugged.

Marco had been trying to keep up, but he was beginning to feel sick. It was like he just stepped off the carnival ride that spins

everyone in a circle so fast they stick to the walls, the inertia holding them back. He was nauseous and dizzy.

Something sprinted through the trees, this time closer. It was so fast all Marco could make out was a blur. He tripped and landed firmly on his face.

Mr. Pikake knelt down and grabbed Marco's hands. "What's happening son?" He sounded worried.

"I don't feel so well." He gulped trying to hold back the vomit. "I saw another one, that way. It's going fast."

"Was it a number?" Mr. Pikake pressed. The group was still walking getting farther and farther away.

"A number?" Marco was confused. Why would there be numbers way out here? There weren't any clocks, or speed limit signs, or jerseys, or any of the other worldly items that held their beady little eyes. "I. I. I don't think so? It was so fast, just a blur. Something running through the woods."

Mr. Pikake helped pull Marco to his feet. "You're phasing. The void is thin up here, it's only going to become more fragile. I believe what you are seeing is the Numberfolk plane."

Great. This was just what he needed. They were following some cryptic map to save the world from a lunatic, and now Marco was hallucinating. The two climbed up a steep hill and found the rest of the group had settled in a small clearing.

Dr. Pound pulled out a tent from her backpack and everyone chipped in to set it up. She ordered them out to collect wood, and they lit a campfire right as the sun set behind the large cliffs to their west. Marco looked around the flames and felt a peace wash over him. His sister and Penelope giggled quietly across from him, Dr. Pound and Mr. Pikake were deep in conversation to his right, and Liam was chatting about something to his left. He realized how lucky he was to have such an amazing group of friends, of family.

"Marc. Marc-O," Liam was poking at him. He snapped out of his head. "Where's Oliver?"

Marco felt horrible, he had forgotten about his friend. He was already mentally preparing for Oliver's move, his removal from their close-knit group. Scanning the tree line, he saw Oliver leaning against a tall birch. Pushing himself up, he followed Liam toward the forest.

"Dude! You've got bars?" Liam called out and Oliver quickly tucked the phone away. "I haven't had any since we left the van. You must have killer service."

Oliver looked nervous. "Oh. *Um.* Yeah. It's a sat-phone. My mom gave it to me to check in and for, you know, emergencies." He gave Liam a playful push, and they returned to the campfire.

"Okay. Penelope believes that to rip apart the formation, you have to divide. How can you divide a polynomial?" Maggie was chattering. Marco was not in the mood for more math lessons, they had been at it since eight that morning. The landscape seemed more fitting for a story.

"How about a scary story? We are camping after all," he pleaded.

"Perhaps two birds one stone?" Dr. Pound raised her right eyebrow. Mr. Pikake responded with a wicked grin.

"Long ago, five children, much like yourselves wandered into this campground," Mr. Pikake began in a deep bellow. "Little did they know, four ghosts haunted the area. The children had to determine the four lines to create a shape that would deter the ghosts forever!" He let out a hearty laugh.

"That's it?" Liam stared. "That's pathetic. Let's do this so I can get some sleep."

"Finding the zeros, the soldiers rooted at ground zero, isn't easy. There are a few tricks some hunters have up their sleeves. I'll show 'em to you. Just know we created computers for this. Polynomials are powerful formations," Dr. Pound began.

"Yeah! Like Death Stars!" Liam interrupted.

"Yes. . . . But much more than that. Polynomials are used to model people, behaviors, money, investments. They govern our entire world in a way. What you know already, about their shape, how they work, is quite strong alone. However, in a bind, you might need to be able to find these key soldiers."

She drew out a long division sign. "You already really know most of it and also don't. It's been hiding right in front of you. The way they teach Slants oft obscures both beauty and simplicity. Look here. If you wanted to divide 1496 by 22, what would you do?"

Marco felt giddy. This was the good math. The math from elementary school that wasn't so hard for him. He raised his hand like a kid in a classroom. Everyone stared. As his cheeks turned a rosy red he answered, "I can do it."

In the light of the campfire, Marco went through the tedious steps of long division. He drew out the 1496, made a little house over it, and placed 22 on the outside. Immediately, he regretted volunteering. He had to figure out how many 22's he could stuff into 149. Wishing he could simply say "Alexa, what's 149 divided by 22?" he allowed his imagination to take over hoping to avoid embarrassing himself.

In fifth grade, he had created the division monster. He was thankful he allowed it refuge in his mindroom rather than throwing it out with all the other elementary topics he'd forgotten. This monster only ate 22's, he needed to know how many to feed it until it was satisfied. Piling them into the mouth of the beast, he skip counted 22, 44, 66, 88, 110, 132, he stopped. No way he could fit another meal in there. Marco had been keeping track on his left hand. That was six meals for the monster. He wrote a six atop the house and 132 below. A full stomach was 149 which meant there was still 17 hunger bars left over. *Bring down the 6.* Next up, stuff more 22s into 176. Starting where he left off, 132, 154, 176. Perfect, no remainder. That was eight 22's that could be forced down the monster's throat. He finished up and revealed his work.

```
     68
22 ⟌1496
   -132
    -176
     176
       0
```

"Great, now what if we changed these numbers into polynomials?" Dr. Pound said with excitement and intrigue.

"Oh!" Maggie gasped. "That's like what Oliver did with the multiplication."

Oliver looked mortified as he flashed back to Maggie's hug and only gave a shrug. Dr. Pound copied Marco's steps in a strange and alien way. "The leading value is going to drive," she started. "First, I need to determine what wizard can turn a 20 into 1000, which is obviously 50." She scribbled the results. "Multiplying 50 by 22 is really just multiplying 50 by 20 and 50 by 2 and adding them together." More marks. "Subtract, you've got 300 left over. Again. Now, I'm looking to turn 20 into 300, I need a 15 for that." She continued her process to the next line. "Alright, last one, I need to turn a 20 into a 60, that's simple, 3." Dr. Pound finished her process. Atop the division house sat 50 + 15 + 3 which was indeed the same 68 as Marco found.

```
           50 + 15 + 3
20 + 2 ⟌1000 + 400 + 90 + 6
       - 1000 + 100
                300 + 90 + 6
              - 300 + 30
                      60 + 6
                    - 60 + 6
                           0
```

"This is *amazing,*" Liam screeched. Everyone nodded along. While Marco was busy trying to force feed 22's to his beast, Dr. Pound had turned everything into very basic problems with lots of friendly zeros.

"How do we do that with a polynomial?" Penelope spoke up.

"Same way." Dr. Pound swiped away her work with her shoe and wrote out the previous polynomial.

"There's your cylon," Oliver whispered poking at Liam. "Super memory."

"If our goal is to factor a formation, rip it apart like a Numberfolk, we need to know at least one factor. Or guess at one. Just as we know 3 is not a factor of 22 because $22 \div 3$ has leftovers, we can determine if $x + 1$ is a factor by attempting to tear it out of the formation."

$$x+1 \overline{)\,2x^4+5x^3-8x^2-17x-6}$$

"Okay . . ." Penelope took the lead. "The 20 led before, so the x should lead here?" She looked for approval before continuing. "We want to turn the x into $2x^4$. To do that, we'd need $2x^3$." As Penelope spoke, Dr. Pound transcribed with her stick in the dirt. "x times $2x^3$ gives us the $2x^4$ we are looking for and $2x^3$ times one is just $2x^3$. Now, we subtract. All the other places are really just zeros, that's why we can 'bring down' the other values."

$$\begin{array}{r} 2x^3 \\ x+1 \overline{)\,2x^4+5x^3-8x^2-17x-6} \\ -(2x^4+2x^3+0+0+0) \\ \hline 0+3x^3-8x^2-17x-6 \end{array}$$

"I guess I never really thought about how you add and subtract these terms." Maggie tilted her head as she watched her friend fill in all the blanks.

"They are like-terms, Bug," Marco replied. "Just like $5x - 3x$ is $2x$, you can just think of them as things. If x^3's are dogs, you have five dogs minus two dogs which gives you three dogs."

"That makes so much sense," Liam added. "I always used to think you took away the mask too. Like $5x - 3x = 2$. But five dogs minus 3 dogs is for sure not 2! Two what?" He laughed at his own joke.

$$\begin{array}{r} 2x^3+3x^2 \\ x+1 \overline{)\,2x^4+5x^3-8x^2-17x-6} \\ -(2x^4+2x^3+0+0+0) \\ \hline 0+3x^3-8x^2-17x-6 \\ -(3x^3+3x^2+0+0) \\ \hline 0-11x^2-17x-6 \end{array}$$

"Alright, keep going," Dr. Pound led Penelope on.

"Well, now we need to turn the

x into $3x^3$ so that is just multiplying by $3x^2$." Multiply, subtract, repeat.

"Now we need $-11x^2$ so we multiply by $-11x$. That will give me $-6x$ so I multiply by -6! That's it!"

Dr. Pound tediously wrote out the last few steps. "There you have it. You have divided $2x^4 + 5x^3 - 8x^2 - 17x - 6$ by $x + 1$. In turn, you have found, since there is none left over, that one of the lines used to make this polynomial is in fact $x + 1$."

$$\begin{array}{r} 2x^3+3x^2-11x-6 \\ x+1\;\overline{)\,2x^4+5x^3-8x^2-17x-6} \\ -(2x^4+2x^3+0+0+0) \\ \hline 0+3x^3-8x^2-17x-6 \\ -(3x^3+3x^2+0+0) \\ \hline 0-11x^2-17x-6 \\ -(-11x^2-11x+0) \\ \hline 0-6x-6 \\ -(-6x-6) \\ \hline 0 \end{array}$$

"I prefer synthetic division," Mr. Pikake announced. "It requires less writing, and my chicken scratch is too difficult to read most of the time," he laughed. "Remind me to show you sometime. I'm quite tired. We have a big day ahead . . . should get some sleep."

Dr. Pound handed out PB&J sandwiches and everyone got settled in. Glad no one had a peanut allergy kicking them out of the domain of the trip, Marco was still smiling as he entered the tent. He was amazed. It was larger than any pop up he had ever seen. It was big enough to fit eight adults, so the five kids and two chaperones had no problem getting comfortable. It had one main area in the middle, and two smaller pull-outs to each side. Liam, Oliver, and Marco took the left, while Penelope and Maggie settled on the right. Mr. Pikake and Dr. Pound sat in the center.

"I get the division but how do you know *what* to divide by? Like did you just pick $x + 1$?" Maggie nagged at Dr. Pound.

"It is a bit guess and check, unfortunately," Dr. Pound began. "If all the coefficients . . ."

"Oh, I know that. Those are the numbers in front of the mask, like in $2x^3$ it is 2," Maggie interrupted.

"Yes . . . if all the coefficients are Integers, when you multiply the linear factors: $(a_1x + b_1)(a_2x + b_2)\cdots$, the product of the a's becomes the leading coefficient and the product of the b's become where Zero is sent. This allows you to make a list of the possibilities to work through, although it can still be tedious."

"Can you show me?" Maggie pushed.

$(8x^2 - 99x + 225) \div (6x - 18)$

"Maybe tomorrow. We need to get some sleep. There is an evil villain to stop in the morning."

Maggie reluctantly agreed and the tent finally quieted down. Liam's snores made it impossible for Marco to fall asleep. He rolled over to face the tent wall. Outside was pitch black. There was no moon at all. Marco felt like he was falling down the infinite throat of the shadow monster. There were no shapes, there was nothing except darkness.

Shifting his gaze towards his feet, the flickering of the campfire was just what he needed to grasp back on to reality. He stared at the flames and watched the evil twins of the trees sway in the distance. Right as he nodded off, a new figure crept into the landscape. He pushed his eyelids tightly together trying to force the phasing away. When he opened them, the figure was gone. *Another Numberfolk*, he thought. Deep inside, doubt slithered into his throat. The shadow had looked strangely . . . human.

EXTEND YOUR POWER

Master the mother, perfect the pupils

DETERMINE THE SHAPE

What do you want your formation to look like?

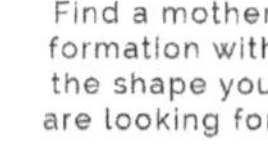

Find a mother formation with the shape you are looking for.

FIND THE MOTHER

MOLD IT

Move it, flip it, stretch it, shrink it, into your perfect battleplan.

A few formations . . .

RADICAL

A jail cell. You are often limited by the soldiers in your dominion.

$m(x) = \sqrt[n]{x}$

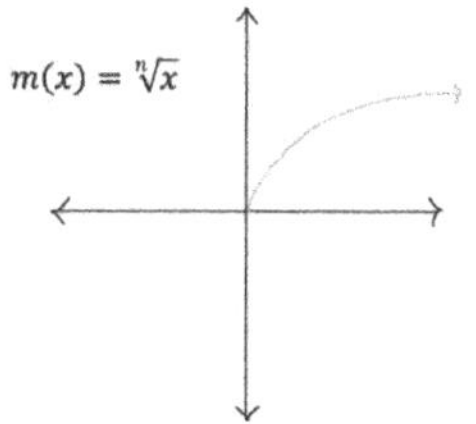

Each value of n has a different mother and thus a different shape. The above formation is for n=2.

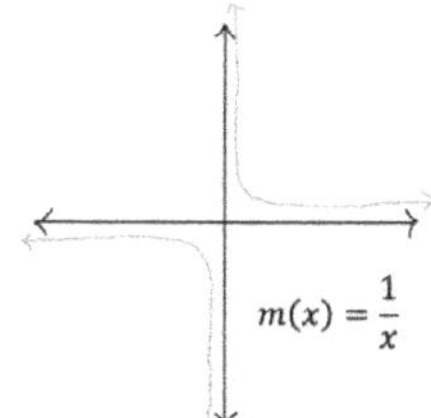

$m(x) = \frac{1}{x}$

RATIONAL

A rational formation is created by dividing two polynomials.

$$r(x) = \frac{p(x)}{q(x)}, \qquad q(x) \neq 0$$

Rational formations have both soldiers who are not allowed in their army and a Station on the field that is impossible to cover.

EXPONENTIAL

A brigade of a certain Numberfolk. Note the number of soldiers in the unit is the value which changes.

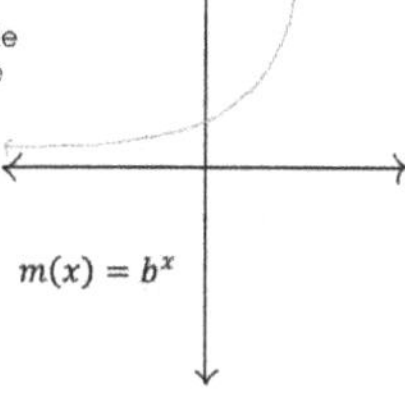

$m(x) = b^x$

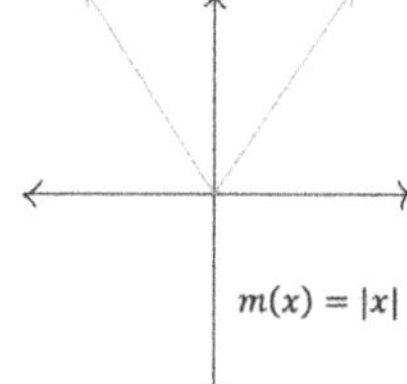

$m(x) = |x|$

ABSOLUTES

The mother of an absolute formation sends everyone to non-negative stations.

Incantations

Your incantations will work with any functional formation! Let f(x) be the existing formation where $k > 0$ and $0 < p < 1$

VERTICAL

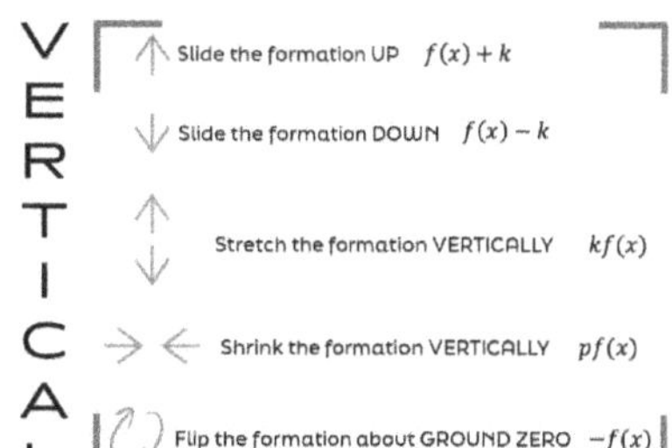

- Slide the formation UP $f(x) + k$
- Slide the formation DOWN $f(x) - k$
- Stretch the formation VERTICALLY $kf(x)$
- Shrink the formation VERTICALLY $pf(x)$
- Flip the formation about GROUND ZERO $-f(x)$

HORIZONTAL

- Slide the formation LEFT $f(x + k)$
- Slide the formation RIGHT $f(x - k)$
- Stretch the formation HORIZONTALLY $f(px)$
- Shrink the formation HORIZONTALLY $f(kx)$
- Flip the formation about CENTER STAGE $f(-x)$

18

THE BOOK OF THE DEAD

"FOR THE SAKE OF BREVITY, WE WILL ALWAYS REPRESENT THE NUMBER 2.718281728459... BY THE LETTER e."

- LEONHARD EULER

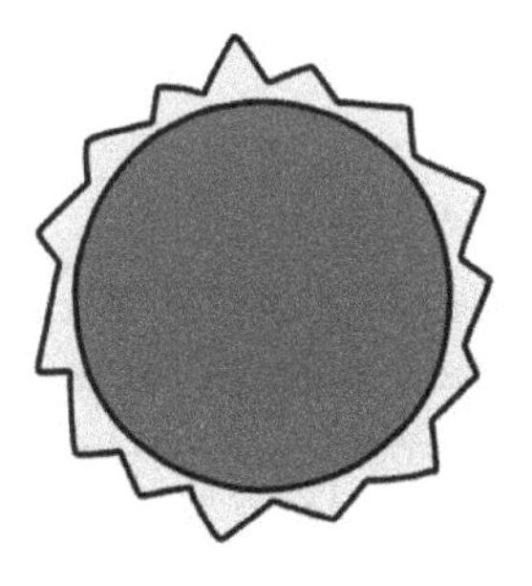

Deep in the trenches of his jail, Three sat and watched. While he had released his prisoners, he kept one, just one, as a trophy. The 6 sat behind the bar breaking itself apart, then gluing the pieces together again and again.

$$\sqrt{6} = \sqrt{2 \cdot 3}.$$

It could force parts of itself into different rooms of the jail, but it could not escape,

$$= \sqrt{2} \cdot \sqrt{3}.$$

What was this strange new creature he had made? The ramifications of throwing a rational being into a cage were quite extraordinary.

He remembered back to when Thirty-six was jailed in the town square. She had gone rabid but was somehow able to turn things around, to remain rational.

Of course. Thirty-six could be split perfectly into identical twins. She was a special Numberfolk who was able to remain rational behind square bars. That was it. It all came down to genetics. Traits. If a Numberfolk was born with traits that could create identical twins, or triplets, or quadruplets, or on and on, they could escape Three's jails.

Thirty-six never had the pleasure of being shoved into a prison General Three created. His weren't any old penitentiaries, but ones designed to break down the creatures inside. He grinned as he realized stuffing the mayor into a $\sqrt[3]{}$ cell would have ended differently. To remain rational, sensible, she'd have to split herself into three identical pieces. Since $36 = 2 \cdot 2 \cdot 3 \cdot 3$, she couldn't escape such a prison, at least not with her sanity intact.

He looked back at his test subject. This poor six, as perfect as it was, simply didn't possess the genetic copies it needed to escape. And as it tried, again and again, it became angrier and angrier, losing all rational thought.

While the threat began as an apparition, a hazard that may or may not exist, Three had unintentionally birthed a real enemy. One that he himself had trained. An army that knew every battleplan. Even worse, this legion was unpredictable. Having stolen their rationality, the beast he was about to face was capable of unthinkable moves and formations that citizens of Rational would never expect. For the first time, Three was terrified.

Irrational Numberfolk were those who could not disguise themselves with a vinculum and two Integers. This was one quality Three was thankful for. He could easily recognize them. Their dot-form was never ending with no repetition. While some Rationalites had very long dot-forms, with over twenty digits, over two-hundred digits, or even more, before they circled back, Irrationals never circled back. While there may be a string like 345 that came again at some point, in one instance it would be followed by a 6 and in another a 1. You just couldn't predict what would happen next, ever.

Three gathered his scientists to complete a threat assessment. The results were petrifying.

The first question was determining just how many irrationals could be out there. While Three had thrown his prisoners out onto the barren continent, it was quickly determined that the population was a complete mystery.

Rational was closed. If any two Rationalites had a child, the child would also be rational. If any two Rationalites dueled, the result would be rational.

Irrationals did not follow these orderly laws. If two $\sqrt{3}$'s had a child, the result would be

$$\sqrt{3} \cdot \sqrt{3} = \left(\sqrt{3}\right)^2 = 3.$$

A Rational! In the same way evil twins could vanquish each other, Irrationals could too! The crazy $\sqrt{2}$ and their twin $-\sqrt{2}$ would create the Rational zero in a duel.

$$\sqrt{2} + \left(-\sqrt{2}\right) = 0.$$

Not only were their infinite digits unpredictable, so was their population. Out in the dark continent, they could have millions, billions, of rational beings on their side.

Attempting to get a hold of their numbers, Three and his scientists began counting how many Irrationals there could be.

The Naturals were countable, after all, they made up the counting Numberfolk (1, 2, 3, …). Not knowing all the Rationals, the team developed a method to tally the citizens, without actually counting them. They developed an ordering system.

If every Natural Numberfolk could be paired with exactly one from another group, the unknown group must be countable.* To do this, they required a functional order that was both one-to-one and onto. One-to-one protected against over-counting. It meant every Natural was assigned

* This is in fact how we count! If given a pile of M&M's to find the total we assign each candy a number, 1, 2, 3, ….

to a unique member of the opposing group, no ambushes. Being onto assured that everyone in the other group was counted, no one left out.

Infinite groups are mind bending. One might believe there *must* be more Integers than Naturals. Afterall, if I counted all the bears in a forest, it would be absurd if there were somehow more brown bears than total bears. Counting the Integers, we have all the Naturals plus their negative twins. It seems like there must be twice as many Integers as Naturals, right? Wrong.

The scientists defined an order that sent every even Natural to the integer half its size. For odd Naturals, they determined how far they were from one and cut that in two.

$$c(x) = \begin{cases} \dfrac{x}{2} \text{ if } x \text{ is even} \\ \dfrac{1-x}{2} \text{ if } x \text{ is odd} \end{cases}.$$

In this order $1 \to 0$ as one is odd, while $3 \to -1$, and $5 \to -2$, and so on. For the evens, $2 \to 1, 4 \to 2, 6 \to 3, \ldots$. Do you see where this is going? They could show the order was one-to-one using a clever trick.

An order that is 1:1 means no two soldiers can be deployed to the same location. In a normal ledger, one could simply check. However, they were dealing with infinite residents. Forcing a check on everyone would take an eternity. They had an idea. Suppose you had a small team of five and assigned each member a color. In ordering their team shirts, you noticed there were two green jerseys. This was not possible! Every player had their own color. Back tracking, you found that it was a simple accounting error. Your assistant had included player 5 two times on the list. In other words, the only way two players could both have green was if they were in fact the same person and someone made the unfortunate error of writing their order twice.

Thus, to test if their commands were 1:1, they needed only to assume two different soldiers were sent to the same station. Consider the first soldier is x_1 and the second is x_2.

If x_1 is even, they would be sent to $x_1/2$. If soldier two is odd, they couldn't possibly be sent to the same location, as all odds were sent outside the Village. Thus, soldier two must also be even and sent to $x_2/2$. Suppose this is the same station then,

$$\frac{x_1}{2} = \frac{x_2}{2}.$$

This led to the conclusion $x_1 = x_2$, meaning if two soldiers were sent to the same location, it was an accounting error, and someone had simply transcribed the same thing twice. Two different soldiers could not be deployed to the same post. After confirming the same results when the soldiers were odd, they knew their order was 1:1.

To confirm their order was onto, they worked backward. For every station, they found the Natural Numberfolk who would be assigned to it. First, they looked at all the positive stations.

Supposing k was a positive station (1, 2, 3, ...), they saw the Numberfolk $2k$ would be deployed to it. Since $2k$ was always an even Natural it traveled to,

$$c(2k) = \frac{2k}{2} = k.$$

As k was a Natural, they found every positive station had some Natural deployed to it.

What if the station was negative? What about station zero? Were they sure some Natural was assigned to these as well? For a station m, where $m \leq 0$, the Numberfolk $1 - 2m$ was the odd Natural assigned to it as,

$$\begin{aligned} c(1-2m) &= \frac{1-(1-2m)}{2} \\ &= \frac{1-1+2m}{2} \\ &= \frac{2m}{2} = m. \end{aligned}$$

They'd done it! Using masks, they could find that every Natural soldier was assigned one station in Integer and that every station in Integer was covered. This was mind blowing.*

They had determined not only that they could count the Integers, but that somehow, if they deployed only the Naturals, they would be able to guard each of their own homes and still have enough of them to guard the home of every evil twin as well. The even Numberfolk would stay in the Village and each protect a location there (2 protected the home of 1, 4 watched over 2's house, 6 was assigned to 3's mansion, and so on) while the odd Naturals would be deployed to watch over TNZ and Zero (as 1 cared for 0, 3 for -1, 5 to -2...). This was simply magnificent!

* As crazy as this seems, this is the power of the infinite. It is in fact true that the Naturals $(1, 2, 3, \ldots)$ and the Integers $(\ldots, -3, -2, -1, 0, 1, 2, 3, \ldots)$ are equivalent in size.

Using their tools, they could also show that everyone in Rational was countable meaning the Naturals alone could protect all Rational citizens.

They turned their focus to the Real continent and discovered a huge problem. It wasn't possible to guard every Real Numberfolk! Assigning each soldier to one station was easy enough, unfortunately they simply couldn't cover them all! While the Naturals could protect everyone in Rational, they couldn't come close to matching the enormity of Real. In fact, they couldn't even guard all the Reals between 0 and 1. How the scientists determined this was based on a favorite pastime, a game called 4's and 5's.

Fours and Fives

A two-player game, Fours and Fives wasn't difficult to play. The first Player was provided with a 6×6 grid to fill out. On their turn, they completed a full row, adding in fours and fives in any manner they picked. They could choose 4-4-4-4-4-4 or 5-5-5-5-5-5 or any of the 2^5 possibilities in between.

Player Two had only a single row of six blanks to complete. In order to win, Player Two must construct a row, a 6-digit number, that was different than all of Player One's creations. If Player Two was unable to create something unique, the first player won.

A lover of games, a seventeen on the science team knew that if she was Player Two, she could win every time. Here's how.

Player One begins and writes some combination of fours and fives before passing the play to Seventeen to fill in her first box.

Player #1					
4	5	5	4	5	5

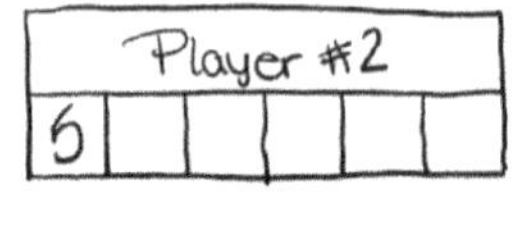

All she needed to do, was pick the opposite of whatever Player One put in their first box. Since they selected a 4, she'd pick 5. This ensured that whatever she ended up with, it would be different than Player One's first row.

Back to Player One. A smart gamer would also start with a 5, attempting to push Seventeen into a corner. They play, 544544. Since Seventeen had

Player #1					
4	5	5	4	5	5
5	4	4	5	4	4

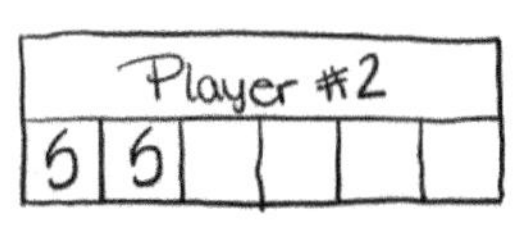

already created a different number than Player One's first row, she no longer cared about it. It was now time to make sure her entry didn't match row two. Since both started with a 5, she needed only to make her second entry the opposite of Player One's second entry.

The game continued. Each round, Seventeen played the defensive. In round *n* she looked at the *n*th row of Player One's board and the *n*th entry and selected the other number. If Player One picked a 4, Seventeen picked a 5 and if Player One picked of 5, Seventeen would swoop in with a 4 to claim the game.

Player #1					
4	5	5	4	5	5
5	4	4	5	4	4
5	5	5	4	5	5
5	5	4	5	4	4
5	5	4	4	5	5
5	5	4	4	4	4

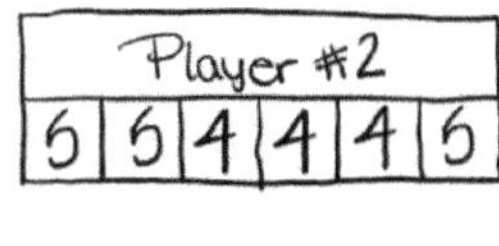

Seventeen won. It wasn't any surprise. Using this strategy, she could win *every* time, no matter what, by simply picking whatever Player One didn't. She could win on any sized board. Seventeen could always find a number which Player One missed on their list.*

This was exactly how the scientist found not only that the Irrationals alone far surpassed everyone in Rational, but that there were more Numberfolk on the tiny nail of the pinky toe of Real than there were in the entire body of Rational Country.

They started by pretending they had some master list of all the Real Numberfolk between 0 and 1. They didn't. They had not a clue who all

* Give it a try! Play with your parents and then tell them you just proved the real numbers are uncountable.

these Numberfolk were as it would contain both Rationals like $1/2$ or 0.5 and Irrationals like $\sqrt{1/2}$. They pretended nonetheless.

From the plague, they knew everyone on their list had some dot-form. Because they were only looking at the Numberfolk between 0 and 1, their dot-form would be,

$$0.anumbersomethingelseanotherthing\ ...$$

Or . . . something like that. It didn't actually matter *what* the dot-form was, just that they had one. Now, if their list really did contain all the Real Numberfolk between 0 and 1 and they could assign a Natural to each of these, then it was countable, and they had a chance. But, if they could find just one, anyone, who didn't make the list, that meant there were more Real Numberfolk just between 0 and 1 than there were in all of Rational Country.

Seventeen found one, a missing Numberfolk on the list. She found it quickly and easily, and she named him Ron. Supposing Ron had some dot-form,

$$\text{Ron} = 0.r_1r_2r_3r_4\ ....$$

All she needed to do was for the nth digit, pick something different from the nth Numberfolk on the list's nth digit. The same strategy she used for Fours and Fives.

If the first Numberfolk on the list was $0.0somethingblah$ she'd make Ron $0.1somethingelse$. Looking to the second Numberfolk on the registry, if it was $0.03blippyblop$ she'd make Ron $0.15cantcatchme$. And that was that.* Seventeen could always create a Ron that wasn't on the list. They came to the devastating conclusion that there were many, many more Real Numberfolk than there were citizens of Rational, they simply couldn't guard them all.

Now, Rationals were Real. Thus, the Rationals had essentially counted themselves in the mix. It made sense to directly compare the Rationals to the Irrationals to determine, if by some miracle, they were large enough to go up against these crazy neighbors. It wasn't good news. If you took two countable groups, the result would be another countable group. If I can count $\{1, 2, 3\}$ and I can count $\{4, 5, 6\}$, I can also count

---

* There was one tricky situation. We know $0.199\overline{9} = 0.2$ for the same reason $0.\overline{9} = 1$. So Seventeen made sure not to pick nines. It was easy since she only needed a *different* number, which meant she still had 8 digits to choose from.

$\{1, 2, 3, 4, 5, 6\}$. It might take longer, but it is doable.

If the Irrationals could be tallied, since the Rationals were countable and Real was made up of only the two groups together (rational and irrational), Real would have to be countable as well. It wasn't. Ron proved that. Thus, the Irrationals alone consisted of an insurmountable army, so big you couldn't even comprehend their infinite size.

Three was in serious trouble. The enemy was much larger. Even if Three made every man, woman, and child fight, Irrational had more. Worse, since the Irrationals weren't closed, they could be out there creating Rational spies to send into the country and attack at any moment. The situation was dire. The only thing protecting them all was the fact that Irrationals couldn't enter Rational. The spies were a problem, but a manageable one. Three realized their only hope was to find a way to truly separate the sets. If he could somehow rip away everything outside of Rational Country, they would be safe.

Working long and hard for many weeks, Three and his team found their solution. It would require every Rational, it demanded perfect timing, and even better precision, but it could be done.

They gathered every Numberfolk in Rational and commanded them to a station. Together they formed a parabola. A Death Star that could harness the power of all their Spogs combined and focus it on one key location: the volcano.

Everything was coming together. When One first discovered the volcano, her timing was impeccable. Had she traveled at any other moment, she would have missed the eruption and never created her mirrors. They later found the eruptions were exceedingly rare. Lava would once again shoot from its mouth, but if they missed it, fifty-two years would pass before they had the chance again. The volcano only cried during harmony. A special moment in which the cosmos was in perfect alignment.

Terrified, every Rational deployed to their station. They held hands to create an unbreakable bond. Beams of light shone down stabbing each and every resident directly in the chest. Protected by their Spogs, the energy reflected and focused on the volcano. It began to rumble. Two Numberfolk were overcome by the force and their palms slipped. They quickly grabbed back on. If they died, at least Numberville would live. If they failed, the entire country would soon be invaded by insane and unforgiving beasts – irrational monsters that would rip them apart and devour them piece-by-piece, stealing away their digits to add on to their infinite tails.

$$\sqrt{\frac{16}{9}x^2 - \frac{88}{3}x + 121}$$

The moment finally came. As the blazing red tears of the volcano began to drip down its face, many Numberfolk cried too. They yelled and they wailed, but they held tight. Just when they began to doubt they could be successful, at that moment when all they wanted to do was let go and give up, an outstanding glow radiated from the volcano's rim directly into the sky. The ground began to shake violently. The Numberfolk were thrown from their formations as a thunderous crack filled the air before everything went dark.

When they awoke, they found the continent was gone. There was no more Real. There was only Rational. They cheered, they screamed, the celebration lasted for fifty-two days and fifty-two nights. The Rationals were finally safe. While they continued to squabble with each other, the negatives and the positives never quite getting along, the Naturals always believing they were better than the Logos, they were at peace.

From far out in the galaxy, the charmed ones watched in dismay. What had become of the children they created? While Natural was formed in their image, the Irrationals were too. Every digit a charmed one, and every Numberfolk vital for completion. The Mirror of Wonders was not concerned.

The irrationals now exist on Counter-earth

An entire world has given birth.

Give it time and you will see,

The power of irrationality

While logic creates formidable borders

'Tis in the chaos where you will find order.

And thus, Counter-earth was formed. It grew. Because Irrationals were not closed, they could create their own complete system. For $\pi/\pi = 1$ and $\sqrt{2} \cdot \sqrt{2} = 2$ and so it expanded. For centuries and centuries, it swelled and became stronger than Numberville, the original Earth, ever could be.

ARE YOU IRRATIONAL?

The thought that killed Hippasus

The Pythagoreans were out to sea. Hippasus thought it would be fun to study one of the sails. It had been made from a 1-cubit square that was sliced down the diagonal to form a right triangle.

As he tried to find the length of the diagonal, something bothered him. Using their leader's famous theorem he calculated,

$$1^2 + 1^2 = d^2$$
$$2 = d^2$$
$$\sqrt{2} = d$$

Who was this strange number? It wasn't anyone Hippasus had ever met before. Having a crisis of faith, he decided to prove for himself √2 was indeed rational.

1 Hippasus begun by assuming in good faith √2 was rational. All rational numbers can be written with two integers and a vinculum. He didn't know what integers these would be, so he gave them masks and named them p and q.

$$\sqrt{2} = \frac{p}{q}$$

2 Rationals are known for the infinite disguises they can wear. This meant p and q couldn't be any integers, they needed to be enemies. If they shared any common traits, Hippasus would need to push them through a mirror and start all over again.

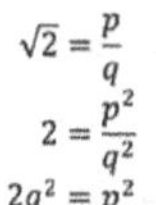

3 He got to work attempting to hunt something, anything, that would provide him the solace he needed.

$$\sqrt{2} = \frac{p}{q}$$
$$2 = \frac{p^2}{q^2}$$
$$2q^2 = p^2$$

This was not the result he was hoping for. The value on the left was even, after all it had a trait of two. But if the two values were equal, this meant the value on the right must be even too!

Any integer with a 2-trait must be even!

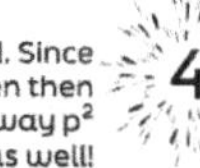

4 If p^2 is even, what does that say about p? Hippasus wondered. Since $p^2 = p \cdot p$ and p was either even or odd, he went to work. If p is even then $p^2 = \text{even} \cdot \text{even} = \text{even}$. If p is odd then $p^2 = \text{odd} \cdot \text{odd} = \text{odd}$. The only way p^2 could be even was if p was even as well!

5 Knowing that p was even meant there was some integer, let's call it k, so that p=2k as this was the definition of an even number – even numbers all have a 2-trait.

$$p = 2k \quad \text{therefore} \quad p^2 = 2k \cdot 2k = 4k^2$$

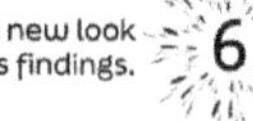

6 Now things were getting interesting. Using his power of substitution, he put this new look of p^2 back into his findings.

$$2q^2 = 4k^2$$
$$q^2 = \frac{4k^2}{2}$$
$$q^2 = 2k^2$$

Wait . . . that means q^2 has a 2-trait too! So q must also be even.

7 Hippasus noticed something remarkable. If both p and q are even, it means they both have a 2-trait: that's something in common. If they have something in common, they can't be enemies! He carefully looked over his work, each step followed logically from the one before it. It was pristine, beautiful, perfect. Alas, something here was rotten. If all his steps were correct, the only thing left was his assumption. Assumptions are the worst.

√2 CANNOT BE RATIONAL, ERGO IT MUST BE IRRATIONAL!

The Pythagoreans believed all numbers are rational and while Hippasus excitedly shared his news with his crewmates, they began plotting. It wasn't long before they banned together and threw Hippasus overboard for his heresy. For in proving an irrational number existed, they thought him to be the irrational one.

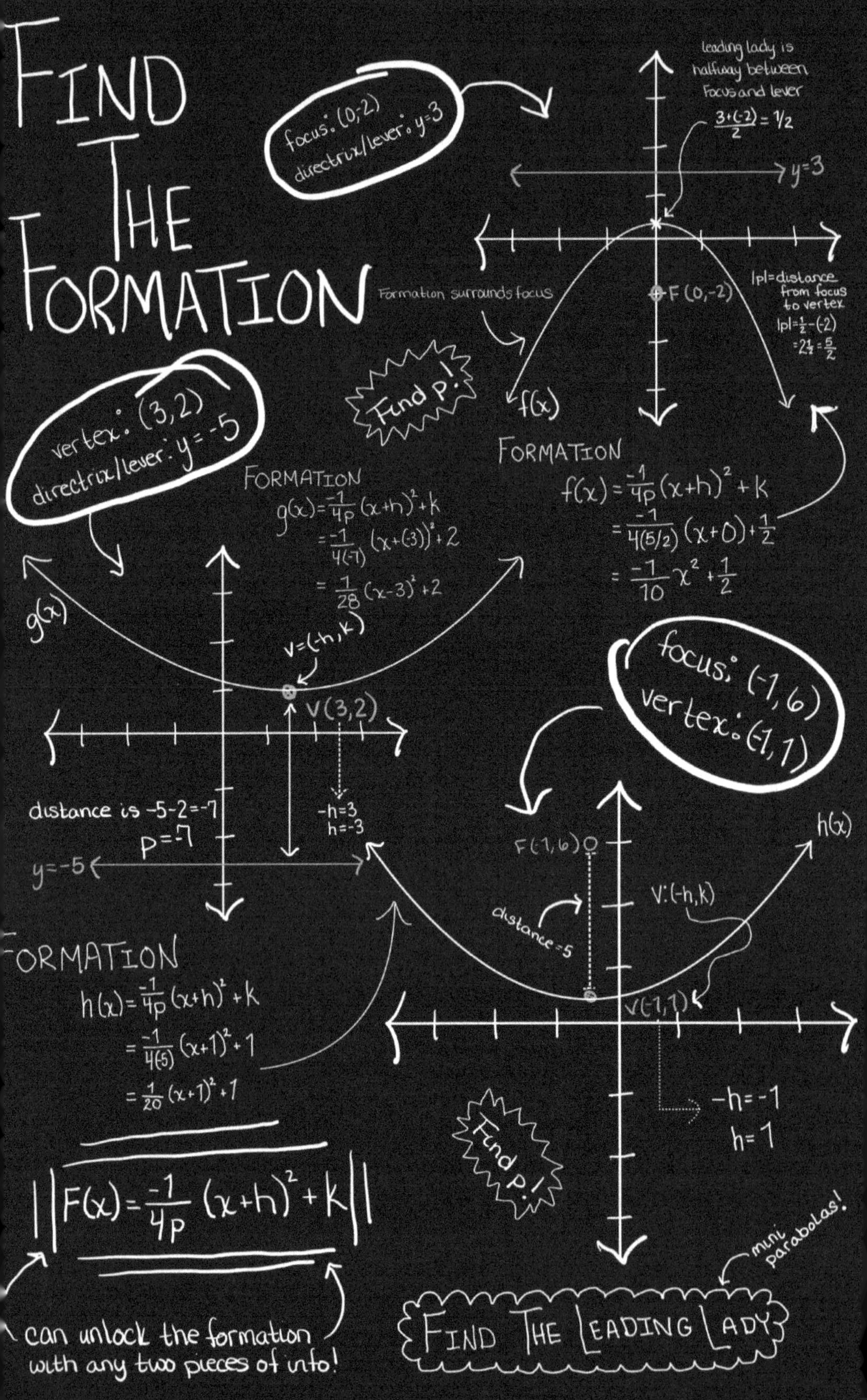

FIND THE FORMATION
focus: (0,2)
directrix/lever: y=3
leading lady is halfway between focus and lever
3+(-2) / 2 = 1/2
y=3
Formation surrounds focus
F (0,-2)
|p|=distance from focus to vertex
|p|=1/2-(-2)
=2 1/2 = 5/2
f(x)
Find p!
FORMATION
f(x) = -1/4p (x+h)² + k
= -1/4(5/2) (x+0) + 1/2
= -1/10 x² + 1/2
vertex: (3,2)
directrix/lever: y = -5
FORMATION
g(x) = -1/4p (x+h)² + k
= -1/4(-7) (x+(-3))² + 2
= 1/28 (x-3)² + 2
g(x)
v=(-h,k)
V(3,2)
distance is -5-2=-7
p=-7
-h=3
h=-3
y=-5
focus: (-1,6)
vertex: (-1,1)
h(x)
F(-1,6)
V:(-h,k)
distance=5
V(-1,1)
FORMATION
h(x) = -1/4p (x+h)² + k
= -1/4(5) (x+1)² + 1
= 1/20 (x+1)² + 1
-h=-1
h=1
Find p!
F(x) = -1/4p (x+h)² + k
can unlock the formation with any two pieces of info!
mini parabolas!
FIND THE LEADING LADY

19

THE END

"JUST BECAUSE WE CAN'T FIND A SOLUTION IT DOESN'T MEAN THAT THERE ISN'T ONE."

-ANDREW WILES

Right as the sun poured in like a morning cup of orange juice, Marco awoke. The tent was roomy. He stretched out and noticed the entire middle section was empty. Climbing over his friends, careful not to wake anyone, he unzipped the door and stuck out his head.

Dr. Pound and Mr. Pikake were huddled over the dead campfire looking at the map. "I think I know where this is leading us," Dr. Pound said in a hushed tone. "It's not good. You are going to have to help the boy, or . . ."

"Or what?" Mr. Pikake replied angrily.

"Or leave him here, Blaise. It's dangerous."

Marco jumped out of the tent. "I'm coming," he said firmly.

"Calm down, son." Mr. Pikake walked to Marco and helped him exit the tent. "She is just worried. The place we must go, the location of the book, it's high in the mountains. It will be challenging for you . . . you will likely be able to see the Numberfolk, but no one else will."

"My granddad left me that map. It's my duty to protect that book, and it's my fault Fredrick is so close. I am going."

The arguing must have woken up the others as one-by-one they exited the tent. "What's going on?" Maggie asked, clearing the sleep from her eyes.

"Nothing. Nothing is going on," Marco barked. "We're leaving."

No one said a word as they packed up their gear and prepared to head out. The horrible taste that lingered in Marco's mouth made him wish he had remembered to pack a toothbrush. Forming a line behind Dr. Pound, they headed towards the hill. It was steep and required grabbing on to tree limbs and rocks to pull themselves up as they went. Maggie was the first to speak.

"Do you guys remember the fly? This situation totally reminds me of this one time there were like a lot of flies around for some reason. Anyway, I had the window open and one got into my room. I really struggled with the decision to close the window. Like, on the one hand, if I closed the window no *more* flies would get in, which would be good. But, on the other hand, if I closed the window, the fly that was already in there driving me crazy would continue to bother me for all of eternity."

"To close or not to close, that is the question," Liam uttered in his best Shakespearean voice.

"Right?" Maggie continued. "It's the perfect analogy for a difficult situation."

"Whoa! What's that?" Oliver interrupted.

"What?" Liam looked around panicked, half expecting a bear or a wolf to jump out.

"The whole world . . . it looks . . . funny?"

Marco saw it too. His nausea kicked into high gear.

"The eclipse." Mr. Pikake pointed to the sky. "It has begun."

As day initiated its transition to night, a strange filter was cast over the entire landscape. It wasn't anything like sunset. Sunset was calm and warm. The sky projected beautiful and vibrant colors of pink, orange, and red. They'd dance and mix to form silvery purples before turning into the deep cobalt of night. This was *nothing* like that.

It was bright, but it wasn't. The world looked almost like a haze. The shadow monster was devouring an unscheduled snack as the familiar deep greens and lush browns of the forest became muted, trapping everyone in a shared dream. Marco felt like he was falling into darkness as the world took on unfamiliar eerie hues. The warm hug the sun normally gave eagerly was being ripped away as the temperature began to drop.

Penelope looked at Minnie Mouse. "It's 10:17," she said. "The eclipse will last near three hours with totality at about 11:42."

"About?" Liam joked.

"We don't have much time." Mr. Pikake pushed the children forward. "We must hurry!"

Marco had been seeing Numberfolk everywhere for the past thirty minutes but hadn't said anything. The giggling 5 was following them. Thankfully, he hadn't yet come across the angry 1 or any other nefarious digits.

The group finally made it to the top of the hill. With one last pull, they landed in a small flat area leaving only a single big climb until they reached the peak. As everyone continued forward, Marco held back.

"What's wrong?" Mr. Pikake bent down beside him.

"Mirrors," he whispered. "They're everywhere, all around the clearing."

"Those are portals. Numberfolk use them to disguise who they are. Nothing to be afraid of." The tutor took his arm and stayed beside him as they crossed the field.

Marco kept his head down, trying not to look directly at them. Suddenly, one of the mirrors began to move. Like a puddle of water, waves expanded across its face forming circles that increased in diameter as they neared the edge. He saw a foot. Someone was coming through.

Clenching onto Mr. Pikake's arm, he watched in awe as the being exited the frame and stood tall to reveal itself. It was 16/4. He let go. "It's just a four," he sighed. Unable to stop himself, he gawked at the number. He couldn't understand why anyone would want to wear such a disguise.

"What are you looking at?" the four spit directly at Marco.

He latched back onto the professor and forced his eyes to the ground trying to make himself invisible to the Numberfolk. Attempting to distract him, Mr. Pikake spoke, "Did I ever tell you that Descartes is the one who introduced the modern masks?"

"Really?" Marco kept his eyes down.

"Really. He was the first to describe a march as $ax + by = c$. Humans have known about quadratics for centuries. The early Babylonians in 400 BC could understand such formations. Al-Khwarizmi determined how to complete the square. Reading these old texts is quite challenging. As hard as it may seem, Descartes' notation is actually quite easier," he chuckled.

Marco thought about Penelope. She had told him the same thing, that even though math looks hard, it is just a way to make complicated things simple. He looked up and smiled at his tutor.

In that moment another mirror sparkled to life. A loud laugh escaped his throat when he saw who it was: 8/22. The eight formed a beehive hairstyle from the sixties. He remembered seeing a photo of his mother wearing one at a costume party before he was born. The 22 looked like two legs, stilts, but curvy and every step made the shape wobble causing Marco to laugh more.

The 8/22 shot him an angry look, "Whatta you laughing at?" she asked starkly, "Least I don't gotta tail." The Numberfolk turned and walked away. Her butt wiggling along like a worm.

"A tail?" Marco asked out loud. He looked behind him to see she was right. He hadn't even noticed it. A long trail of numbers were following him, they seemed to be random: threes and fives and sixes and eights. He saw a one far back in line followed by a nine then a two. He squinted trying to see the end of the row but failed, it was too far away. He tugged on Mr. Pikake's sleeve.

"They're following me," he whispered pointing behind him. Mr.

Pikake looked but couldn't see anything.

"What do you mean?" he asked.

"Behind me, like a tail. There's a bunch of numbers."

A concerned look slid onto the professor's face, he grasped his chin with his left hand. "Tell me Marco, do the numbers repeat?"

"Repeat? You want me to talk to them and see if they echo back?"

"No, no. The numbers you see, do they repeat? Like in 1111 or 123123. Do they repeat themselves?"

Marco glanced behind quickly before snapping his head back to Mr. Pikake, "I don't think so. They seem random."

"Interesting."

Interesting? Yeah, I know. Big help you are.

Looking around at all the Numberfolk that had invaded the clearing, Marco could see Maggie and Penelope up ahead, already beginning the next climb. Liam followed closely behind. Where was Oliver? He swiveled around. His friend wasn't anywhere.

"Oliver!" Marco screamed manically. "Oliver!"

Everyone looked back. "Where is Oliver?" he yelled toward them.

"Did he fall behind?" Maggie called back.

"I thought he was up ahead," Liam answered.

"What should we do?" A horrible and frightening feeling took over Marco's chest. He looked to Mr. Pikake for answers.

Before he could speak, Dr. Pound shouted from the lead, "We have to keep moving. The eclipse draws near. We'll find your friend, I promise, but we don't have time to search now."

A strong gust of wind blew through the mountainside. Marco held Mr. Pikake tighter. Maggie and Penelope huddled together.

"What about me?!" Liam yelped before grabbing onto Maggie's arm.

They all put their heads down and pushed through the wind. An invisible wall was trying to keep them back, making each step harder than the last.

"The cave! It's here! I see it!" Dr. Pound shrieked.

One-by-one the group entered the safety of the structure giving a collective sigh of relief. The cave seemed impossibly quiet next to the wind's roar outside. A shadow flashed in the distance. "There!" Marco pointed. Nobody moved.

"Marco, we can't *see* your numbers," Maggie said obnoxiously.

"It's not a number, it's a person!" Marco snapped back. They all turned.

Fredrick emerged from deep in the cave. He stood straighter and taller than ever. Each step was a dramatic lunge forward like a king on coronation day. His arms were directly in front of him, they sat at right angles. He was carrying something.

"*The Book!*" Liam screamed.

"Yes, young one. It is the book. The Book of the Dead. The Book of Us."

"Does he mean like us us or him us?" Maggie whispered.

"Harmony is near. You cannot stop me. I have read the text and now know how the Numberfolk split the worlds. Philolaus was wrong. Counter-earth is not on the other side of the central fire. Counter-earth is *here*!"

"Wait like here-here? This isn't really Earth?" Liam exclaimed, "Battlestar Galactica *was* right!"

"Come!" Fredrick bellowed.

"Who is he talking to?" Maggie asked.

Another shadow appeared in the arm of the cave. They were considerably smaller than Fredrick, tall, but smaller. Everyone watched as the apprentice entered the room.

"Oliver?!" Marco cried.

"I didn't see that coming," Maggie whispered to Penelope. "And I'm generally very good at predicting the ending."

"He's crazy!" Liam shouted at his friend, "What are you doing?"

"Plenty of people think you're crazy Liam!" Oliver shot back. "But I know better. And I know better here too. The Numberfolk separated the worlds, and now, now, they want to destroy our side. Fredrick is going to set things right! No more irrationality. It makes us do horrible things. It causes anger and jealousy and bitterness. It's why I'm moving to Berlin, because my parents couldn't get along, they made all sorts of bad, irrational choices. Let him fix it. It will all be better. I won't have to move. I can stay here with you!"

Marco hesitated. *Maybe Oliver is right,* he thought. Emotions were horrible. The last few months showed him that. The grief from losing his grandfather was an ugly monster that never went away. If Oliver could fix that, if that was Fredrick's plan, it could be a good thing. He looked at his sister. She was abrasive and annoying. She drove him crazy and too often made him want to defenestrate her. But Maggie also made him so proud. He had an absurd need to protect her and help her and care for her at all costs. It was all irrational. The

$$-\frac{1}{6}(8x^2 - 137x + 570), (6, 4)$$

good and the bad.*

"Oliver!" Marco cried. "You're right. Emotions are irrational. They make you angry, and sad, and helpless. The rest of them though, the rest of them are good. It means you can love. Love is the most irrational of all. Love is caring about someone so much you'll do anything for them, no matter how crazy. It's all the Great Scale. You can't get rid of the bad without getting rid of the good too. But we need to feel. It sucks sometimes, I know. I feel sad every day that my grandad is gone . . . that's a good thing. It means I love him. Without irrationality we won't be friends. We'd all be robots!"

"Cylons!" Liam shouted, "He's turning us into cylons!"

Mr. Pikake stood, staring at Fredrick. He hadn't said a word. He hadn't moved.

"Mr. Pikake!" Marco demanded, "Talk to him! He's *your* brother!"

Fredrick grabbed Oliver's arm and led him out the side of the cave. No one followed. The seven of them all stood in shock. Maggie was the first to break the silence, "Why does the bad guy *always* tell the good guys their evil plan? They always think no one can stop them, then the good guys stop 'em every time. So typical."

"This isn't a movie, Maggie," Marco scuffed. He walked to his tutor, "Why didn't you do something?"

"I. I haven't seen Fredrick since I was a boy. I . . . I figured he was dead all these years. And all these years he never called, sent a postcard, nothing. He obviously doesn't care much about me or what I have to say." The professor collapsed to the ground and put his head between his knees.

"So that's it? We are just going to let him do whatever he's trying to do?" Maggie yelled. "No way! We are the good guys!"

Dr. Pound was leaning against the cave wall, she pushed herself up. "From what that young man said, it seems as though his plan, whatever it is, is to remove irrational numbers from existence. To rationalize everything. But that isn't possible. Irrational numbers are the glue that holds everything together."

"The glue?" Penelope asked.

"Without π, you can't have a circle, and without $\sqrt{2}$, you can't construct a right triangle with 1×1 sides. The rational numbers are

---

* Defenestrate is an excellent word that means "to throw out a window." It is also the only word Marco remembers from his 6th grade vocabulary test.

an incomplete system, there would be holes everywhere. Everything would break down," Dr. Pound explained.

Little crescents danced on the cave floor. They grew larger and larger. Around them, shadow bands waved back and forth alternating between light and dark making the ground look like an ocean. The eclipse continued to progress outside the cave and its effects were getting closer. The crescents grew into a large, plump silhouette.

"I may be of some assistance," a voice bellowed.

They turned to see Maxwell standing at the entrance of the cave.

"Maxwell?" Mr. Pikake was surprised, "What are you doing here?"

"I called him," Dr. Pound stated bluntly.

"I know you think I am unaware of this little club, but on the contrary. You are not fooling anyone," he began.

Fear surged down Marco's spine. He spun his head to Mr. Pikake who nodded furiously indicating he hadn't shared their secret.

"I believe I have some explaining to do . . ." he continued. Speaking rapidly and with incredible strength, Maxwell told his story. "I have allowed this to continue because I understand. As a boy, I too loved the numbers. I made promises to change things, to get back to a society of peace." He looked at Marco with regret in his eyes. "When my father explained what things really were, I knew I couldn't. Everything began in Numberville. As the numbers evolved and irrational beings formed, they felt threatened and used their powers to cast all irrationality into its own world, this world. Our world was birthed from chaos."

"Wait? We are numbers?" Maggie squealed.

"We are born from the Numberfolk, they are our ancestors. Think about it, everything around us, inside us, it is all numbers. It is a wicked and fragile house of cards, and any wrong move can cause everything to come tumbling down. Think of the deck. There are 52 cards, 52 weeks per year, 4 suits, 4 seasons. The 12 face cards for 12 months and should you sum the value of the entire deck with Jacks as 11, Queens as 12, and Kings as the lucky 13, you will find it is 364, with two jokers to readjust for the irrationality of our orbit."

"Irrationality of our orbit?" Penelope wondered out loud.

"Our world is irrational, that is how it was created and what it remains. It currently takes approximately 365.25 days to rotate about the sun, which means four rotations are near 1,461 days. Our years are but 365 totaling 1,460 every fifth of a score. A leap year attempts to catch us up. It is more complex than this. While we attempt to rationalize things, keep them nice and clean and make our days precisely 24 hours, this too is off. Thus, we have the leap second. A total of 27 seconds have been added thus far, each trying to adjust. It is in our nature as humans to strive for rationality, for the world where we were born, to find order in the chaos."

"Yeah, that's why people see patterns in clouds!" Liam added.

"Exactly, we are programmed to see patterns, some say they aren't really there, but they are, we just have not yet determined *what* they are and thus can only claim a cloud is an elephant in a clown hat because that is what we know." Maxwell glanced outside, it was getting darker and gloomier. "We don't have much time. Like it or not, the Numberfolk are at war. That is why we could not go back to the days of adoring them, we must continue to work to understand and gain authority, or we risk annihilation. Remove one card, one day, one season, one month, and all will spiral out of control. Fredrick doesn't care about this world, he wants to return to the original Earth, to Numberville, to the Rationals. He must be stopped. Only a certain type of person can cross over into Numberville, if another tries, they could rip apart both realms. It would certainly tear them to pieces."

Dr. Pound stepped up. "What do we need to do?"

The group huddled together in the cave and formed a plan. Maxwell explained that crossing over could only occur at special places during key times. The eclipse ensured alignment and harmony, creating a bridge which addressed the time. Phaseville was one of the few locations on Earth where the void was thin enough to make the journey.[*] Fredrick would need to harness the power and focus it on the bridge to unveil a hole between the two realities. The problem was, no one knew how he could do this or exactly where the bridge was located. They would have to search, look for clues, and move quickly before it was too late.

As they began to exit the cave, Marco glanced behind. "Hey! Less Numberfolk are following me!" He could see the end of the line now;

---

[*] Predictably, another place where the void is thin is, of course, the Bermuda Triangle.

$$-\frac{1}{3}(4x^2 - 65x + 264), (6, -2)$$

it wasn't long at all.

"That's it!" Mr. Pikake yelped. "That is how he is doing it. He is truncating, pulling away all the infinite digits to power his bridge."

"What's truncating?" Marco asked.

"Think of square roots." Mr. Pikake speech was frenzied. "The square root of 2 is a never-ending number that goes on and on forever. If you consider every digit as a trait or an experience, then as humans we are each a combination of everything we have lived before but also everything of everyone that has lived before us. Every invention, every adaptation, they are all a part of who we are. If we erase some of those, we begin to lose ourselves."

Marco imagined how different his life would be without his experiences. If his dad hadn't died, Maggie wouldn't exist. If he hadn't sat at that table with Oliver and Liam in kindergarten, he would be an entirely different person. He remembered the chalkboard in the warehouse. That's what Fredrick was doing. He was trying to document himself, each number representing a key point in his life. He was figuring out which he could change or remove safely.

"Truncating is taking away the digits. Trying to rationalize the irrational. Approximate it. Instead of $\sqrt{2}$, we use 1.41 or instead of π, we use 3.14, it seems harmless, but it is disastrous. Every digit is key. No matter how much we may believe certain experiences shaped us, in truth, they all shape us. Removing even one can be catastrophic."

They ran out of the cave to the dark, muted world. Marco saw Numberfolk everywhere, but they were different. Before, they seemed like people, living their lives, chatting, traveling. Now, they were like soldiers. Expressionless zombies marching toward the west.

"Something's changed!" he yelled. "The Numberfolk, they're marching, that way."

A strong force hit Marco from the side. Stunned, he tumbled to the ground. His eyes shot up to see what had hit him. A 1 lay next to him dazed and confused.

"I'm so sorry, sir!" the number spoke. "9 and I were ordered to Station 7. But I simply cannot find it, and I am already late."

Marco opened his mouth to answer but the number jumped up and scurried away, falling back in line.

Liam rushed over and pulled Marco up. "What happened? Looked

like you were hit by the Invisible Man. You fell out of nowhere."

"It was a number, a one. It spoke to me."

"What did it say, son?" Mr. Pikake pressed.

"That him and nine were sent to Station 7, and he was lost." Marco felt crazy even saying the words.

"Wait!" Penelope cried. "We can use that. If we know that both one and nine are being sent to station seven, and Maxwell said Fredrick had to focus the power to create the bridge, he must be calling them to a parabolic formation."

"Yeah! She's right." Maggie pulled her notebook from her backpack and crouched down. "The order must be $a(x + h)^2 + k$, and parabolas are symmetric, which means if one and nine are sent to the same station, the leading lady has to be halfway between them."

$$\frac{9-1}{2} = \frac{8}{2} = 4.$$

"So, the leader must be four units from each of them, four from 1 is 5 and four from 9 is 5. The leader is a 5," Penelope added.

"How can we find the station the leader is at?" Liam questioned.

"I think if we use what we already know, we can find the order." Maggie substituted the information in.

$$7 = a(1-5)^2 + k$$
$$7 = a(9-5)^2 + k.$$

"I get it!" Penelope smiled. "You put in the soldiers 1 and 9 because you know they are being sent to station $y = 7$."

"Yeah. And, since we know the leader is 5, that leaves us with two equations and two masks to find."

"It's a system," Marco pointed. "Simplify, then subtract the two."

Maggie obeyed. "Okay, the first is $16a + k$ because $1 - 5 = -4$ and squaring that gives 16."

"The second is $16a + k$ too," Liam whined. "If you subtract those, you'll get zero, all the masks will be gone."

"You need more information!" Mr. Pikake yelled. "A line, a march requires two ordered pairs, a quadradic needs three!"

"Marco." Maggie grabbed onto her brother's shoulders, "Can you find zero? If you can figure out where zero is sent, we can solve this."

He scanned the area, they were beginning to form their shape, some Numberfolk were marching but others stood still, already at their station. Smiling as he saw one had found his place, he looked to

$$-\frac{1}{3}(4x^2 - 60x + 216), (6, 0)$$

the left, there was zero. "I see him, but I don't have any clue what station he is at."

"Parlay!" Liam shouted.

"*What* are you talking about?" Maggie snapped.

"Parlay. When you ask someone, a soldier, their leader, they have to tell you.

"Liam, we have all seen *Pirates of the Caribbean*. I doubt that's our best move," Maggie moaned.

"Can't hurt." Marco shrugged and approached the 1. The number was looking straight forward at full attention. "*Um.* Excuse me? Could you possibly tell me where your leader is located?"

"Leader was deployed to Station 5," the one snapped. A surge of joy sprung up Marco's chest. He felt like a tourist who finally succeeded at getting one of the Royal Guards in the red coats and big fluffy black hats to speak.

"He says the leader is at Station 5!" Marco ran back to the group.

"Now both equations are just $16a + 5 = 7$, I don't think that helped," Liam questioned. "Also, if that one's so chatty, you probably should've just asked him the order . . . saved us the work."

"That's all I need!" Maggie insisted. "We don't need to find the order, we are just looking for the focus. Remember, we can use,"

$$y = \frac{1}{4(-p)}(x + h)^2 + k.$$

"Oh, yeah! And since $a = \frac{1}{4(-p)}$, that means if you find a, you can find p!" Liam exclaimed.

"It's 1/8!" Marco screamed. Using his x-ray vision, he knew right away who was hiding in $16x + 5 = 7$."

$$\begin{aligned} 16a + 5 &= 7 \\ 16a &= 2 \\ a &= \frac{2}{16} = \frac{1}{8}. \end{aligned}$$

"Right, that means p is -2. And . . ." Maggie continued.

"Since the leader is at $(5,5)$, the Death Star has been shifted. If $p = -2$, the lever is at $y = k + p = 5 + (-2) = 3$. That is two away from Station 5, so the focus is 2 stations above 5. The focus must be at $(5,7)$!" Penelope finished. Maggie looked mildly perturbed her

$-\frac{1}{2}(2x^2 - 29x + 104), (6,2)$

thunder was stolen. Pushing her feelings aside, she sprinted with the others toward the formation.

They couldn't tell where each station was, the terrain wasn't marked like a football field, but luckily, since One was at Station 7, they needed only to walk in a straight line toward 9 and look for the leader, 5, to find the focus.

Marco saw Fredrick. The parabola was taking shape, and the sun was seconds away from slipping entirely behind the moon. Beams of light began entering the Death Star and focusing directly onto Fredrick's location. Oliver stood next to the evil man, holding the book.

Squinting, Marco could make out the bridge. He saw a three on the other side. It was primal, like an animal ready to attack his prey.

"We've gotta get in there!" Dr. Pound screamed. The wind was roaring out of control and there was a high-pitched buzzing sound that whistled beneath the gusts.

"No! It will obliterate you!" Maxwell screamed. "It has to be Marco!"

"Liam!" Marco cried. "You get Oliver. Do whatever you have to do but get him out of the blast area!"

The two ran toward the focus. They dipped and dodged the rays entering the formation and the beams that reflected into the center of the shape. Liam got there first.

"You have to come with me Oliver."

"No! You need to trust me. This is the right thing."

"You will die!" Liam shrieked. "You are the pansy! It's like all the movies. This thing's gonna rip a hole in reality!"

"He said if I did this, I wouldn't have to move! Things would get better!"

"He lied! Well, I guess technically he didn't. You won't have to move if you're dead!"

"Why would he do that? Why would he lie?"

"What am I? Answer man?" Liam shot back. "How could I possibly know that? I do know he doesn't really care about you. I do! We do! Please listen to me."

Oliver pushed him away. Marco and Fredrick stood face-to-face. Fredrick looked down and revealed a wicked grin.

"What has happened will happen again," he smirked. "You know, you look just like him. Long ago, he stood right where you are. Tried to stop me too."

$$-\frac{1}{2}(2x^2 - 29x + 104), (6.25, -0.25)$$

He killed my father. Marco realized. *That's why my dad hid the map. He was trying to stop Fredrick and paid the ultimate price.* Every irrational feeling surged through Marco's body. He was infuriated. He was devastated. He was embarrassed he had made such a horrible mistake.

Marco lunged. The element of surprise caught Fredrick off guard, and they tumbled to the ground. As the final sliver of sun disappeared, the parabolic mirror amped up to full force. The formation was closed. A gigantic U-shaped ring of light burst towards the sky. Liam staggered back to Maggie and Penelope just in time. The beams twinkled with soft yellow and silvery light. Distracted by the brightness, Fredrick bested him. With a swift kick, Marco rolled toward the 9. His arm grazed a beam of light and a burning seer shot through his body. He could feel it everywhere, the pain was overwhelming. Like a lightning bolt struck his side, the sting tingled causing his right eye to fill with tears.

He was stuck. Caught between two rays of death, he searched for an exit. Then, he saw him. Oliver. His best friend. They locked eyes. Marco didn't see a madman. He didn't see anger. He didn't see someone who had betrayed his trust. What he saw was fear.

Oliver dropped the book and pounced towards Marco. It was too late. Fredrick grabbed him by the shoulders, lifting the boy at least two feet off the ground. Oliver was strong. He fought with everything he could, but Fredrick's arms were too long.

Click.

Click.

Click.

Click.

Marco began piecing the puzzle pieces together. What was Fredrick trying to do? He had opened a bridge between old Earth, Numberville, and here. Maxwell said he wanted to cross over, but the power could obliterate a human. *He's using Oliver.* As the full picture formed, a pain struck his gut like a knife. Fredrick was using Oliver like a doorstop. The energy was too much, if he forced Oliver to take the brunt of it, Fredrick could sneak through the bridge unharmed, destroying this Earth in the process.

Everything stopped. In the time between seconds, Marco looked past the chaos. Maggie and Penelope were huddled together, Liam stood over them. Mr. Pikake was screaming at Dr. Pound and

$$-\frac{1}{3}(8x^2 - 90x + 243), (6, 1)$$

Maxwell, insisting they save him. They looked bewildered, helpless. No one was coming. There was nothing they could do. The parabola was an impenetrable wall.

A final thought ran through his mind. He was back in the library with Mr. Pikake. "You will become a great master, Marco. One day, you will protect not only yourself, you will have the power to protect many." This was his destiny. He had found something remarkable, that no margin could contain. He was Marco the Great.

Taking a deep breath, he leapt. With every ounce of power inside him, he charged toward Oliver. The beams pierced his body, like a whip of fire, they clutched onto him.

"NOOOOO!" he could hear Maggie screaming.

He took one step and then another, pushing himself harder than he thought possible. Every tiny movement felt like an eternity as the beams seared through him. When you die, they say your life flashes before your eyes. What Marco saw was much worse. He saw the friends he would let down. He saw the evil villain grinning in victory. He saw his sister in terrible danger. He saw Oliver's face frozen in fear. And then he saw darkness. As the black began fading in, the pain was all encompassing. Then suddenly, it was gone. Having made it through the beam, still propelling forward, he tumbled onto Oliver releasing Fredrick's grip.

It hit.

As the massive burst of energy soared through the Death Star, he thought he could hear the Numberfolk singing. It was like a chorus all holding a single note as the entire formation gave off an incredible pulse. A stream of pure, white light glistened through the focus. Marco quickly grabbed onto Oliver and rolled behind Fredrick shielding him from the laser, throwing his entire body over his best friend.

Fredrick turned and let out a sharp scream. The beam had pierced his back. He looked down to see the shimmer glowing through a hole in his stomach.

"I can see them!" Maggie shrieked. "The Numberfolk!"

"It's the harmony," Dr. Pound yelled. "It is disintegrating the fabric of the void."

"Disrupt the formation!" Mr. Pikake ordered. They all ran to one Numberfolk or another trying desperately to pull them away from their stations. They couldn't. The hold was too strong. Dr. Pound signaled for everyone to pull together. Clasping onto the one, they each wrapped their hands around a part of its thin frame, tugging with all their might.

The beams shattered as the number tumbled to the ground. The last thing Marco saw was Fredrick, allowing the gravity to take over, his body crumbled. Like a distorted mirror, he watched as Mr. Pikake, his evil twin in a way, collapsed in the distance. Fredrick was no longer a variable in his life, there was no guessing, his brother had been vanquished.

With no strength left to fight, unable to hold on any longer, Marco finally allowed the darkness to take him.

* * *

Disoriented, he opened his eyes and saw an angel. The sun was reemerging, and a beautiful golden halo surrounded her head. "Marco? Are you okay?" she sang. As the blur lessened, he saw her. It was Penelope. A dopey smile spread across his face before he passed out again.

When Marco awoke, he was in the cave. Night had fallen and a fire raged next to him. He heard Maggie.

"Here's my question. That parabola, no one was at ground zero. Think about it. The whole formation was above the x-axis. They said every quadratic has exactly two soldiers assigned to hold it down, the soldier can be the same, but there are always two. So how is that possible?"

"Bug?" Marco coughed.

"Hey!" She ran to his side and cradled his head in her arms.

"I had a dream and you, and you, and you were there!" Marco theatrically pointed to his friends.

"He's fine," Maggie dropped his head with a thump on the cave floor.

"Wizard of Oz," Liam whispered to Penelope.

"Well done, son," Mr. Pikake came over and grabbed his hand. "You did it, you stopped it all. Your grandfather would be proud." He smiled softly. "You had a hard fall and all that energy was running through you. Get some rest, we head back in the morning." He turned and raised his voice, "All of you to bed! We've got a long trek tomorrow."

Marco lay there and watched as his sister, Penelope, Liam, and Oliver smiled, and laughed, and joked. Liam teased Maggie and she got angry, then she got even, then they all smiled again. The feelings

he had inside were the best kind. They were messy, and dirty, and ugly, and out of place, and all over the place, and as irrational as they could be. Inside, he was so happy, and he was also so sad. And it was wonderful.

a new number?

$\sqrt{-1} = i$

The Complex Plane

Imaginary Isle

Real Continent

INTEGER STATE

Village of Natural

THE DARK CONTINENT

OUTSKIRTS OF RATIONAL

Country of Rational

$\sqrt{2}+0i$

$7-ei$

$1-i$

$10^2+10^6 i$

$6+0i$

$42+42i$

$\pi+\pi i$

$3-6i$

$7-\frac{i}{2}$

$4+0i$

$\frac{1}{2}+2i$

$16-i\sqrt{3}$

COMPLEX NUMBERS

$a+bi$

Real → a, imaginary → bi

all Real numbers are complex!

2 is really $2+0i$ ← no imagination

others like $3-4i$ have an imaginary part.

GROUND ZERO IS A PLANE!

SYNTHETIC DIVISION

same thing . . . less writing

$x^3-4x+6 \div (x+1)$

$1x^3+0x^2-4x+6$

make sure to take the evil twin!

$$\begin{array}{r|rrrr} -1 & 1 & 0 & -4 & 6 \\ & & -1 & 1 & 3 \\ \hline & 1 & -1 & -3 & 9 \end{array}$$

add the columns!

$x^2-x-3 \text{ R } 9$

$x^3-4x+6 \div (x+1) = x^2-x-3+\frac{9}{(x+1)}$

20

AFTERMATH

"MATHEMATICS IS NOT ONLY REAL, BUT IT IS THE ONLY REALITY."

-MARTIN GARDNER

Yuck. It had been two days since Marco last brushed his teeth. He remembered on *Survivor* watching the contestants use some kind of stick as a toothbrush. It was crazy they didn't let them bring toothbrushes. *One million dollars and permanent tooth decay!* What a prize. He foraged around and found a stick. Doing his best to scrape each tooth it did nothing for the rotten taste.

Maxwell had disappeared during the night and only KFUN and Dr. Pound headed toward the van. Down was much easier than up. She insisted that if they kept their speed, they could make it back in a few hours.

"Mr. Pikake," Maggie called back.

"Yes dear?" he responded kindly.

"You promised you'd tell us about synthetic division."

Everyone groaned.

"Oh, don't act like you don't want to know too." She stuck out her tongue.

When they stopped at a stream to gather more water, Mr. Pikake fulfilled his promise.

"Suppose you want to divide $4x^3 + 2x^2 - 16x - 8$ by $x - 2$. Synthetic division allows you to avoid writing all the steps out, it's the same as any other division *except*." He held his chin up in suspense.

"Yeah, yeah, go on, except . . ." Maggie pushed.

"Except, it can be tricky when the polynomial you are dividing by is not monic, that is, when the leading coefficient isn't one. I'll show you how when dividing by a line, other polynomials are more complex. Synthetic division essentially ignores the masks and uses only the coefficients."

"That makes sense. You aren't really using the masks anyway. When we divided before, it was like x times what makes $4x^3$. So, it is always a brigade of one less mask. The only thing that really mattered was the coefficients anyhow," Maggie said matter-of-factly. Penelope was the only other person listening.

"For each step, just like in normal division, you multiply but rather than subtract, when dividing synthetically you add." Mr. Pikake pulled out a notepad, on one side he wrote the traditional division symbols, on the other, he wrote it upside down and without any x-masks.

"Why's that one have 2 when it's x *minus* 2?" Maggie pointed to the upside-down division house.

"When the leading coefficient is 1, like in $x - 2 = 1x - 2$, you will always first multiply by whatever is leading the polynomial you are dividing, in this case it is 4. When you multiply that 4 by the negative two, you will of course produce -8. You then subtract negative 8 which is synonymous to adding 8. Synthetic division takes out all that ugly subtraction and negative negative mess, so you just add."

$$x-2\,\overline{)\,4x^3+2x^2-16x-8} \qquad 2\,|\;4\;\;2\;\;-16\;\;-8$$

"That's really why you like it," Maggie teased. "Because you hate subtraction."

Mr. Pikake's Cheshire grin took over his face. "You may know me too well." He winked.

"Okay, so what do I do?"

"To start, bring down the first coefficient, then the same as any division, you multiply. Lastly, rather than subtract, you add, again because of the negative negative." He slowly wrote out the first step in each.

$$\begin{array}{r} 4x^2 \qquad\qquad\quad \\ x-2\,\overline{)\,4x^3+2x^2-16x-8} \\ -(4x^3-8x^2+0x+0) \\ \hline 10x^2-16x-8 \end{array} \qquad \begin{array}{r|rrrr} 2 & 4 & 2 & -16 & -8 \\ & \downarrow & 8 & & \\ \hline & 4 & 10 & & \end{array}$$

"Oh, I see. You brought down the 4, then multiplied $2 \cdot 4$ and put that under the other 2. So next you'd multiply $2 \cdot 10$ and put that under the -16?" Maggie pointed as she spoke.

"Precisely! And we continue in the same manner all the way through. One thing to remember is you must include every brigade. If we had $3x^2 + 2$, synthetically, you would need to utilize 3 0 2 to include the invisible $0x$ term." He completed the process in long smooth strokes.

$$\begin{array}{r} 4x^2+10x+4 \qquad \\ x-2\,\overline{)\,4x^3+2x^2-16x-8} \\ -(4x^3-8x^2+0x+0) \\ \hline 10x^2-16x-8 \\ -(10x^2-20x+0) \\ \hline 4x-8 \\ -(4x-8) \\ \hline 0 \end{array} \qquad \begin{array}{r|rrrr} 2 & 4 & 2 & -16 & -8 \\ & \downarrow & 8 & 20 & 8 \\ \hline & 4 & 10 & 4 & 0 \end{array}$$

"Now we have found the coefficients to be 4, 10, and 4 with no remainder. Thus, $4x^3 + 2x^2 - 16x - 8$ divided by $x - 2$ produces $4x^2 + 10x + 4$."

Maggie's eyes widened. "That's *amazing*! And, it is so much less writing!"

"Yeah, but you could do it without division too," Oliver whimpered. "Look there, a wizard cast a doubling spell, reverse time to see before it was $2(2x^3 + x^2 - 8x - 4)$. Now, look at the pairs. The first two terms have an x^2 in common and the second two share -2. That gives you $2\big(x^2(2x+1) - 2(2x+1)\big)$. Another wizard and you know this is really, $2(2x+1)(x^2-2)$." He pushed an awkward smile on his face.

Maggie would have been thrilled. Oliver had ripped the polynomial to shreds. But she wasn't. A growl grew in her gut as she flashed back to her brother nearly tearing himself apart to save his traitorous friend. She let out a grunt and narrowed her eyes before turning back to the professor and pleading, "Give me one to try."

"Alright, how about $3x^3 - 2x + 5$ divided by $x + 2$?"

"We've gotta keep moving!" Dr. Pound had already reattached her backpack and continued down the hill. Maggie grabbed her notebook and pen and worked as she walked.

"First, I set it up. It needs to be $x - c$ and I have $x + 2$. But that's really $x - (-2)$ which gives me . . ."

"Did you remember the x^2 term?" the professor called from behind her.

"Of course I did," Maggie whined. "Now, bring down the 3, multiply, add, repeat."

She looked at her result, confused. "I got $3x^2 - 6x + 10$ and then another -15?"

-2 | 3 0 -2 5
↓ -6 12 -20
3 -6 10 -15

"Haha," the tutor bellowed. "And what's that mean?"

"It has to be the remainder. That's what happens when you divide, whatever is left is the remainder."

"Excellent. And what does that mean?"

Maggie felt like the professor was a broken record. *What's that mean? What's that mean?* She remembered his speech about the power of figuring things out and tried to appreciate the freedom he was providing rather than be annoyed at all the open-ended questions. "Well, if I divided say 72 by 10, I'd get 7 remainder 2 which means I can split 72 cookies into ten bags by placing 7 in each bag and have two leftover for me. Which if I was being nice, and wanted to also split the 2 leftovers, I'd have to break each cookie into 10 pieces and place a piece from each cookie in a bag making the total amount per bag 7 and 2/10."

"Those would be scraps!" Liam interjected. "Better off splitting each cookie into fifths and putting one of those larger chunks in the bags. Nobody wants crumbs."

So that's how remainders work. Marco felt amazing. Dividing had been one of his strong suits, and he was thankful dividing polynomials worked the same way. He never did understand why you put the leftovers after the letter R. Now it made sense.

He looked around at the scenery and felt good. They had, after all, saved the world. *This is how superheroes must feel,* he thought. *Always saving the world but no one else even knows.* A strange feeling began bubbling inside him. It was almost like he was lighter than he used to be. Panic struck for a moment when he wondered if Fredrick had truncated him, taken away some of his life experiences. Looking back, a long line of Numberfolk were still following along behind him. He smiled, *Must be my imagination.*

"That means," Maggie continued, "$3x^3 - 2x + 5$ divided by $x + 2$ is really $3x^2 - 6x + 10$ remainder -15. If I apply the same logic as the cookies that's . . ."

$$3x^2 - 6x + 10 - \frac{15}{x+2}?$$

"Impressive! That it is. That last bit is what we call a rational order. Those are orders with a vinculum. The domains of such a command can be tricky, they are always leaving someone out." Mr. Pikake shook his head in disdain.

"Is that what we will learn next? Do the incantations work on those too?" Maggie urged.

"No and yes. Incantations work on all functions, so I imagine you won't struggle much with the other orders. Master the mother. . ."

"Perfect the pupils," Liam, Oliver, Penelope, and Maggie all chimed in.

Oliver had been selecting his words carefully. Mainly keeping quiet, he had replaced his class-clown mask with a kind, soft-spoken, and regretful demeanor. Last night, everyone was on a high after preventing the end of the world. Today, as reality set back in, a lingering taste of anger over his betrayal emerged.

Not for Marco. Sure, his friend almost destroyed Earth . . . well, counter-earth, but he couldn't really blame him. Oliver was being torn away from his friends, his school, his whole life. He was just trying to hold onto everything he loved. Marco understood that more than most, he had made the same mistake in trusting Fredrick. Plus, when it mattered, Oliver had tried to save him.

Teenage boys weren't really known for their deep heart-to-hearts, and Marco wasn't sure what he'd say anyway. In a show of forgiveness, he threw his arm around his best friend's waist and they walked side-by-side.

"Okay, one last question!" Maggie shouted. "The formation, with Fredrick, something was wrong with it. It didn't have anyone at ground zero, and you said there were *always* two soldiers assigned to ground zero in a quadratic. So, I have been doing some research."

"How can you do research? There is literally no Google up here," Liam spurted.

"With my *brain,*" she whined back. "When you complete the square, there is always the square root of $b^2 - 4ac$. Now, when you have a quadratic where the $4ac$ is more than the b^2, you end up with a negative. Like if it was $4x^2 + x + 10$, then $b^2 - 4ac$ would be $1^2 - 4(4)(10)$, which is 1 minus 160 or -159. This is *exactly* what happened with Fredrick's formation. It is impossible to take the square root of a negative, any number times itself is positive."

"Or zero!" Penelope found a loophole.

"Oh yeah, or zero. The point is, it can't be negative. You were obviously wrong. Not every quadratic has two soldiers assigned to Station 0," Maggie concluded with an umph.

Marco was starting to feel dizzy, he held tight to Oliver.

"Bruh, you alright?" he heard his friend ask.

"Outstanding observation!" Mr. Pikake boomed. "There must have been an anchor, otherwise the formation would float away. You must widen your mind, invoke your imagination! Recall ground zero is not only a line in the sand, but evolves into another dimension, a miraculous wall that extends to a new plane."

"Are you saying there are anchors in the sky?" Liam scoffed before

nervously glancing upward to check for a Death Star above them.

"We have studied the continent of Real. But," he popped, "there is one type of Numberfolk you have not yet been formally introduced to. The *imaginary*." The professor widened his hands and traced a circle in the air.

"That's cheating!" Maggie squealed. "You can't make up pretend numbers just because you're wrong."

"*Ahh*, the name is misleading. Not-real would be a horrible moniker. Upon their discovery they were misunderstood and the tag stuck. For when ground zero shoots up, it presents to us the complex plane. Complex Numberfolk are in the form $a + bi$ where a and b are from the continent of Real."

"If I really am a Numberfolk, I'd think I'd be complex rather than irrational," Penelope quipped.

"All Numberfolk are complex creatures! The irrationals as well," Mr. Pikake declared. "They possess a real and an imaginary component. It just turns out the Reals lack imagination. For a 3 is nothing other than $3 + 0i$. But there is $2 + 0.5i$ and $\pi + 2i$ and on and on. When a polynomial formation appears to have no anchors, or missing anchors, they are there, hiding on the complex plane."

"I . . . I . . ." Marco grunted quietly.

"What's i?" Maggie yelled back.

"i is exactly what you have found, it is the square root of -1. That is $i = \sqrt{-1}$."

"Wait." Maggie stopped in her tracks. "So you are saying in those cases, where $4ac$ is bigger than b^2, where you get a negative under the square root, there is a solution, a Numberfolk at Station 0?"

"Perfection!" Mr. Pikake purred bouncing along. The lineup had changed as they walked. With the danger behind them, Dr. Pound led, followed by Penelope, Maggie, and Mr. Pikake. The boys trailing behind. "You see, there *were* soldiers assigned to ground zero, you just couldn't see them as they lay on the wall. All polynomials of degree n have exactly n anchors."

Suddenly Oliver fell. He tumbled down the hill knocking over Liam in the process. Dr. Pound was already waiting at the bottom, and Maggie and Penelope jumped out of the way just in time. Everyone ran to the base.

"What happened?" Mr. Pikake asked concerned.

"I don't know. I was walking with Marco. He . . . he had his arm around me, and I had mine across his shoulder. Then out of nowhere he ducked out or something and it threw me off balance," Oliver explained as he picked himself up and dusted the dirt and leaves from his clothes.

Liam looked around, "Where *is* Marco?"

"Marco? Marco?" Maggie started screaming. The others quickly joined in.

"Marco! Marco!" Maggie burst into tears. "Marco!" She was running around franticly. "Where is my brother?!" she cried. Tears streamed down her face. She was shivering beneath the hot sunlight.

Dr. Pound stood still, a statue. "It was always a possibility," she whispered. Mr. Pikake grabbed onto Maggie and squeezed her tight. "It will be okay. It will be okay."

She pushed away; but he held strong. She banged on his chest. "Let go of me! Let go of me! We have to find my brother!" Maggie felt like she was dying, there was no air to suck in. Marco was always there, her whole life, no matter what happened, her brother was there.

Holding her tightly, he bent down. "Maggie. Listen to me," he said firmly. "Marco was hit by the beams. He is okay. I promise. He is just not *here*."

"What do you mean he's not here?" she yelped. "Where is he?"

"He's in Numberville. But I swear to you, we will get him back."

* * *

The entire hike Marco had one strange sensation after another. First came the dizziness. Next, he was lighter as if he might float away. It reminded him of *Back to the Future*, the way Marty McFly must have felt as he was erased from history. Then came the glitches. That was best described as when you are on a Zoom call and someone's internet isn't great; they hiccup in and out. If you could feel glitchy, that's what it would be like, your entire body hiccupping in and out. The last thing he remembered felt like the blip; like Thanos snapped and he was snatched from reality. Then everything went dark, and he felt better. He chuckled to himself as he realized the whole experience read as a lineup of his favorite movies.

Marco raised his hand to block the harsh light of the sun. *Where is everyone*? He was still in the forest, but he was alone. In the distance he saw a shadow approaching. *Must be Mr. Pikake.*

Slowly walking towards the form, it wasn't until it was right on top of him that he could make out their face.

"Dad?"

Stay tuned for the final installment of *Marco the Great*!

GLOSSARY

The Charmed Ones: Magical creatures first created by Zil (the nothing) Each charmed one contains enchanted particles called Spogs. They created Natural in their image and are what we know today as the digits.

Numberville: A fantastical world where Numberfolk live.

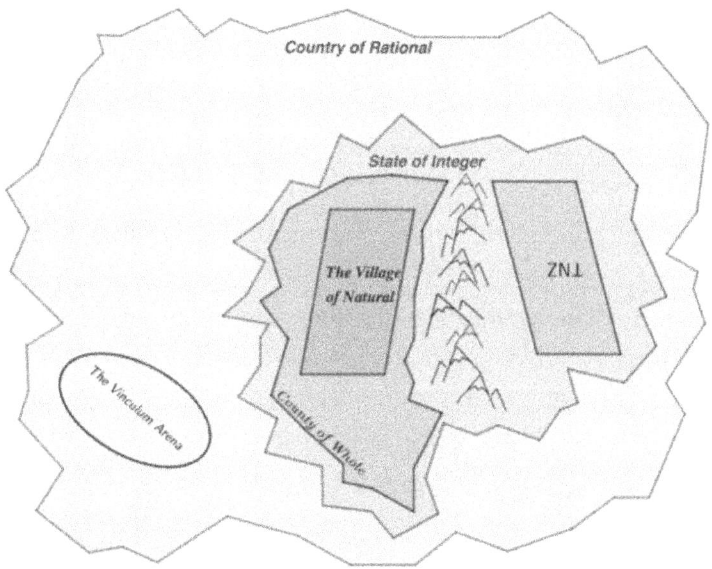

Village of Natural: The creation of the charmed ones. Contains only Numberfolk made of whole and positive particles. $1, 2, 3, \ldots$

County of Whole: The Village of Natural lies within Whole County. All Natural Numberfolk are also Whole. Zero is the only resident who lives in Whole County outside the Village.

TNZ: aka The Negative Zone. The reflected image of Natural created by the Mirror of Wonders to balance the Great Scale. Only Numberfolk made of whole and negative particles may reside in TNZ: $\ldots, -3, -2, -1$.

Integer: The State made up of Whole County and TNZ. Residents include, $\ldots, -3, -2, -1, 0, 1, 2, 3, \ldots$.

Rational: The country that contains Integer.

Rationalite: A protected class of Numberfolk created by the Vinculum. Composed by tearing the particles of Integers, a Rationalite is formed when two Integers are divided by a vinculum.

Logos: Rationalites that are not Integers such as $1/2$, $2/3$, or $5/4$.

Duels: A way to settle a disagreement. When two Numberfolk duel, the result is the combination of their Spogs. If a 2 duels a 3, the result is $2 + 3 = 5$. Dueling in Natural always results in a larger, more powerful resolution. Dueling in TNZ, $-2+(-3)=-5$, always results in a smaller, more negative resolution.

Evil Twin: A doppelganger. Evil Twins are identical but made of opposite particles. For instance, the evil twin of 2 is -2 and the evil twin of $2/3$ is $-2/3$.

Mask: A mask is simply an unknown Numberfolk hiding from us. We utilize letters such as $a, b,$ or x to represent these tricksters.

Hunting: The art of identifying an unknown Numberfolk and ripping off their mask.

Vanquishing: The annihilation of a Numberfolk through a forced duel with an evil twin. To vanquish the 3 in $x+3=5$, force a duel with the evil twin, -3, keeping the scales balanced.

$$\begin{aligned} x+3+(-3) &= 5+(-3) \\ x+0 &= 2 \\ x &= 2. \end{aligned}$$

Vanquishing is vital in hunting as duels with evil twins always produce a 0.

Partner: The partner of a Numberfolk is the only one in existence who can make the Numberfolk whole. The partner of $2/3$ is $3/2$ as,

$$\frac{2}{3}\cdot\frac{3}{2}=\frac{6}{6}=1.$$

Proliferation: The act of increasing or suppressing a Numberfolk's size. In the hunt $3x=6$, the x-mask has been enlarged to three times its magnitude. We require shrinking it by casting a suppression spell with the partner of 3, $1/3$.

$$\begin{aligned} 3x &= 6 \\ \frac{1}{3}(3x &= 6) \\ 1x &= 2 \\ x &= 2. \end{aligned}$$

Proliferation is vital in hunting. As partners always make a whole, we can use this spell to obtain a single mask or bring a mask back to their original size in order to identify them.

Obliteration: The act of casting a proliferation spell to obtain a single mask. For instance, we can cast a Proliferation spell of 2,

$$\frac{3}{2}x = 12$$
$$2\left(\frac{3}{2}x = 12\right)$$
$$3x = 24.$$

However, we still have a mask increased to three times it size. If we instead obliterate using the partner of $3/2$, we can hunt the mask in a single step:

$$\frac{3}{2}x = 12$$
$$\frac{2}{3}\left(\frac{3}{2}x = 12\right)$$
$$x = 8.$$

Mirror: Numberfolk often disguise themselves using the reflection power of 1. By casting a Proliferation spell with a 1, or using a mirror, a Numberfolk can become unrecognizable.

$$\frac{1}{2} = \frac{1}{2} \cdot \frac{2}{2} = \frac{2}{4} = \frac{1}{2} \cdot \frac{3}{3} = \frac{3}{6} = \frac{1}{2} \cdot \frac{4}{4} = \frac{4}{8} \cdots$$

Traits: Numberfolk DNA. Traits determine what a Numberfolk is made of; they represent everything within the value. To find a Numberfolk's traits, you must rip them apart. It is often beneficial to rip them into their prime components.

$$36 = 6 \cdot 6 = (2 \cdot 3)(2 \cdot 3) = 2^2 \cdot 3^2.$$

The Numberfolk 36 is composed only of twos and threes, two of each to be precise.

Ripping: The act of tearing apart a Numberfolk to identify their traits. This can be helpful in many situations including hunting, working with or identifying Rationalites, evaluating the strength of a jail cell, and finding the linear factors of a polynomial formation.

Recta Defensive: When hunting a Numberfolk hiding themselves in the form $ax + b$, the Recta Defensive may be used by first Vanquishing the collaborator b followed by Obliterating the a.

$$3x + 2 = 8$$
$$3x + 2 + (-2) = 8 + (-2)$$
$$3x + 0 = 6$$
$$\frac{1}{3}(3x = 6)$$

$$x = 2.$$

Relation: A set of ordered pairs that associates a Numberfolk from group A called the ***domain*** to a Numberfolk from group B called the ***range***. For instance, a group of Numberfolk soldiers and their location.

Function: A relation in which every soldier in the domain is assigned exactly one location in the range.

One-to-one: A special type of relation in which no soldiers are assigned to the same location (no ambushes).

Onto: A special type of relation in which every location is guarded; no locations are left unattended.

Order: A rule that shows where every soldier has been deployed to. For example, the order $f(x) = 3x + 2$ sends soldier 1 to $f(1) = 5$.

Composition: When an order is stuffed into another, such as $f\big(g(x)\big)$.

March: When Numberfolk soldiers form a line. Orders for such a formation are often expressed as $y = ax + b$, where x is the soldier.

Center Stage: Also called the y-axis, center stage is the line that cuts the real plane in half vertically. To the right of center stage are all the positive Numberfolk soldiers, and to the left of center stage are all the negative Numberfolk soldiers.

Ground Zero: Often called the x-axis, ground zero is the line that cuts the real plane in half horizontally as well as the plane that extends vertically to the heavens. Above ground zero on the battlefield lies all the positive stations; below ground zero lies all the negative stations.

Slope: The rate of change of a formation. For a march, the rate can be found by identifying two soldiers and their positions. The difference of their position divided by the difference of the soldiers will uncover the slope. If soldier 1 is sent to location 3 and soldier 5 is sent to location 11, then the rate of change of the march (how one must travel to get from one station to the next) is,

$$\frac{11 - 3}{5 - 1} = \frac{8}{4} = \frac{2}{1}.$$

In a march, the slope is the a in $y = ax + b$.

Intercepts: The locations or soldiers on the axis. The y-intercept represents the station where soldier 0 has been sent in an order. The x-

intercept represents the soldiers assigned to ground zero, or location zero. In a march, the y-intercept is the b in $y = ax + b$.

Warning Sign (factorial): A group of Numberfolk using multiplicative power. The leader is the largest Numberfolk and is displayed followed by an !. Such a group requires one of each Natural less than the leader.

$$4! = 4 \cdot 3 \cdot 2 \cdot 1.$$

Majors (exponents): A Numberfolk soldier who has been appointed to lead other identical soldiers. The notation 3^6 indicates a 3 is leading a brigade of 5 other soldiers to total six in the unit. Every soldier in the brigade must also be a 3, and the leader is included in the count.

$$3^6 = 3 \cdot 3 \cdot 3 \cdot 3 \cdot 3 \cdot 3.$$

Sand Worm: A strategy in which majors create a brigade beneath a vinculum.

$$\frac{1}{3^6} = \frac{1}{3 \cdot 3 \cdot 3 \cdot 3 \cdot 3 \cdot 3} = 3^{-6}.$$

Scarecrow: An empty brigade meant to scare off intruders. As no soldiers are in the brigade, a scarecrow has no multiplicative power and thus is equivalent to the Numberfolk with no multiplicative power, 1.

$$3^0 = 1.$$

Permutation: A way of counting items when different arrangements should be counted. For instance, 123 Elm differs from 321 Elm.

$$P(n, r) = \frac{n!}{(n - r)!}.$$

Here, n is the number of items and r is how many you wish to pick.

Combination: A way of counting items when different arrangements should not be counted. For instance, if picking marbles from a bag, selecting blue, green, red is the same result as selecting green, red, blue.

$$C(n, r) = \frac{n!}{r!\,(n - r)!}.$$

Jail: A Numberfolk prison, designed by General Three, to rip apart brigades. A $\sqrt{}$ cell rips a brigade into two, while a $\sqrt[3]{}$ rips a brigade into three. Thus, $\sqrt{x} = x^{1/2}$ and $\sqrt[3]{x} = x^{1/3}$. Numberfolk whose traits can be split into an equal number of identical groups as the cell can escape and maintain their rationality but are forever torn. Numberfolk who cannot create such groups within their traits cannot escape and the confinement drives them crazy, creating irrational Numberfolk.

Polynomial: From "poly", meaning many, a polynomial is a Numberfolk order with a finite number of brigades containing masks of each size.

$$3x^4 + 5x^3 + 0x^2 + 2x - 3$$

is an example of a polynomial order. These create curvy formations and we write them starting with the largest brigade and moving down.

Degree (of a polynomial): The size of the largest brigade. In the polynomial

$$3x^4 + 5x^3 + 2x - 3,$$

the degree is 4.

Anchors: An anchor of a polynomial is a Numberfolk soldier assigned to ground zero. These hold down the formation. More identical soldiers creates a heavier pull or a stronger gravity.

Multiplicity: The number of identical soldiers assigned to ground zero. For example, in the polynomial,

$$y = x(x - 1)^2(x + 5)^3,$$

the x occurs only once and thus the soldier 0 has multiplicity one. While the factor $(x - 1)^2$ tells us there are two 1-soldiers anchoring the formation and thus this root has a multiplicity of two. Similarly, the soldier -5 has a multiplicity of three. Even multiplicities "bounce" off ground zero, while odd multiplicities "cross". The larger the multiplicity, the more gravity at said soldier's location causing the formation to take on a flatter appearance.

Quadradic: A polynomial where the largest brigade is made of two soldiers.

$$x^2 + 6x + 5$$

is an example of a quadratic. These orders can take on many forms. The standard form is $y = ax^2 + bx + c$; vertex form is,

$$y = a(x + h)^2 + k.$$

It is called vertex form because you can easily identify the leader in the center of the formation and their location at $(-h, k)$.

Transformations: A transformation is a way to start with the most basic formation and move it or change it to create something new. You can move the formation left, right, up, or down, and you can stretch or squish the formation vertically and horizontally. It is also possible to flip the formation over center stage and/or ground zero.

Incantations: The act of applying transformations to an order to change the shape of the formation.

Insurrection: An attempt to break free of a jail cell. To complete an insurrection, first determine the number of rooms of the cell. A $\sqrt{\ }$ cell has two rooms while a $\sqrt[3]{\ }$ cell has three. Rip a Numberfolk into their traits creating identical groups based on the number of rooms in the cell. Only one trait from each group can safely escape.

Complex Plane: Another dimension above the Real Plane. The extension of Ground Zero to the heavens.

Imaginary Numberfolk: A Numberfolk with an imagination who must live on the Complex Plane. For instance, $i, 3i,$ and $-\pi i$ are all imaginary Numberfolk. Imaginary Numberfolk contain $\sqrt{-1} = i$.

Complex Numberfolk: All Numberfolk are complex and can be seen as $a + bi$, where a and b are real values. Numberfolk with no imagination such as 2, $\sqrt{2}$, or -100 have a b value of 0. For example, $3 = 3 + 0i$. Numberfolk with an imagination may have both real and imaginary components such as $-2 + 5i$ or $\pi - \frac{i}{2}$.

HINTS AND TIPS

To fold easily and without ripping or damage, the slopes are restricted to the easy-to-graph values: $\pm\frac{1}{2}, \pm\frac{3}{4}, \pm 1, \pm\frac{4}{3}, \pm 2$, and ± 3.

Chapter 1
The location of two soldiers close to center stage.

Example
$(1, -2)$ and $(-3, 2)$

> The change in soldiers is $1 - (-3) = 4$ and the change in their locations is $-2 - 2 = -4$. This means each soldier is one station less than their neighbor. As -3 was sent to Station 2, this means -2 is sent to Station 1, -1 is sent to Station 0, and 0 is sent to Station -1.
>
> The order for the march is $y = -x - 1$.

Chapter 2
The order is disguised with a vinculum.

Example
$-{}^{3}/_{4}x + {}^{23}/_{2}$

> From the order we can see the change is $-3/4$ and 0 was sent to 23/2. Locate 11.5 on the right and move up 3 and left 4 to identify a second point.

Chapter 3
The change and the location of one soldier.

Example
$\frac{\Delta y}{\Delta x} = -1$ and $(8, 2)$

> If soldier 8 was sent to Station 2 with a change of -1, then (moving backward) 7 was sent to 3, 6 to 4, 5 to 5, 4 to 6, 3 to 7, 2 to 8, 1 to 9, and 0 to 10.
>
> The order for the march is $y = -x + 10$.

Chapter 4

Mapping pattern.

Example

$1 \to -3, 2 \to -2, \ldots$

This tells us 1 was sent to Station −3, or $(1, -3)$, and 2 was sent to −2, or $(2, -2)$. The pattern shows us each soldier is 1 station above their neighbor. This means 0 must be one station below where 1 was sent. Hence, $0 \to -4$.

The order for the march is $y = x - 4$.

Chapter 5

The order as a function.

Example

$f(x) = -2x + 11$

From the order, we can directly find the change is −2 and 0 is sent to $f(0) = -2(0) + 11 = 11$. Have you found the hidden message in Chapter 5?

Chapter 6

The location of two soldiers far from center stage.

Example

$(57, 48)$ and $(92, 83)$

The change is $(83 - 48)/(92 - 57) = 35/35 = 1$. Since the rate, position to soldier, is 1, and Zero is 57 soldiers away from the closest command, the location of Zero is also 57 units away from the location of soldier 57. As 57 is at location 48, that places Zero at $48 - 57 = -9$.

The order for the march is $y = x - 9$.

Chapter 7
An order providing the change and the command of a single soldier.

Example
$3(y+3) = 4(x-6)$

The form of this order is $y - y_1 = c(x - x_1)$, therefore we can read this as $y + 3 = 4/3(x-6)$. The change is $4/3$ and we know -3 was sent to 6. Hunting y gives us,

$$\begin{aligned} y &= \frac{4}{3}x - 8 - 3 \\ &= \frac{4}{3}x - 11. \end{aligned}$$

Thus, Zero was sent to Station -11.

Chapter 8
Orders with exponents.

Example
$2^2y + 7 \cdot 2^3 = 3(x + 2^2)$

Determining each power, we have $4y + 7 \cdot 8 = 3(x+4)$. Hunting y we find,

$$\begin{aligned} 4y + 56 &= 3x + 12 \\ 4y &= 3x + 12 - 56 \\ 4y &= 3x - 44 \\ y &= \frac{3}{4}x - 11. \end{aligned}$$

Chapter 9
Orders with factorials.

Example
$y = 3(3! - x) + x$

Since $3! = 3 \cdot 2 \cdot 1 = 6$, this order is really,

$$\begin{aligned} y &= 3(6 - x) + x \\ &= 18 - 3x + x \\ &= -2x + 18. \end{aligned}$$

Chapter 10
Dot-forms.

Example
$y = 0.\overline{9}x - 15.24\overline{9}$

As $0.\overline{9} = 1$, this means $0.0\overline{9} = 0.1$ and $0.00\overline{9} = 0.01$. Therefore, $0.\overline{9}x = 1x = x$ and $15.24\overline{9} = 15.25$. This order is really $y = x - 15.25$.

Chapter 11
Transformations.

Example
$y = \frac{1}{2}\left(x - \frac{59}{2}\right)$

These orders are written as transformations of the parent, $y = x$. Here we see $y = x$ was first shifted right by 59/2 units, followed by a vertical shrink by a factor of 1/2. We could also write this order as $y = \frac{x}{2} - 14.75$.

Chapter 12
A different perspective.

Example
$x = 17 - y$

We normally view an order from the perspective of y, the location. Now we are given the viewpoint of the soldiers. Hunt y to determine the order. If $x = 17 - y$ then $y = -x + 17$.

Chapter 13
The Pythagorean distance.

Example
$c^2 = (12 - 8)^2 + (-4 + 8)^2$

In Chapter 13 we see the distance from two points (x, y) and (z, w) can be found using the Pythagorean Theorem, $a^2 + b^2 = c^2$. The clues in this chapter provide the ordered pairs. For $c^2 = (12 - 8)^2 + (-4 + 8)^2$, comparing to the theorem $c^2 = (z - x)^2 + (w - y)^2$ we find $(x, y) = (8, -8)$ and $(z, w) = (12, -4)$. The change between these points is $(-4 + 8)/(12 - 8) = 4/4 = 1$. We can use any of the tools previously explored to identify Zero's location.

$$y + 8 = x - 8$$
$$y = x - 16.$$

Note, all clues in this chapter are in this form with the difference of the x values presented first.

Chapter 14

Focus on the order.

Example

$4(4 + \sqrt{y})(4 - \sqrt{y}) - 2x = x$

Can you find the order hidden in the mess? Simplify each equation to isolate y.

$$\begin{aligned} 4(4 + \sqrt{y})(4 - \sqrt{y}) - 2x &= x \\ 4(4 + \sqrt{y})(4 - \sqrt{y}) &= 3x \\ 4\left(16 - 4\sqrt{y} + 4\sqrt{y} - \left(\sqrt{y}\right)^2\right) &= 3x \\ 4(16 - y) &= 3x \\ 16 - y &= \frac{3}{4}x \\ y &= -\frac{3}{4}x + 16. \end{aligned}$$

For a bonus challenge, consider the domain.

Chapter 15

Fill in the blank.

Example

$(ax + b)(4x + 8) = -3x^2 + 58x + 128$

Multiplying we have,

$$(ax + b)(4x + 8) = 4ax^2 + 8ax + 4bx + 8b.$$

Equating the parts, notice $4ax^2 = -3x^2$ so that $4a = -3$ and $a = -3/4$. We also see that $8b = 128$, making $b = 16$. Thus, the line we are looking for is,

$$y = -\frac{3}{4}x + 16.$$

Chapter 16

A bit of everything.

Dissonance is when something doesn't quite fit. Thus, these clues are a mix of the previous chapters!

Chapter 17
Polynomial Division.

Example
$(2x^2 - 33x + 81) \div (2x - 6)$

$$\begin{array}{r|l} & x - \frac{27}{2} \\ 2x - 6 & 2x^2 - 33x + 81 \\ & \underline{2x^2 - 6x} \\ & \quad -27x + 81 \\ & \quad \underline{-27x + 81} \\ & \qquad\qquad 0 \end{array}$$

The order is $y = x - 27/2$ or $y = x - 13.5$.

Chapter 18
Perfect squares.

Example
$\sqrt{\frac{9}{16}x^2 - \frac{123}{8}x + \frac{1681}{16}}$

Each radical is a perfect square, can you find it?

$$\sqrt{\frac{9}{16}x^2 - \frac{123}{8}x + \frac{1681}{16}} = \sqrt{\frac{9}{16}x^2 - \frac{246}{16}x + \frac{1681}{16}}$$
$$= \frac{\sqrt{9x^2 - 246x + 1681}}{\sqrt{16}}.$$

Now we must find a and b such that $(ax + b)(ax + b)$ gives us $9x^2 - 246x + 1681$. As $9 = 3 \cdot 3$, we know $a = \pm 3$.

If $a = 3$, then $(3x + b)^2 = 9x^2 + 6bx + b^2$. As $\sqrt{1681} = 41$, we could have $(3x + 41)^2$ or $(3x - 41)^2$. Notice $(3x + 41)^2 = 9x^2 + 246x + 1681$, which doesn't contain a $-246x$ term. Therefore, it must be $(3x - 41)^2$.

If $a = -3$, then $(-3x + b)^2 = 9x^2 - 6bx + b^2$. This gives us $b = 41$ and $(-3x + 41)$.

Our solutions can be,

$$\frac{\sqrt{9x^2 - 246x + 1681}}{\sqrt{16}} = \frac{\sqrt{(3x - 41)^2}}{4}$$
$$= \frac{3}{4}x - 10.25$$

or,

$$\frac{\sqrt{9x^2 - 246x + 1681}}{\sqrt{16}} = \frac{\sqrt{(-3x+41)^2}}{4}$$
$$= -\frac{3}{4}x + 10.25.$$

To "dog ear" the page, any equation on the left must have a negative slope while the right pages all have a positive slope. As this example is on a right page (page 312), the order is,

$$y = \frac{3}{4}x - 10.25.$$

Chapter 19

Factoring and one order.

Example

$-\frac{1}{6}(8x^2 - 137x + 570)$, (6,4)

Factor the polynomial to uncover the order for both folds! Select a good method based on the polynomial. Using the quadratic recipe for $8x^2 - 137x + 570$, we have,

$$x = \frac{137 \pm \sqrt{137^2 - 4(8)(570)}}{2(8)}$$
$$= \frac{137 \pm \sqrt{18769 - 18240}}{16}$$
$$= \frac{137 \pm \sqrt{529}}{16}$$
$$= \frac{137 \pm 23}{16}.$$

The first value is $\frac{137+23}{16} = \frac{160}{16} = 10$. The second value is, $\frac{137-23}{16} = \frac{114}{16} = \frac{57}{8}$.

When $x = 10$, we have $x - 10 = 0$, giving us $x - 10$ as a factor. When $x = \frac{57}{8}$, we have $x - \frac{57}{8} = 0$ as a factor. We can also write this as $8x - 57 = 0$.

Placing this back into our equation we have,

$$-\frac{1}{6}(8x^2 - 137x + 570) = -\frac{1}{6}(x - 10)(8x - 57).$$

We are in a pickle. There are a few options as we don't know which linear factor to distribute the $-1/6$ to.

Knowing the slope values are limited along with the clue (6,4) will help us. The first two possibilities for our lines are,

$$-\frac{x}{6}+\frac{10}{6} \text{ and } 8x-57,$$

or,

$$x-10 \; and \; -\frac{8}{6}x+\frac{57}{6}.$$

Since we were given (6,4), we need the order that sends 6 to Station 4. None of these will fit. Noticing $x-10$ will send 6 to Station -4, we can instead distribute the negative to the first factor and the 1/6 to the second to find,

$$-x+10 \text{ and } \frac{8}{6}x-57/6.$$

Use both pages and clues to help you in Chapter 19! Both the left and right page contain the same polynomial to factor but with different clues as to one of the ordered pairs on the line.

Chapter 20

Things just got complex.

Example

$y = i^2x + (3 + 0i)$

Knowing $\sqrt{-1} = i$, this means $i^2 = \left(\sqrt{-1}\right)^2 = -1$.

We can simplify the order into $y = -x + (3 + 0i)$. Although we are not used to writing numbers like 3 with an imaginary component, all numbers are complex! This means $3 = 3 + 0i$, and we find the order to be $y = -x + 3$.

Visit www.MathBait.com/folds for additional hints and tips, how-to videos and more!

ACKNOWLEDGEMENTS

As crazy as it may seem, this tale is based in truth. The Pythagoreans were a real number-cult, and many did believe in what they called counter-earth as well as limited and unlimiteds.

Recently, researchers "debunked" a theory of Pythagoras regarding musical harmonies. He believed humans preferred certain combinations of notes in which the distribution was rational, such as 3:4. The study found the participants in fact liked combinations that were slightly different from the perfect rational combinations, what is called dissonance. Perhaps this is a sign that humans are related to the irrational Numberfolk?

In 2023, two high school students disproved another belief by mathematicians: proving the Pythagorean Theorem using trigonometry was an impossible task. But that didn't stop Calcea Johnson and Ne'Kiya Jackson from trying and succeeding. I hope this inspires you to realize you are powerful at any age. Question everything. Dig deeper. Numbers and mathematics are limitless, offering a wild playground to explore.

Thus, I acknowledge you. The ones reading this book. The ones who want to know more. The ones brave enough to embark on a new journey. The ones creative enough to imagine mathematics in a new light. To my wonderful students who inspire me every day and force me to question every assumption and prove every statement,

Thank you.

Topics Covered in this Book

- The Coordinate Plane
- Graphing
- Relations
- Functions
- One-To-One and Onto
- Linear Functions
- Linear Equations
- Slope
- Intercepts
- Group Theory
- Set Theory
- Combinations and Permutations
- Arithmetic Sequences
- Exponents
- Decimals
- Radical Expressions
- The Pythagorean Theorem
- Quadratic Functions
- Quadratic Equations
- Focus and Directrix
- Polynomials
- Polynomial End Behavior
- Multiplicity
- Polynomial and Synthetic Division
- Cardinality
- Complex Numbers

ABOUT THE AUTHOR

SK Bennett is an award-winning educator and now an award-winning author as *The History of Numberville* has won multiple awards in the categories of Children's Education and fiction. The majority of her time is dedicated to her wonderful husband, her five amazing children, and the hundreds of outstanding students who love her unique and creative approach to learning math. She believes deeply that education should be an adventure full of exploration, excitement, and enjoyment. Bennett sees all children as "math kids" and hopes every student can find a bit of themselves in Marco, Maggie, Oliver, Liam, and Penelope to discover and perfect their own magical powers.

THE ADVENTURE DOESN'T STOP HERE!

The Society of the Abolishment of Numbers is real.

Learn from the best by joining *The Kryptografima*. Embark on missions, gather items, and play your way to success in this one-of-a-kind learning platform.

Head to www.MathBait.com to learn more.

www.ingramcontent.com/pod-product-compliance
Lightning Source LLC
Chambersburg PA
CBHW021917190326
41519CB00009B/820

* 9 7 9 8 9 8 8 0 8 6 1 2 3 *